Economic activity and land use:
The changing information base for local and regional studies

Edited by

Michael J. Healey

Routledge
Taylor & Francis Group

LONDON AND NEW YORK

First published 1991 by Longman Group UK Limited

Published 2014 by Routledge
2 Park Square, Milton Park, Abingdon, Oxon OX14 4RN
711 Third Avenue, New York, NY 10017, USA

First issued in hardback 2017

Routledge is an imprint of the Taylor & Francis Group, an informa business

ISBN 13: 978-1-138-41698-7 (hbk)
ISBN 13: 978-0-582-05724-1 (pbk)

British Library Cataloguing-in-Publication Data

Economic activity and land use: The changing information base for local and regional studies.
  I. Healey, Michael J.
  333.0028

Library of Congress Cataloging-in-Publication Data

Economic activity and land use: the changing information base for local and
  regional studies / edited by Michael J. Healey.
    p.   cm.
    Includes bibliographical references and index.
    ISBN 0-470-21794-4
    1. Land use—Great Britain—Data processing. I. Healey, Michael.
  HD596.E26   1991
  333.73′13′0285—dc20   91-4015
                                                                            CIP

Set in 10/12 Linotron 202 Times Roman

Contents

Contents

List of figures

List of tables

Editor's preface

In order to understand how and why economic and social conditions in the UK vary from place to place and change over time an accurate, up-to-date, and accessible information base is a necessity. The marked deterioration in the state of the UK database over the last few years is thus a source of concern. Previously the protagonists in the debate might hold different views on the interpretation and meaning of the figures, but there was a general acceptance that the data were a reasonable representation of the situation. Now increasingly the debate is over the reliability and validity of the statistics themselves. Most of these discussions have revolved around the state of national statistics; this book extends the debate to local and regional data.

The increasing dissatisfaction with the official statistics available for subnational analyses of economic activity and land use has resulted in greater use being made of unofficial statistics and more surveys being undertaken and commissioned. Both trends increase the need for guidance for users as to the strengths and weaknesses of the different kinds of information available and the advantages and disadvantages of different methods of collecting information about businesses and land uses. This book is primarily about *statistical* information, including not only processed statistical data, but also raw data collected from businesses and individuals and by direct observation or using remote sensing techniques.

Ironically, as the quality of the database has deteriorated the growing importance of computerized information systems has improved the ability to store and analyse the information that does exist. Interest in the development of specifically *geographical* information systems was stimulated by the publication in 1987 of the Chorley Report, *Handling Geographic Information*. These developments are having a significant effect on the way in which information is used by those who have access to the systems, though their numbers are often restricted by the charges imposed. However, the real potential of geographical information systems (GIS) to become an effective decision-making tool for public policy planning depends not only on ready access to information, but also on the availability of high-quality, nationally available statistical series which are geographically disaggregated. This emphasizes that discussions about the handling of data should not be separated from those concerned with data collection, quality and availability.

The improvements in the handling of information, brought about by computerization and on-line access, increase not only the possibility of misinterpretation of data, but also the danger of misuse of information. In the first case the availability of more 'raw' data enhances the need for guidance for unwary users as to the limitations and pitfalls of the data; while the second emphasizes the need for improved security systems to restrict access to confidential information and to prevent infringement of copyright.

The issues involved are explored in this specially commissioned collection of essays. The aims of the book are:

1. To provide a guide to the nature, uses, availability and limitations of the main data sources for interpreting and undertaking local and regional studies of economic activity and land use in the UK; and
2. To examine the methods of collecting information on these topics, the GIS used for storing and analysing the data, and the range of techniques used to forecast local and regional economic change.

The book is divided into three sections: Part A examines data on different *measures of economic activity and land use*, including employment; unemployment, vacancies and redundancies; output, costs and profitability; floorspace and commercial developments; and cartographic, statistical and remote sensing sources on urban and rural land use. Part B analyses methods of *monitoring economic activity and land use*, including GIS on manpower, and land and property; local and regional information systems for public policy; obtaining information from businesses; land use surveys; and forecasting local and regional economic change. Part C assesses the data available on different *economic sectors*: agriculture, forestry and fishing; mining, utility and construction industries; manufacturing; producer services; consumer services; and transport, trade and tourism. Whereas the chapters in Parts A and C are mainly concerned with *data sources*, the chapters in Part B are more concerned with the *techniques* for collecting, storing, monitoring, presenting and forecasting data. An introductory chapter provides an overview of the nature and recent changes in the information base and makes recommendations for its improvement. Annex 1 gives a list of useful addresses.

The book has been written to appeal to a wide ranging 'semi-professional' audience, including consultants, market research workers and business people concerned with local and regional economic development and land use; planners and economic development officers in local and central government; research workers and undergraduates in business, economics, geography and planning; and reference librarians and others wanting a critical guide to the nature and uses of data sources on economic activity and land use in the UK. It should also be a useful complement to the many recent texts which have analysed the changing pattern of economic activity and land use in the UK.

The contributors write about data sources which they have personal experience of using in their research, teaching and consultancy activities. The authors come from a variety of backgrounds, including geography, planning, economics and surveying. All of them have experience of working in and/or undertaking consultancy work for local and central government, research organizations and/or business. Each chapter can be read independently of the others, but to avoid excessive overlap on the discussion of some sources and techniques cross-references between chapters have been added to help the reader identify further discussion of particular sources and techniques.

The book is mainly about local and regional data sources and information systems in the UK. However, many of the techniques and issues discussed are widely applicable outside the UK. Reference is also made to the national and international data sources, such as publications from the Central Statistical Office, the United Nations and Eurostat, which provide a context in which to interpret local patterns and changes. Variations in the data sources available in England, Northern Ireland, Scotland and Wales are noted, and examples of some of the uses to which the various data sources and techniques have been put are included. Considerable effort has been expended in attempting to provide an up-to-date (i.e. 1989/90) statement of the current nature of these information sources and the way in which they have changed over the last decade or two. However, the nature of the information base alters rapidly and several changes have occurred in the few months since the manuscript for this book was written. For example, with the reorganization of training in the UK the Training Agency has ceased to exist and many of its functions have been taken over by the Training, Enterprise and Employment Division of the Employment Department. North of the border the Scottish Development Agency has become Scottish Enterprise. This reorganization has affected the collection and availability of labour market information. For example, the quarterly *Skills Bulletin* has ceased publication and the Computer-Assisted Local Labour Market Information System (CALLMI) has been wound up (Chs. 3 and 8). Another body which has been reorganized is the Scottish Development Department, which is now the Scottish Office Environment Department. The continuing trend towards privatization is also affecting the organizational structure of various industries. The latest is the electricity industry (Ch 15). The Central Electricity Generating Board is being replaced by National Power PLC and Powergen plc, while Scottish Power plc and Scottish Hydro–Electric plc are to takeover the two Scottish Electricity Boards. Technological advances are also continuing to affect the form in which the information is stored and available. For example, the Ordnance Survey (OS) is in the process of abolishing its master survey drawings (Ch 12). It is moving to direct data transmission from field surveyors to OS headquarters so that the digital record is always correct. A further example stems from the increased sophistication of computer analyses. This has increased the possibility of identifying individual agricultural holdings from the parish summaries of the Agricultural Census (Chs. 7 and 14). The last parish summaries to be released were for 1988, since when the Ministry of Agriculture, Fisheries and Food have been investigating the issue and considering alternative ways in which sub-county information might be released. A list of publications in which new developments are regularly announced is included in Chapter 1 (section 1.4).

Numerous people have helped in the preparation of this book. The role of some are acknowledged at the end of particular chapters. Others have helped in a more general capacity. I am particularly grateful to the many individuals who provided information and answered questions about the data sources emanating from their organizations, and to the librarians at Coventry Polytechnic and the University of Warwick who helped trace the answers to many queries. Special thanks, of course, are due to the authors and their readiness to respond to suggestions quickly and positively. They will feel their endeavours worth while if they lead to an increased use of the data sources discussed; a greater critical appreciation of their strengths and limitations; and a more informed debate on the urgent need for improvement in the information base.

Michael Healey
Solihull
January 1 1991

Acknowledgements

We are indebted to Oxford University Press for permission to reproduce Fig. 16.1 from Fig. 16.2 (Healey and Ilbery 1990).
Although every effort has been made to trace and contact copyright holders, this has not always been possible. We apologize for any apparent negligence.

Abbreviations and acronyms

AA	Automobile Association
AGB	Audits of Great Britain
AGI	Association for Geographic Information
AH	activity heading
ANA	Article Number Association
BACMI	British Aggregate Construction Materials Industries
BERT	Tayside Business Establishment Register
BPA	British Port Association
BPF	British Ports Federation
BQSF	British Quarrying and Slag Federation
BR	British Rail
BSO	Business Statistics Office
BT	British Telecommunications
BTA	British Tourist Authority
BTS–M	*British Tourism Survey – monthly*
BTS–Y	*British Tourism Survey – yearly*
BURISA	British Urban and Regional Information Systems Association
CAA	Civil Aviation Authority
CALLMI	Computer-Assisted Local Labour Market Information System
CBI	Confederation of British Industry
CCS	Countryside Commission for Scotland
CE	Cambridge Econometrics
CEC	Commission of the European Communities
CEGB	Central Electricity Generating Board
CEPG	Cambridge Economic Policy Group
CGP	Cambridge Growth Project
CIFS	*Commercial and Industrial Floorspace Statistics*
CIG	Corporate Intelligence Group
CIPFA	Chartered Institute of Public Finance and Accounting

COBIS	Tyne and Wear Countywide Business Information Service
CODOT	Classification of Occupations and Directory of Occupational Titles
CoE	Census of Employment
CoP	Census of Population
CORINE	Co-ordinated Information on the European Environment
COSLA	Convention of Scottish Local Authorities
CSO	Central Statistical Office
CSV	comma separated value
CWS	Co-operative Wholesale Services
D&B	Dun and Bradstreet
DAFS	Department of Agriculture and Fisheries for Scotland
DANI	Department of Agriculture in Northern Ireland
DE	Department of Employment
DoE	Department of the Environment
DSS	Department of Social Security
DTI	Department of Trade and Industry
DTp	Department of Transport
EC	European Community
ED	enumeration district
EDC	Economic Development Committee
EFTA	European Free Trade Association
EOSAT	Earth Observation Satellite Company
ERI/II	Employment Record I/II
ESRC	Economic and Social Research Council
ESU	European size unit
ETIC	Employment and Training Consortium
FC	Forestry Commission
FES	*Family Expenditure Survey*
FICGB	Forestry Industry Committee of Great Britain
FPC	Family Practitioner Committee
GATT	General Agreement on Tariffs and Trade
GB	Great Britain
GDP	gross domestic product
GHS	*General Household Survey*
GIS	geographical information system
GISP	general information system for planning
GNP	gross national product
GSS	Government Statistical Service
ha	hectares
HIB	Herring Industry Board
HIDB	Highlands and Islands Development Board
HMLR	Her Majesty's Land Revenue
ICAO	International Civil Aviation Organization
ICC	Inter Company Comparisons
IEA	International Energy Agency

IER	Institute for Employment Research
IGD	Institute of Grocery Distribution
I–O	input–output
ILAG	Inquiry into Locational Attitudes Group
ILO	International Labour Organization
IMF	International Monetary Fund
IPD	Investment Property Databank
IPF	iterative proportionate fitting
IPS	*International Passenger Survey*
IRVO	Inland Revenue Valuation Office
IRS	Institute for Retail Studies
ISIC	International Standard Industrial Classification
IT	information technology
ITE	Institute of Terrestrial Ecology
ITT	invitation to tender
JANET	Joint Academic Network
JUVOS	Joint Unemployment Vacancies and Operating System
KO	key occupations
KRIOS	Keele Regional Input–Output System
LAMIS	Local Authority Management Information System
LAMSAC	Local Authority Management Service and Computing
LAN	local area network
LEA	local education authority
LEC	Local Enterprise Company
LFS	*Labour Force Survey*
LHA	local highway authority
LIS	land information system
LLMA	local labour market area
LMS	labour market statistics
MAF	Ministry of Agriculture and Fisheries
MAFF	Ministry of Agriculture, Fisheries and Food
MDS	maritime and distribution systems
MISEP	mutual information system on employment policies
ML	Market Location Ltd
MLH	minimum list heading
MLURI	Macaulay Land Use Research Institute
MSS	multispectral scanner
NACE	Nomenclature Genérale des Activités Economiques dans les Communautés Européennes
NATLAC	National Land Characteristics Database
NBC	National Bus Company
NCC	Nature Conservancy Council
NEDO	National Economic Development Office
NERC	Natural Environment Research Council
NES	*New Earnings Survey*
NFS	*National Food Survey*

NFU	National Farmers' Union
NHBC	National House Building Council
NHS	National Health Service
NHSCR	National Health Service Central Register
NI	Northern Ireland
NIEC	Northern Ireland Economic Council
NIERC	Northern Ireland Economic Research Centre
NITB	Northern Ireland Tourist Board
NJUG	National Joint Utility Group
NLUC	National Land Use Classification
NLUSS	National Land Use Stock Survey
NMWG	National Managed Workspace Group
NNDR	national non-domestic rates
NTF	national transfer format
NOMIS	National On-line Manpower Information System
NPC	National Ports Council
NTS	*National Travel Survey*
O/D	origin-destination
OECD	Organization for Economic Co-operation and Development
OPCS	Office of Population Censuses and Surveys
OS	Ordnance Survey
OXIRM	Oxford Institute of Retail Management
PACEC	P A Cambridge Economic Consultants
PAYE	pay as you earn
PC	personal computer
PLC	public limited company
PPRU	Policy Planning and Research Unit
PSMRU	Public Sector Management Research Unit
PSS	Packet Switching Service
PTE	Passenger Transport Executive
PYO	pick your own
RAC	Royal Automobile Club
RAD	Rural Areas Database
R&D	research and development
RCF	relative change forecasting
RES	relative employment shortfall
RPI	Retail Price Index
RSI	Retail Sales Index
RSS	Royal Statistical Society
SAGA	Sand and Gravel Association
SAM	social accounting matrix
SAMI	Sales and Marketing Information Ltd
SDD	Scottish Development Department
SEG	socio-economic group
SFIA	Sea Fish Industry Authority
SFST	*Sea Fisheries Statistical Tables*

SIC	Standard Industrial Classification
SMD	standard man-day
SOC	Standard Occupational Classification
SPA	Sales Performance Analysis
SPOT	Satellite Probatoire d'Observation de la Terre/Le Système Pour l'Observation de la Terre
SQL	structured query language
SSEW	Soil Survey of England and Wales
SSF	Social Science Forum
SSFST	*Scottish Sea Fisheries Statistical Tables*
SSRC	Social Science Research Council
TA	Training Agency
TB	tourist board
TEC	Training and Enterprise Council
TEED	Training, Enterprise and Education Division
TEST	Transport and Environmental Studies
TM	thematic mapper
TMS	Textile Marketing Services
TRRL	Transport and Road Research Laboratory
TSGB	*Transport Statistics Great Britain*
TTWA	travel-to-work area
UBO	Unemployment Benefit Office
UK	United Kingdom
UKSPA	UK Science Park Association
UNCTAD	United Nations Conference on Trade and Development
URBED	Urban and Economic Development Group
URPI	Unit for Retail Planning Information
USA	United States of America
UU	Unemployment Unit
U/V	unemployment/vacancy
VAN	value added network
VAT	value added tax
VPS	Voluntary Population Survey
WAA	Water Authorities' Association
WALTER	Wales Terrestrial Database Project
WAN	wide area network
WFA	White Fish Authority
WOAD	Welsh Office for Agricultural Development

List of contributors

Blakemore, Michael Lecturer in Geography, University of Durham; Co-Director of NOMIS; author of many articles on GIS.

Coppock, Terry Emeritus Professor of Geography, University of Edinburgh; Secretary and Treasurer of the Carnegie Trust for the Universities of Scotland; author of numerous articles and several books on land use, recreation and agriculture including *An Agricultural Geography of Great Britain* (Bell, London, 1971); *Agricultural Atlas of England and Wales* (Faber and Faber, London, 1976); author of 'Land use' in *Reviews of United Kingdom Statistical Sources* vol. VIII: 1–102 (Pergamon, Oxford, 1978); joint author of *Review of Approaches and Sources for Monitoring Change in the Landscape of Scotland* with R.P. Kirby (SDD, Edinburgh, 1987); joint editor with P. Kivell of *Geography, Planning and Policy* (Geo Books, Norwich, 1986); editor of *International Journal of Geographical Information Systems*.

Coull, James Senior Lecturer in Geography, University of Aberdeen; author of *The Fisheries of Europe* (Bell, London, 1972); author of many papers on fisheries in the UK and other countries around the North Atlantic; consultant on fishery matters to the White Fish Authority, Shetland Islands Council and Orkney Islands Council.

Dale, Peter Professor, Department of Land Surveying, Polytechnic of East London; author of numerous articles on GIS and LIS; co-author with J.D. McLaughlin of *Land Information Management* (Oxford University Press, Oxford, 1988); land use and information systems consultant; member of Editorial Board of *International Journal of Geographical Information Systems, Cities* and *Mapping Awareness*; Chairman of the International Federation of Surveyors working group on Land Information Systems in Developing Countries.

Dawson, John Professor of Marketing, University of Edinburgh; formerly Secretary of the Institute of British Geographers; author of numerous articles and several books on retailing and services including *The Marketing

Environment (ed.) (Croom Helm, London, 1979); *Retail Geography* (ed.) (Croom Helm, London, 1980); *Shopping Centre Development* (Longman, London, 1983); retail consultant to businesses, local authorities and central government.

Green, Anne Senior Research Fellow, Institute for Employment Research, University of Warwick; consultant on economic development; author of many articles on employment and unemployment; co-author of *Technological Innovation, Structural Change and Location in UK Services* with J. Howells (Gower, Aldershot, 1988) and *Changing Places: Britain's Demographic, Economic and Social Complexion* with A.G. Champion *et al.* (Edward Arnold, London, 1987).

Healey, Michael Principal Lecturer in Geography and Deputy Director, Centre for Local Economic Development, Coventry Polytechnic; author of several articles on industrial change in the UK; editor of *Urban and Regional Industrial Research: The Changing UK Data Base* (Geo Books, Norwich, 1983) and *Industrialization of the Countryside* with B.W. Ilbery (Geo Books, Norwich, 1985), co-author with B.W. Ilbery of *Location and Change: Perspectives on Economic Geography* (Oxford University Press, Oxford, 1990), and *People and Work in Coventry: A Survey of Employers* with P. Elias (Coventry City Council, Coventry, 1991).

Hoare, Tony Senior Lecturer in Geography, University of Bristol; author of many articles on industrial change, transport and trade; author of *The Location of Industry in Britain* (Cambridge University Press, Cambridge, 1983).

Ilbery, Brian Senior Lecturer in Geography, Coventry Polytechnic; author of many articles on agricultural change and the geography of Europe; author of several books including *Agricultural Geography* (Oxford University Press, Oxford, 1985) and *Western Europe* (Oxford University Press, Oxford, 1986); joint editor with M.J. Healey of *Industrialization of the Countryside* (Geo Books, Norwich, 1985); co-author of *Agricultural Change: France and the EEC* with H. Winchester (John Murray, London, 1988); *Location and Change: Perspectives on Economic Geography* with M.J. Healey (Oxford University Press, Oxford, 1990); consultant to MAFF.

Kivell, Philip Senior Lecturer in Geography and Director of Centre for Regional Information Research, University of Keele; author of many articles on the use and misuse of land; joint editor with J.T. Coppock of *Geography, Planning and Policy Making* (Geo Books, Norwich, 1986); joint author of *Inner City Wasteland* with M. Chisholm (Institute of Economic Affairs, London, 1987).

Marshall, Neill Lecturer, Centre for Urban and Regional Development Studies, University of Newcastle upon Tyne; consultant to local and central government; author of many articles on services and regional development, the links between services and UK manufacturing industry; co-author of *Services and Uneven Development* (Oxford University Press, Oxford, 1988).

Mather, Alexander Senior Lecturer in Geography, University of Aberdeen; author of many articles on forestry in the UK; author of *Land Use* (Longman,

London, 1986) and *Global Forest Resources* (Belhaven, London, 1990); editor of *Scottish Geographical Magazine*.

Owen, David Research Officer, ESRC Wales and South West Regional Research Laboratory, Department of Town Planning, University College of Wales; author of many articles on modelling economic and employment change in the UK; co-author of *Changing Places: Britain's Demographic, Economic and Social Complexion* with A.G. Champion *et al*. (Edward Arnold, London, 1987); consultant on economic development.

Perry, Martin Visiting fellow in Geography, National University of Singapore; author of *Small Factories and Economic Development* (Gower, Aldershot, 1986); co-author of *Property and Industrial Development* (Hutchinson, London, 1987).

Roberts, Peter Professor, Department of Urban Planning, Leeds Polytechnic; formerly Senior Consultant at Ecotech and Chairman of the Regional Studies Association; consultant to several local authorities on mining developments and economic change; co-author of *Mineral Resources in Regional and Strategic Planning* with T. Shaw (Gower, Aldershot, 1982); joint editor of *Regional Planning at the Crossroads: European Perspectives* (Jessica Kingsley, London, 1988).

Senior, Derek Senior Lecturer in Urban Planning, Leeds Polytechnic; formerly worked in planning departments of Liverpool City Council, Cheshire County Council and SDD; previously Group Planner South Yorkshire County Council responsible for minerals, waste disposal and land use.

Townsend, Alan Reader in Geography, University of Durham; Co-Director of NOMIS; author of many articles on employment and industrial change in the UK; author of *The Impact of Recession* (Croom Helm, London, 1983); joint editor with J. Lewis of *The North–South Divide: Regional Change in Britain in the 1980s* (Paul Chapman, London, 1989); joint author with A.G. Champion of *Contemporary Britain: A Geographical Perspective* (Edward Arnold, London, 1990).

Tyler, Peter Research Associate, Department of Land Economy, University of Cambridge; consultant PA Cambridge Economic Consultants; author of many articles on industrial change and regional policy; co-author of *Geographical Variations in Costs and Productivity* (HMSO, London, 1988).

Worrall, Les Principal Policy Planner, Wrekin Council, Telford and Research Fellow, Department of Civic Design, University of Liverpool; author of many articles on information systems for urban and regional planning; Vice-Chairman Standing Committee of Regional and Urban Statistics, International Statistical Institute; Chairman, Local Authorities Research and Intelligence Association (LARIA).

1

The information base: nature, changes and recommendations for improvement

Michael Healey

1.1 Main themes

In recent years the nature, quality and accessibility of the information base in the UK has become a key issue of debate among research workers and decision-makers in business and local and central government. Concern about deterioration in the information base, particularly official statistics, was fuelled by the implementation in the early 1980s of the Rayner *Review of the Government Statistical Services* (Rayner 1980). In the late 1980s and early 1990s discussion over the accuracy and integrity of government statistics became more widespread. The Social Science Forum (SSF 1989) established a campaign for improvements to be made to government statistics and the Royal Statistical Society (RSS 1990) published an influential report entitled *Official Statistics: Counting with Confidence*. Debates also took place in Parliament (e.g. on the problems with statistics on unemployment and the National Accounts), on television (e.g. Channel 4 1989), and in the press (e.g. Waterhouse 1989a, b). The issues raised in these debates also apply to the important subset of UK information sources with which this book is concerned, namely those available for local and regional studies of economic activity and land use. This introductory chapter attempts to give a context for the subsequent chapters by reviewing the main features of the information base, the ways in which it is changing, and recommendations to improve it.

The paucity of appropriate, accurate, up-to-date information on variations from place to place in the nature of economic activity and land use has been a theme developed in earlier reviews (Coppock 1978; England *et al.* 1985; Healey 1983; Owen 1988). Unfortunately, as most local and regional statistics are derived from disaggregations of national official data sets, any deterioration in the quality of the latter is magnified the smaller the area under consideration. Increasing dissatisfaction with the official statistics available for subnational analysis has encouraged two trends. First, many research workers have been forced to undertake or commission their own surveys of businesses and land uses. This raises questions about the optimal allocation of the nation's resources concerned with data collection because in many cases it has led to several different organizations collecting similar information. A second trend has been for a growth to occur in the number of private-sector organizations providing local data. As access to this information is determined by the ability to pay it is leading to distinctions between information-rich and information-poor groups and regions (Openshaw and Goddard 1987). Charges for official statistics are also becoming more common.

A further trend has been the growing importance of information systems used to store and analyse the information used by local and regional research workers. Whereas the quality of 'Database UK' has decreased over recent years, the quality of the information systems developed to handle the data has improved significantly. The rapid growth of interest in geographical information systems (GIS), in particular, is witnessed by the Committee of Enquiry set up by the government under Lord Chorley, and its report *Handling Geographic Information* (Department of the Environment (DoE) 1987); the initiative by the Economic and Social Research Council (ESRC) in establishing eight Regional Research Laboratories in 1988; and the launch the following year of the Association for Geographic Information. Although GIS can now be regarded as 'having reached the age of maturity' (Newby 1988: 8), concern has been expressed that 'without high quality, nationally available statistical series covering the key issues of social concern, GIS will never graduate from being a sophisticated toy to an effective decision-making tool' (Worrall 1990: 7). This emphasizes that discussions concerned with the handling of data should not be separated from those concerned with data collection, quality and availability. The need for good-quality data applies even more so to forecasting local and regional economic change because any inaccuracies are amplified. Existing social and economic statistics fall far short of what is needed to implement an adequate working model of a complex spatial economy. As one commentator recently summed it up: 'nice model, pity about the data' (Hunter 1989: 26). A further issue, which is raised by the increased ease of handling information brought about by computerization and on-line access, is that of confidentiality and copyright. Security systems need to be developed to restrict access to confidential information and to prevent infringement of copyright.

1.2 Nature of the information base

The information base for local and regional studies of economic activity and land use is highly selective. The information is structured by: what questions are asked, of whom, and when; how accurate and representative the methods of data collection are; what information is analysed; how the data are held; whether the information can be manipulated by the user; how the data are aggregated (e.g. spatially and sectorally); what information is published; what access there is to unpublished data; and how much it costs. Attention in this section is focused on three factors which underlie the nature of most of the data available. They are: how the information is classified; what statistical and spatial units are used for collecting, storing and presenting the data; and the way the information is collected.

Classifications of economic activity and land use

The UK economy consists of a range of *economic activities* 'through which goods are produced or services rendered by firms and other organisations' (Central Statistical Office (CSO) 1979: 2). Similar economic activities may be grouped together into 'industries', for example into agriculture, coal extraction, motor vehicle manufacture, banking and finance, retail distribution, air transport. These industries occupy land and may therefore also be considered as *land uses*, but the use of land by human activities is not necessarily always for financial profit or gain. A basic distinction may be made between urban (e.g. industry, offices, housing) and rural (e.g. agriculture, forestry, recreation) land uses. For analytical purposes various classifications of economic activities and land uses have been developed. They provide a uniform and comparable framework for the collection, presentation and analysis of data. Most are based on the 'major' activity or use and exclude minor activities and uses.

The first comprehensive *Standard Industrial Classification* (SIC) for the UK was issued in 1948. It was subsequently revised in 1958, 1968 and 1980 to take account of changes in the organization and relative importance of a number of industries, and to distinguish new industries not separately identified previously. The 1980 revision brought the SIC in line as closely as was practicable with the 'Nomenclature Générale des Activités Economiques dans les Communautés Européennes' (NACE) classification used by the Statistical Office of the European Community (EC). It can also be rearranged to agree with the United Nations International Standard Industrial Classification (ISIC) at aggregated levels (CSO 1979). Unlike the 1968 SIC which has 27 orders, each divided into a number of minimum list headings (MLH) denoted by three-digit numbers (181 in all), the 1980 SIC has a hierarchical structure based on 10 divisions (denoted by a single digit), 60 classes (two digits), 222 groups (three digits) and 334 activity headings (four digits). A summary of the relationship between the two classifications is shown in Table 1.1.

There are several other classifications which may be used in conjunction with the industrial classification. These include, for example, the classification of occupations, which categorizes workers on the basis of the kind of job they do rather than on the basis of the industry in which they work. A revised *Standard Occupational Classification* (SOC) was published in 1990 (Employment Department Group and Office of Population Censuses and Surveys (OPCS) 1990). An example of its application is given in Appendix 11.4. It was designed as a single up-to-date classification to replace CODOT

Table 1.1 Comparison of 1980 and 1968 SICs. (*Source:* CSO 1979:3)

Present divisions	Former orders
0 Agriculture, forestry and fishing	I
1 Energy and water supply industries	II (MLH 101 and 104),IV, XXI
2 Extraction of minerals and ores other than fuels; manufacture of metals, mineral products and chemicals	II (MLH 102, 103, 109), V, VI, XVI
3 Metal goods, engineering and vehicles industries	VII – XII inclusive
4 Other manufacturing industries	III, XIII–XV, XVII–XIX
5 Construction	XX
6 Distribution, hotels and catering; repairs	XXIII, XXVI, (MLH 884–888, 894, 895)
7 Transport and communication	XXII
8 Banking, finance, insurance, business services and leasing	XXIV, XXV (MLH 871, 873, 879)
9 Other services	XXV (remainder), XXVI (remainder), XXVII

(Classification of Occupations and Directory of Occupational Titles) and the 1980 version of the *Classification of Occupations* (OPCS 1980, and also Ch. 2). Another classification is that of products. An example of this is the alphabetical list of industries and their typical products which accompanies the 1980 SIC (CSO 1980). A further one is the Use Classes Order which local authorities employ in deciding on whether or not to grant planning permission to proposed developments. The Town and Country Planning (Use Classes) Order 1987 introduced a new business use class (B1) which involved a combination of the previous office (II), light industrial (III) and warehouse (X) use classes. This was intended to remove restrictions on the supply of multi-use accommodation appropriate for high-technology industry (Henneberry 1988).

There are also several different ways of classifying land use. An important distinction may be made between *formal* and *functional* classifications. The first kind is based on the land form or cover, for example types of building or open space, whereas the second kind is based on the land function or activity and concerns what land is actually used for. One of the simplest land use classifications is the sixfold division used by the *First Land Utilization Survey of Britain* into: arable; heath and rough pasture; orchards and nurseries; meadowland; forest and woodland; urban areas. The more recent *Second Land Utilization Survey* is more comprehensive in that it identified 70 land use types, grouped into 13 major classes (Ch. 7). Neither survey, however, provided much detail for urban areas. A more balanced classification is used by the DoE in the annual reports it has published on land use changes in England since 1986. Urban and rural land uses are each divided into 5 groups, which are further subdivided into 24 categories (see Fig. 6.1). An earlier taxonomy which provided much more detail on urban land uses was the *National Land Use Classification* (NLUC) which was devised to serve the various purposes of planning throughout the country (NLUC 1975). It is a hierarchical system with 15 major orders, 78 groups, 150 subgroups and over 600 classes. Interestingly, an attempt was made to relate the NLUC to the 1968 SIC which meant that some limited connection between land use and economic activity was forged (Ch. 6).

Statistical and spatial units

The ways in which the statistical and spatial units used in collecting economic activity and land use statistics are defined is important because they can affect how the information is interpreted. For instance, the use of different statistical units for different enquiries can result in different findings. An example is given in Chapter 16 of how the numbers employed in manufacturing industries in Great Britain differ by up to 9 per cent, according to which source is used, and by up to 50 per cent for individual industries (two-digit classes) (see Tables 16.1 and 16.2). Different findings may also occur between studies using different spatial units. It is an ecological fallacy to assume that a description or relationship found using one spatial unit (e.g. local authority areas) applies to another (e.g. households or standard regions) (Openshaw 1987). The problems of statistical and spatial data collection units may be illustrated with the cases of the SIC and land use surveys respectively.

The SIC is applied to units on the basis of their principal activity (CSO 1979). Frequently there is no problem in defining the relevant unit. This applies, for example, where the unit is at a single address, all its activities occur within the same heading of the classification, and it is operated as a separate unit for record-keeping and accounting purposes. However, in many cases the situation is more complex. Sometimes a unit produces, sells or provides a variety of goods or services, in which case the unit as a whole is allocated to the industry in which the greater part of its production, sales or services are found. Another difficulty can occur when dealing with large organizations operating at several locations. Frequently some of the data required cannot be provided for each site, because the organizations have centralized information systems. For example, many retail chains operate centralized purchasing and stockholding arrangements. Sometimes in order to secure compatibility between different enquiries some units may be allocated to more than one classification. For instance, a unit at one address may be able to provide a limited amount of information, such as employment, but for more comprehensive information it may have to be combined with other units under the same ownership which are mainly engaged in a different activity. Where units located in different places are combined, as happens sometimes with data in the annual Census of Production, the potential for accurate geographical analyses is limited (Ch. 4).

Conceptual and practical difficulties also occur in defining spatial units for the collection and presentation of land use data. As Rhind and Hudson (1980: 20) point out:

> Choice of the 'geographical individual' – the section of ground which is considered as a unit for subsequent purposes – has considerable implications for analysis and even more for the level of effort needed for any ground survey. Frequently, however, choice of spatial unit is outside the scope of the end-user of the data and, if particular technology is used, of even the data collector.

Land use data may be collected by using point samples, line samples and zones or areas. The last is much the most common way of treating land use both in surveys and in storing the data in map form. The principle is that within each zone, for example a field in a cultivated rural area, land use should be homogeneous or as near this as makes no difference for the intended use of the data. The amount of detail is, however, limited by the human resources available for a survey, the spatial resolution of the technology being used, or, most sensibly, the purpose for which the data are collected. For example, the 1969 *Survey of Developed Areas in England and Wales* excluded any discrete areas of built-up land less than 5 ha in extent. Land use mapping based on remote sensing techniques vary widely from scales of about 1 : 2500 for low-altitude photography of urban areas to 1 : 250 000 upwards for Landsat imagery. A scale of 1 : 25 000 may be a suitable compromise when both urban and rural areas are being photographed (Ch. 12).

Sometimes the statistical and spatial units used to collect the data are the same as those used to store and present it, but often they differ. A major problem restricting the integration of different data sets is the variation in the system of locational referencing and the spatial units used for analysing and reporting information. For example, within the energy and utilities sector, the coal, electricity, gas and water industries all use different statistical regions (Ch. 15). There are a multiplicity of spatial units (e.g. enumeration districts (EDs), local authority districts, travel-to-work areas (TTWAs), standard economic regions, assisted areas, district health authority areas, a variety of retail catchment and drive–time areas) many of which are incompatible in that they do not 'nest' into each other. With the deterioration in the quality and frequency of information collection, research workers are increasingly faced with integrating data from a variety of sources and spatial frameworks. The finer the level of spatial disaggregation the greater the potential for manipulation to other spatial units.

There is no one universal method of referencing locational information. Grid references, addresses, grid-squares, unique property reference numbers, and postcodes are all in common usage (Openshaw 1987, 1990). Unfortunately the same data are sometimes presented using different spatial units for different parts of the UK and different spatial units are used over time making examination of changes over space and time very difficult. The Census of Population (CoP) is the classic case. In 1971 data were made available in computer form on a grid-square basis. In 1981 the base in Scotland was the postcode sector, while in England and Wales it was the ED (Ch. 8).

Methods of information collection

The nature and quality of the information available for local and regional studies of economic activity and land use depend to a large extent on the way in which the information is collected in the first place, because the strengths and limitations of the different methods are reflected in the published and unpublished statistics and maps. Inadequacies and deteriorations in the database are increasingly making it necessary for research workers to undertake or commission their own business and land use surveys.

The majority of information on economic activities comes originally from direct surveys of businesses. Different types of survey technique may be distinguished, including: observation, self-administered questionnaires, standardized interviews and non-standardized interviews. The suitability of different methods varies with the situation and is dependent on a range of factors, including the research design, the kind and amount of information required, the resources available, and the size, organizational structure, sector and location of the businesses to be approached. Often a mixture of methods is used within the same study. The close links between theory and method mean that the same method may be used in different ways in different research designs. For example, in extensive research an interview may be used to try to quantify and generalize about aspects of business behaviour, whereas in intensive research the aim of the interview may be to uncover the causal mechanisms underlying a particular process and to refine and test previously used categorizations. Unfortunately the lack of guidance in the literature on undertaking business surveys means that most research workers have no choice in designing their surveys but to rely on a combination of common sense and previous experience (Ch. 11).

Different methods of collecting information on land use may also give varying results (Ch. 12). Remote sensing techniques are best at identifying land cover, as errors can arise when trying to identify land function. For example, houses may be converted to offices without any change in the external form of the building. Accuracies of 80–90 per cent have, however, consistently been reported from trained personnel interpreting air photography where the relationship between form and function is unambiguous (Rhind and Hudson 1980). Ground survey or documentary evidence is required to establish an accurate picture of land activities (Ch. 6). Sometimes the different methods are used in combination. For example, as part of the *Countryside Survey 1990* the Institute of Terrestrial Ecology (ITE) is using images from Landsat to map the proportions of up to 20 major land cover categories in each of the 250 000 kilometre squares into which Great Britain has been divided. Detailed field surveys of 512 sample kilometre squares will then be used not only to check the accuracy of the interpretation of the remote sensing imagery but also to predict which habitats and vegetation types are present within each square. Hence, if the satellite images define 'woodland' or 'grassland' in a square, then the field survey results may be used to predict the type of woodland or grassland. This is the first time that extensive satellite data and detailed probabilistic field information have been integrated in this way at a national level (Barr 1990).

1.3 The changing information base

> Data are . . . conceived of as social products: statistics are not *collected* but *produced*. . . . Their production is a social process which is carried out for specific reasons, and in specific ways (Irvine *et al.* 1979: 3).

Given that statistics are related to the nature of the society which produces them, the characteristics of the information base, including methods of assemblage, storage and presentation, are under continual pressure to respond to changes in society. Four of the principal pressures which have affected the nature of the information base over the last 10–15 years are: changes in demand; government expenditure cuts and privatization programmes; developments in information technology (IT); and changes in attitudes to the commercial value of information. However, the information base has not always changed in response to these pressures. Sometimes the pressures may lead in opposite directions. For example, a demand for new data may be impeded by government expenditure cuts. Where changes have occurred this can lead to problems of comparability over time.

Changes in demand

These stem from a variety of sources. Perhaps the most fundamental are changes in the nature of economic activity and land use. Trends such as deindustrialization and the growth of the service sector, the growth of large organizations and the revival of small businesses, the extension of urban land uses into rural areas, farm diversification and the industrialization of agriculture, skill shortages and increased labour flexibility, all create demands for changes in the information base used to examine them. However, the supply of new information sources often lags well behind demand. One of the clearest examples of this is with the availability of information on the service sector. Over the last few decades there has been a marked expansion in the range, variety and complexity of the service sector (see e.g. Ch. 18). Despite the service sector accounting for twice as many workers as manufacturing, and output from services exceeding that from production industries, the official statistical database is focused on measuring the goods sector. Although some modifications are being made, the statistical information on the service sector is weak in quality and limited in extent (Ch. 17). Data inadequacies are particularly apparent in new and rapidly changing sectors such as IT. For instance, *Computer Weekly* of 10 March 1988 (p. 6) commented cynically:

> The government does not know how many computer discs, tapes and cartridges were sold in the UK and how many were exported, according to written answers from the Department of Trade and Industry. However the DTI does know how many ballpoint pens, pencils and parts have been sold. . .

Nevertheless, Miles (1990: 29) argues that, although suffering from limitations, 'a large number of potential data sources can be used to gain some insights into the "information economy"'.

Another reason for alterations in the demand for information is changes in the theoretical frameworks used for examining economic activity and land use (Healey and Ilbery 1990). In particular, the growth of interest in structuralist and realist approaches has increased demand for information sources on the national and international context of change. Continued interest in behavioural approaches and the emphasis on specific outcomes of wider economic processes in the structuralist and realist approaches, have put a premium on the ability to identify the responses of individual organizations to economic changes. These theoretical shifts have also led to a growth of interest in intensive research methodologies and their associated non-standardized interviewing techniques for collecting information (Ch. 11).

Policy changes also affect the demand for information. The shift in the latter part of the 1970s and 1980s away from regional economic policy to local economic policy emphasizing, for example, inner urban areas and rural areas, enterprise zones and urban development corporations, increased the need for small area data to monitor and evaluate their impact. Unfortunately, such data have, for the most part, not been forthcoming. Over the same period the extent of economic intervention by local authorities increased. The information requirements for implementing policies concerned with statutory planning and industrial promotion differ from those for regenerating the local manufacturing sector or managing the local economy. Employment, land and premises data are not sufficient for implementing policies aimed at higher levels of intervention where issues of business management, technological innovation and the operation of the labour market become crucial (Breese 1986). The introduction of structure plans in 1971 provides a further example of a change in the kind of information demanded. The preparation of these plans required the integration of different data sources, such as land use and employment.

The demand for information depends on the purpose for which it is to be used and, as this alters over time with the changes in the nature of economic activity and land use, shifts in theoretical frameworks, and policy changes, it is difficult to identify in advance either precise information needs or the uses to which information will be put. There are also many cases where information is used in a variety of ways not envisaged initially. Data on household energy consumption may, for instance, be used as an indicator of wealth; while the electoral register may be used to identify local population shifts. In other cases, information may only become useful when it is integrated with another data set; for example, combining information on the employment and floorspace of manufacturing establishments in an area allows employment densities to be calculated (Ch. 5).

Government expenditure cuts and privatization programmes

Continuing reorganization of the Government Statistical Service (GSS) following implementation of the Rayner review in the early 1980s has led to significant cuts in the amount of official information collected and published. Some aspects of the debate that this deterioration in the official database has stimulated were touched on in the introduction to this chapter. An indication of the extent of these cuts is that employment in the GSS fell by over half between 1979 and 1989 (Table 1.2). The cuts to the statistical services have decreased the frequency and coverage of many official surveys and frequently increased the cost of the resulting statistics to the users. One of the clearest examples of the effect of expenditure cuts on information availability is the changes which have occurred to the Census of Employment (CoE). Between 1971 and 1978 a census was taken annually; from 1978 to 1987 it changed to a triennial census, since when a census has been taken every other year. Consequently during a period in which the British economy was undergoing a dramatic change the latest employment data for local authority areas were sometimes 5–6 years out of date by the time they became available. Of just as much concern for local and regional research was the introduction of sampling in 1984, which reduced the reliability of industry data below divisional level for small area analyses (Chs 2 and 16). Charges were also introduced in

Table 1.2 Employment in the GSS, 1979–89.
(*Source:* RSS 1990:16.)

Department	1979	1984	1989–90
Central Statistical Office	263	196	145
Inland Revenue	608	313	166
Customs and Excise	1311	983	526
Employment, etc.	1430	960	314
Environment and Transport	527	360	287
Health and Social Security	517	322	307
Office of Population Censuses and Surveys	1100	791	495
Industry and Trade	1399	1054	837
Others	1846	1472	1151
Total	9001	6451	4228

the 1980s to obtain a notice giving access to the CoE. The cost in 1990 for bona-fide research workers to acquire a notice was £55 + VAT (£27.50 for postgraduates with ESRC awards).

A key principle established by Rayner and confirmed by subsequent governmental reviews of statistical provision is that it is the needs of government which should determine what data are collected (Hoinville and Smith 1982). This has led to a reduction in the amount of detailed spatial referencing of information in official statistical sources because of a feeling in government that its needs could be met adequately by nationally based information (Ch. 18). This ignoring of the needs of other users of official statistics stems in part from the decentralized nature of the GSS and the close political control exercised over the service by individual ministries (RSS 1990). One of the consequences of the pre-eminence given to the needs of government and of the decentralization of control is that increased reliance has been placed on using data which are a by-product of administrative systems. The best-known instance of this concerns the unemployment figures. Between 1979 and 1989 this series underwent 29 downward revisions (Ch. 3). These changes not only reduced the total number officially unemployed, they also led to shifts in the relative position of different local economies. For example, in 1982 when the basis of the monthly count switched from the number registering as unemployed to the number receiving unemployment benefit, Coventry changed overnight from having the highest unemployment rate of the seven district councils in the West Midlands County to having the lowest. The reason for this is that Coventry has a high female economic activity rate and the largest group to be excluded from the change in counting procedures were married women whose husbands were employed. Among the many other official statistical series dependent on the use of administrative sources are those which present data on retail trade, small businesses and farm output.

A related policy to that of cutting expenditure has been the privatization of previously state-run businesses and the switching of some functions previously undertaken centrally, such as employment training, to the private sector. Both have consequences for the quality, availability and extent of information provision. On the one hand, not only has the private sector less obligation to make information available, but where functions have been decentralized to a number of different bodies incompatibility in the nature of the information collected and released can become a major problem. The latter is of particular concern for the provision of local labour market information with the development of the Training and Enterprise Councils (TECs and LECs – Local Enterprise Companies in Scotland), which are taking over some of the functions previously under the control of the Training Agency. Such changes are leading to a greater diversity of information sources from one part of the country to another. Some data sources consequently have a patchy coverage. On the other hand, a short-term bonus of privatization has been the provision of previously unpublished information in the share prospectuses of many of the organizations involved, such as those in some of the utility industries (Ch. 14).

Developments in information technology

These have had a revolutionary effect on methods of collecting, storing, analysing, accessing and presenting information. The clearest examples concern developments in remote sensing techniques and GIS.

The application of remote sensing techniques for analysing land use dates back to the use of aerial photographs in the mid nineteenth century, but the main developments were stimulated by military needs during the First and Second World Wars. The first complete photocoverage of Britain was undertaken by the RAF at the end of the Second World War (Rhind and Hudson 1980). The use of aerial photography for large-scale land use studies was complemented by satellite imagery with the launching by the USA of the first Landsat satellite in 1972, followed 14 years later by the launching of the first French SPOT (Satellite Pour L'Observation de la Terre) satellite (Mather 1990, and also Ch. 7). Satellite imagery is most useful for rural land use studies; its resolution (10 m for Spot 2) is insufficient for detailed analysis of urban areas. Aerial photography can provide more detailed imagery and is generally the preferred approach for land use surveys in the UK, despite the fact that ground survey, when undertaken by skilled and well-trained staff, remains the most reliable procedure for the recording of the uses to which land is put (Ch. 12). When suitably processed, remote sensing data can form an input into GIS, although the technical problems of combining 'raster' data, derived for example from scanning a photographic image, with 'vector' data, such as may be obtained from digitizing information from a map, have yet to be resolved (Mather 1990, and also Ch. 9).

The Chorley Report highlights the significance of recent developments in geographical information handling technology. In its view the development of GIS is 'the biggest step forward in the handling of geographical information since the invention of the map' (DoE 1987: 8). The term 'GIS' is used here in its generic sense to include land information systems. GIS are essentially information systems whereby the data can be related to a specific location on the earth's surface. GIS add value to the data, first, by allowing more information to be stored and faster retrieval than is possible by manual methods, and, second, by permitting different data sets relating to the same geographical areas to be integrated. It is this interlinking of different data for the same location which makes GIS 'an analytical and decision making tool fundamentally different from a paper map' (DoE 1987: 84). For example, the capability of linking different data sets has made the development of integrated forecasting easier (Ch. 13).

An excellent example of the ability of GIS to integrate different types of information is provided by the National On-line Manpower Information System (NOMIS) which combines population, migration, employment, vacancy and unemployment data for various geographical areas from EDs upwards. The system allows simple analysis of the data on-line, supports interfaces into mapping systems, and assists the supply of raw data into statistical packages (Ch. 8). A major feature of NOMIS is that it integrates a series of *national* data sets. It is much more difficult to organize a system when the inputs consist of a large number of *local* data sets which vary widely in their quality and coverage, as witnessed by the demise of the ESRC Rural Areas Database.

Despite the importance of national systems the majority of applications of GIS are in-house, used primarily to improve the handling and presentation of data. For example, many local authorities have constructed land and property information systems (Ch. 9). These systems have been made possible particularly by developments in digital mapping, which allow maps to be converted into digital form, and advances in database management systems, which allow large volumes of textual material to be stored, retrieved and processed. The principal advantages of GIS are apparent when information is treated as a corporate resource, and different departments, and sometimes also organizations, share the data. For example, the Wales Terrestrial Database Project (WALTER) has adopted a collaborative approach to data integration and management (Bracken and Higgs 1990). There are, however, both technical and institutional constraints to be overcome before such schemes can be implemented (Ch. 9). Particular care has to be taken not to infringe confidentiality or copyright restrictions.

Changes in attitudes to the commercial value of information

The increased demand for information, which the development of GIS has helped to stimulate, is one factor which has encouraged government and other holders of data to recognize that information is a resource which is a tradable commodity. However, this 'commodification' of information, i.e. 'the transformation of knowledge into information which can be exchanged, owned, manipulated, and traded in a multitude of ways among a multitude of users' (Openshaw and Goddard 1987: 1423), means that access is being determined more and more by the ability to pay. Indeed as more information is computerized, those without the necessary technology – computer and telecommunication links – and money will increasingly be denied access. Webster and Robins (1989) argue that the conventional conception of the 'information revolution' as a question of technology and technological innovation is inadequate. Rather it is better understood as a matter of differential (and unequal) access to, and control over, information resources.

The potential value of information has been recognized most clearly by the private sector. In the last 10–20 years several companies have been established to exploit the potential of packaging, repackaging and selling geographical information. Some companies, such as CACI and Pinpoint Analysis, have used the electoral register and the small area statistics from the CoP to bring together information about people's social and economic characteristics with information about where they live. Such 'geodemographics' are based on a multidimensional classification of people according to the type of areas in which they live rather than the more conventional unidimensional criteria of social class or income (Brown 1990). Data Consultancy (formerly the

Unit for Retail Planning Information) maintains databases on hypermarkets and superstores, retail warehouses, managed shopping schemes, new retail proposals, planning appeal decisions and company information (Ch. 18). The Investment Property Databank contains details of the property holdings and new investments made by most insurance companies, pension funds, property companies and financial institutions in the UK (Ch. 5). Market Location has data on 151 000 establishments in Great Britain (approximately half the manufacturing locations – see Ch. 16); while SPA Database Marketing provides a similar listing for 178 000 farms in Great Britain (Ch. 14). Some of these applications simply package data in an accessible and useful form, others undertake analyses of the data, thereby adding further value to the information. Perhaps the best examples of the latter come from various organizations which undertake local and regional forecasts of economic change. The private sector is becoming increasingly involved in spatial forecasting and small area labour market forecasts, often combining their own databases with those from official sources (Ch. 13). Unfortunately the charges for accessing most of these private-sector information bases and obtaining details of the forecasts are beyond the budgets of most research workers, as they reflect what the commercial market will withstand. Two examples will suffice. In November 1990 Market Location charged £390 per 1000 locations on their database; while the June 1990 update of *Regional Economic Prospects* by Cambridge Econometrics cost £1500.

By contrast the public sector has been far slower in exploiting the commercial potential of its vast information holdings. Despite the Tradeable Information Initiative (DTI 1986/1990) the widening in the availability of data has not been significant, largely because the Treasury rules mean that any profits go to the Treasury (Ch. 8). On the other hand, slow progress in implementing the initiative may mean that some data sets to which research workers already have access may continue, for a short time at least, to be available at a reasonable cost. Not that all government data are cheap, as witnessed, for example, by the cost of obtaining digital maps from the Ordnance Survey (OS) or special runs from the CoP or the Census of Production.

1.4 Recommendations for improving the information base

Ironically, during a period characterized by revolutionary improvements in the ability to store, manipulate, analyse and display information, the quality of 'Database UK', which was already inadequate in many respects for subnational analyses of economic activity and land use, underwent a marked deterioration, largely due to cut-backs in the collection and provision of official statistics. Concern over the quality of the information base has stimulated a wide-ranging debate which was referred to in the introduction to this chapter and is continued in many of the subsequent chapters of this book. A large number of recommendations on how the limitations of the information base may be overcome or reduced arise from this debate. Many of the proposals refer to ways of improving the provision of information, others touch on ways in which users may help themselves. Many organizations are, of course, both providers and users of information.

Improving the provision of information

Official statistics

As central government is the main provider of information the essential need is to improve the quality, availability and extent of official statistics. The crucial change required is that government becomes more responsive to the needs of the wide community of users of the information it collects. To meet the needs of users, including Parliament, local authorities, industry and commerce, academia, the media and the general public, would mean reversing the doctrine, promulgated ever since the Rayner review, that only the needs of government should determine what data are collected. The so-called 'savings' that have resulted from the implementation of this policy are apparent rather than real, because they ignore the hidden costs incurred by users allocating additional resources to collect their own data and the costs of making poor decisions in the absence of good-quality suitable information. The Rayner principle also belies the requirement that systematic and consistent information 'must be available if people are to be enabled to participate fully in a democratic society' (SSF 1989:1).

A wide range of recommendations has been made to improve the *quality* of official statistics. The Working Party of the RSS (RSS 1990), for example, make four main proposals for reversing the decline in public confidence and improving the quality of the statistical product. They are for:

1. Greater centalization of the major activities of data definition, collection, processing, primary analysis and publication;
2. The establishment of an advisory National Statistical Commission;
3. Legislation in the form of a United Kingdom Statistics Act; and
4. The setting-up of a research unit to strengthen evaluation and method-ological research.

These suggestions are largely concerned with improving the organizational and operational environment of the GSS. Others are more specific. For instance, to avoid dependence of some statistical series on administrative procedures, the Working Party of the RSS recommends that they are supplemented by surveys, as already occurs in many countries. This is especially important if the differences between the administrative source and the required estimates are changing over time, as has happened, for example, with the unemployment count (Ch. 3). It would also be very helpful if government departments were to adopt a comparable set of statistical units and definitions. A step in the right direction is the proposal to establish a common register of businesses. At present one of the reasons why differences occur in the findings of the CoE and the Census of Production is that different registers are used (Ch. 16). Pressures to improve and standardize statistics and produce new series are also coming from the EC, as witnessed by the production of regional gross domestic product (GDP) figures when the UK joined the Community (Ch. 17).

The proposals mentioned so far apply generally to all official statistics. One essential proposal for improving the quality of local and regional information concerns the system adopted for locational referencing. A common problem at present is the incompatibility in spatial units used for different data sets and the general lack of spatially disaggregated data. Ideally data should be locationally referenced so that a high degree of aggregational flexibility is possible. This means using the smallest practical basic spatial unit. The Chorley Report recommended that all geographical information, including remotely sensed data, relating to the land areas of the UK should be referenced directly or indirectly to the National Grid or Irish Grid as appropriate. The preferred bases for holding and/or releasing socio-economic data, according to Chorley, should be addresses and unit postcodes, which should themselves be grid referenced (DoE 1987). The smaller the basic spatial unit the greater the possibility of aggregating data accurately to other larger spatial units, although, if the units do not nest exactly, there will be incorrect assignments along the boundaries. For example, postcodes do not aggregate precisely to local authority wards and this has caused the misallocation of some CoE data units since they started using postcodes for locational referencing (Ch. 16). The standardization of spatial units would greatly ease the integration of different data sets and would enable some of the enormous potential provided by GIS to be realized. However, the support of Chorley's recommendations by the government (DoE 1988) is only advisory. It is left to users to define their own standards for GIS (Openshaw 1990).

A further way of improving the official statistical base is to increase the *availability* of the vast amount of unpublished information which central government collects. The SSF (1989) argues that all statistics and data sets collected with public funds should be in the public domain and available, subject to confidentiality constraints, for analysis by interested parties from the National Data Archive (i.e. the ESRC Data Archive at the University of Essex) and related bodies. These points also apply to the increasingly important data sets collated by the EC and other international bodies of which the UK is a member. The SSF further proposes that there should be no charge for access to data collected at the taxpayers' expense, though it recognizes that there may be a need to charge for the costs of extracting, packaging and documenting data sets for specific enquiries. These are important principles which need to be debated widely. However, the trend in the UK, at least as far as charging is concerned, is moving in the opposite direction, towards what the market will pay. This contrasts with, for example, the massive and cheaply available data outpourings of the USA Department of Commerce (Ch. 8). Where charges are introduced or increased it is important that a differential pricing system is operated, such as already occurs with NOMIS (Ch. 8), so as not to exclude *bona-fide* research workers.

To improve the availability of unpublished data better information is also needed about which statistics are collected. A properly documented register of official data sources, which is kept up to date and maintained in an on-line computerized form, is required. At the commencement of the Tradeable Information Initiative the government published a list of some of the sources available from different departments, but this list has not been updated since a second edition was published in 1988 (DTI 1988). An on-line register of information sources should also include material about published sources currently included in the *Guide to Official Statistics* (CSO 1990) and *Guide to Northern Ireland Statistics* (Policy Planning and Research Unit (PPRU) 1988), which quickly become out of date because of long gaps between editions.

A frequent reason given by government for restricting access to data is that to release it may infringe the rules of confidentiality. It is important in a democratic society that the rights of individuals and businesses to privacy are adhered to and this is reflected in the Data Protection Act 1984. The potential to store information on individuals on computer databases which may then be linked together have raised fears of 'big brother'. Unfortunately the rules are frequently used to prevent access to data which are not confidential or only of historic interest. Often data can be made anonymous simply by removing names and addresses. Social scientists can learn much of value from such data sets and their availability can save resources by avoiding duplication of data collection. For example, many social surveys would be largely unnecessary if the OPCS would release anonymized local, regional and national samples from the CoP. Maintaining the anonymity of businesses by removing their names and addresses from data sets would be more difficult in local and regional analyses because the fewer numbers involved would make identification easier. However, again many resources would be saved and much of value would be learnt if the Department of Employment (DE) would release the CoE tapes to *bona-fide* research workers. Confidentiality (if it is still thought necessary to apply to employment figures several years out of date) could be maintained by research workers agreeing not to publish data which could be attributable to individual businesses, in the same way that those given access to the local area CoE data already promise. Such a change would require an alteration to the Statistics of Trade Act 1947 (and the Statistics of Trade Act (Northern Ireland) 1949), which empowers the DE to collect the information, and the Employment and Training Act 1973, which determines the release of unpublished CoE information. Given the political will legislative changes are possible. For example, pressures for the public to have access to HM Land Registry data in England and Wales, similar to that which they already have in Northern Ireland and Scotland, led to appropriate clauses being included in the 1988 Land Registration Act (Ch. 6). A further initiative which would greatly help research workers would be the passing of a Freedom of Information Act, such as already operates in the USA. Enshrining in legislation the principle of the 'right to know' non-confidential information would help overcome the tendency for excessive secrecy by government.

Other proposals concern improving the *extent* of the official database, including increasing the frequency of surveys, such as the CoE (recently every 2 or 3 years) and CoP (currently every 10 years), and collecting additional information. The following chapters highlight the generally poor state and numerous gaps in the information base for local and regional studies of economic activity and land use in the UK. The official information shopping list of the local and regional research worker is a long one. To mention just a few examples of the gaps: self-employment is measured only by a sampled household survey (*Labour Force Survey (LFS)*) in between the dates of the CoP (Ch. 2) – this is particularly serious in economic sectors, such as fishing (Ch. 14), dominated by the self-employed; output and investment statistics by sector (particularly the private services sector) and for local areas are almost non-existent (Ch. 4); commercial floorspace data are no longer published for England (Ch. 5); the coverage of urban land use statistics is inadequate and patchy (Ch. 6); no comprehensive records exist of patterns of land ownership and occupation (Ch. 7); the SIC needs to be overhauled to provide substantially more information on services (Ch. 17); and the Census of Production has lost interest in interregional transport cost patterns (Ch. 19).

Unofficial statistics

Central government is the prime source of statistical information, but a growing proportion is coming from other sources (Key Note Publications Ltd 1990; Mort 1990). A characteristic of the post-industrial economy is the growth in the number of companies whose business is concerned with obtaining, packaging and selling information. The privatization of previously state-run businesses and the development of organizations, such as the TECs and LECs, has shifted the origin of important sources of data from the public to the private sector. Local authorities, and other bodies, such as chambers of commerce, industrial organizations and trade unions, remain significant other sources of unofficial statistics. Given the disparate nature of this group, it is difficult to make realistic proposals for improving the quality, availability and extent of information they provide. However, encouraging the setting of minimum standards for the collection, presentation and dissemination of information and extolling the virtues of best practice are reasonable targets. It would also be helpful for academic bodies to negotiate differential charging policies with private-sector organizations selling information for the provision of data for non-profit-making research purposes. Furthermore, there is an urgent need to ensure that these organizations keep archive copies of their data files at regular intervals. This is particularly important where computer databases are continuously updated and old data are destroyed. Out-of-date information has little commercial value, but can be invaluable for research workers wishing to examine trends. Once old data files have no commercial value a system should be established for them to be deposited at the National Data Archive.

Improving the use of information

The generally poor and deteriorating state of the information base for local and regional studies of economic activity and land use means that research workers and policy-makers need to examine how they can make the best possible use of the information that does exist and devise means for improving their own procedures for collecting, handling and analysing information. There is much that users of information can do to improve their use of existing sources. The starting-point is to ensure that they are aware of the nature and limitations of the main data sources and how these have changed over time and that they keep up to date with new developments in the information base. This book aims to do the first, while periodicals such as *Statistical News*, *Employment Gazette*, *Business Briefing*, *BURISA Newsletter*, *Business Information Review*, *Mapping Awareness* and *Association for Geographic Information Yearbook* should help with the second. Establishing and maintaining good relationships with the 'guardians' or 'gatekeepers' of unpublished data is another important strategy. This process has been made easier over recent years with the growth in the number of academics undertaking research projects and consultancy work for central and local government. Increasingly strong links with the private sector are also apparent (Breheny 1989). One practical step that research workers can take where there are grounds for concern about the quality of the statistics is, if possible, to verify the data against other sources. This procedure often brings to light additional information, and discrepancies may be resolvable through further checks. This strategy is recommended, for example, when compiling lists of businesses in an area (Chs 11 and 16). Such checks also encourage a cautious attitude in the nature of the conclusions which may be drawn from analyses using these sources.

Often research workers and policy-makers have no choice but to collect their own information or to commission special surveys. However, undertaking surveys of businesses and land uses is a skilled and often expensive procedure which should only be attempted after all other sources of information have been exhausted (Chs 11 and 12). In the case of business surveys, the lack of guidance in the literature on how to obtain information means that there is an urgent need for a greater priority to be given to experimenting with different survey methods and reporting on specific ways which have increased response rates, speeded up replies, obtained clearer answers, improved the flow of interviews and so on. The cut-backs in official surveys have increased the need for research workers to undertake their own surveys, but there is a danger of over-saturation, particularly of larger businesses (Ch. 11). There is a case for making the data derived from surveys more widely available by, for example, depositing the data in the ESRC Data

Archive, or collaborating with other organizations in the collection of data of common interest. Among the advantages of data pooling are the reduction in the number of requests for information, opportunities to increase survey coverage, access to more up-to-date information, and the development of closer co-operation and understanding between organizations (Foley 1990). However, individual organizations have less control over what information is collected and the issue of data confidentiality needs to be sorted out before commencing a data-pooling arrangement.

The last decade has witnessed marked improvements in the ability of research workers and policy-makers to handle geographical data. With technological developments and an expanding market the cost of GIS hardware has fallen and there are now a bewildering variety of systems from which to choose (Pearce 1990). There is also a significant amount of 'off the shelf' software available. However, the technical sophistication of the hardware has progressed faster than has the development of *appropriate* software. Recent studies have shown that the critical factor in successful applications is the way in which computers and computing are used in organizations, i.e. the design of the 'orgware' as it has come to be called (Batty 1988). More worryingly though, the pace of technological development has far exceeded that of the quality of the data. Such technical advances have also increased the potential for inappropriate use of data, particularly where the data are themselves available on-line or on computer disc. If local and regional analyses and planning are to be effective relatively more attention needs to be given to the usability and quality of the underlying data than to the sophistication of the technology used to handle the information. In some applications, such as land use data, the acquisition of the data can account for at least 80 per cent of the cost of initiating a spatial information system (Ch. 9). One consequence of these different rates of development is that most applications of GIS have been concerned with relatively low-level routine data processing rather than with the forecasting, design and decision-making functions which are essential for public policy-making. Unless more consideration is given to the organizational context of GIS applications and improving the quality of the data sets used and their potential for integration, the same fate may lay in store for GIS as befell the large-scale models of the 1960s and 1970s (Ch. 10). Decision-makers are not geographers and they will see GIS as no more than another tool. The basic problem is to get them to use this tool (Grimshaw 1989). If more attention was given to integrating different data sets the value of GIS to decision-makers would be more apparent. One development which would help local policy-makers is the linking of local data sets to NOMIS (Ch. 10).

Exciting though developments in GIS are, the principal new opportunities, according to Openshaw and Goddard (1987), increasingly lie elsewhere in the application of techniques of geographical analysis to add value to information by repackaging it and make it more useful and usable. They note that (1987: 1434):

two decades of planners flirting with computer models for various spatial subsystems has resulted in few working examples. If anything, the gap between the academic modeller of urban and regional systems and the practising planner, and more recently, the market researcher, is still immense.

There is a need for more research into how decision-makers can make better use of spatial information (Chorley 1988). Clarke (1990) argues that to demonstrate how conventional regional science methods can be adapted for use in business, commerce and public-sector planning rests not in developing new methods, but rather in gaining a better understanding of the needs of potential users and the environment in which they operate.

One possibility for overcoming the general lack of small area data is to adapt quantitative forecasts made at one spatial scale to smaller scales, often with the addition of subjective judgements about the likely trend in employment (Ch. 13). Another is to adopt an *a priori* based accounting framework. An example of this approach, which synthesizes a wide range of data, is described in Chapter 4 to assess how costs and profits vary between existing establishments. A further possibility is to produce synthetic data sets using missing data estimation procedures and microsimulation. This technique has been used to generate individual and household incomes (Birkin and Clarke 1989) and to generate updated estimates of the characteristics of the population living in small areas (Clarke 1990). The main problem with such exercises is the lack of base data to validate the results, as the very reason for the procedure is to generate data which do not exist.

1.5 Conclusion

There is a bewildering variety of official and unofficial data sources available to the local and regional research worker. However, most are limited by their spatial coverage, level of spatial disaggregation, frequency and completeness. One function of this book is to provide a guide through this minefield. A reliable, accurate, up-to-date and accessible information base, which is available free (or at a minimum charge), is a prerequisite for a full understanding of the nature and reasons for spatial variations in economic activity and land use. The poor and deteriorating state of the existing database affects not only the ability of academics to analyse and explain these important geographical patterns, but also the capacity of local and central government planners, market researchers and business managers to monitor local and regional changes in economic activity and land use. This in turn affects the potential for local and central government to develop appropriate policies and to monitor the impacts of these policies. Inadequate information impedes effective decision-making and can lead to the inefficient allocation of resources. It needs to be remembered, however, that as Hasluck (1989: 10) notes:

Even if a high quality detailed information base was to exist this would not 'solve' the problems of local decision-makers, because 'the facts do not speak for themselves' but have to be interpreted and analysed. This adds both a theoretical and political dimension to intelligence gathering.

It is important that discussions of data requirements are not divorced from the theoretical and political frameworks used to inform the analysis of the problems being examined.

Pressures are building up to improve the organization of the GSS and the provision of official statistics, but with a few notable exceptions the perspectives of research workers and policy-makers concerned with the quality of local and regional information have not figured strongly in the discussions. It is sad that issues involving local and regional data were hardly touched on in any of the verbal or written contributions to the widely publicized RSS debate on 'Public confidence in the integrity and validity of official statistics' (Hibbert 1990). It is hoped that this book will contribute to correcting the balance.

Note: Further examples of recent changes in the information base are referred to in the Editor's preface (p ix).

Acknowledgements

The author is indebted to Anne Green, Philip Kivell, Andy Pratt and Doug Watts for their helpful comments on an earlier draft of this chapter.

References

Barr C 1990 Mapping the changing face of Britain. *Geographical Magazine* Oct: 44–7

Batty M 1988 Editorial: the cult of information. *Environment and Planning B: Planning and Design* **15**: 375–82

Birkin M, Clarke M 1990 The generation of individual and household incomes at the small area level using synthesis. *Regional Studies* **23**: 535–48

Bracken I, Higgs G 1990 The role of GIS in data integration for rural environments. *Mapping Awareness* **4**(8): 51–6

Breese R 1986 Information requirements for local economic intervention. *BURISA Newsletter* **72**: 10–13

Breheny M J 1989 Chalkface to coalface: a review of the academic–practice interface. *Environment and Planning B: Planning and Design* **16**: 451–68

Brown P J N 1990 *Geodemographics: A Review of Recent Developments and Emerging Issues – Towards an RRL Research Data* Regional Research Laboratory Initiative Discussion Paper 5, Department of Town and Regional Planning, University of Sheffield

CSO 1979 *Standard Industrial Classification Revised 1980* HMSO, London

CSO 1980 *Indexes to the Standard Industrial Classification Revised 1980* HMSO, London

CSO 1990 *Guide to Official Statistics* No 5 HMSO, London

Channel 4 1989 *Dispatches: Is the Government Falsifying Official Figures?* 18 Jan

Chorley R 1988 Some reflections on the handling of geographic information. *International Journal of Geographical Information Systems* 2: 3–9

Clarke M 1990 Regional science and industry: from consultancy to technology transfer. *Environment and Planning B: Planning and Design* 17: 257–68

Coppock J T 1978 Land use. In Maunder W F (ed.) *Reviews of UK Statistical Sources* vol VIII Pergamon, Oxford pp 3–101

DoE 1987 *Handling Geographic Information, Report of the Committee of Enquiry Chaired by Lord Chorley* HMSO, London

DoE 1988 *Handling Geographic Information. The Government's Response to the Report of the Committee of Enquiry Chaired by Lord Chorley* HMSO, London

DTI 1986/1990 *Government-held Tradeable Information: Guidelines for Government Departments in Dealing with the Private Sector* DTI, London

DTI 1988 *Government Data Resources* DTI, London

Employment Department Group and Office of Population Censuses and Surveys 1990 *Standard Occupational Classification* vol I HMSO, London

England J R, Hudson K I, Masters R J, Powell K S, Shortridge J (eds) 1985 *Information Systems for Policy Planning in Local Government* Longman, Harlow, Essex

Foley P D 1990 Growth in the collection of labour market information. *Regional Studies* 24: 367–71

Grimshaw D J 1989 Geographical information systems: a tool for business and industry? *International Journal of Information Management* 9: 119–26

Hasluck C 1989 Local labour markets: information needs and issues. *Local Government Studies* 15: 8–18

Healey M J (ed.) 1983 *Urban and Regional Industrial Research: The Changing UK Data Base* Geo Books, Norwich

Healey M J, Ilbery B W 1990 *Location and Change: Perspectives on Economic Geography* Oxford University Press, Oxford

Henneberry J 1988 Conflict in the industrial property market. *Town Planning Review* 59: 241–62

Hibbert J 1990 Public confidence in the integrity and validity of official statistics. *Journal of the Royal Statistical Society Series A* 153: 123–50

Hoinville G, Smith T M F 1982 The Rayner review of government statistical services. *Journal of the Royal Statistical Society Series A* 145: 195–207

Hunter P 1989 *Computer Weekly* March: 26

Irvine J, Miles I, Evans J 1979 Introduction: demystifying social statistics. In Irvine J, Miles I, Evans J (eds) *Demystifying Social Statistics* Pluto Press, London pp 1–7

Key Note Publications Ltd 1990 *The Source Book* Key Note Publications Ltd, Hampton, Middlesex

Mather P M 1990 Earth observation: a resource for GIS. *Mapping Awareness* 4(5): 17–18

Miles I 1990 *Mapping and Measuring the Information Economy* Library and Information Research Report 77, British Library

Mort D 1990 *Sources of Unofficial UK Statistics* The University of Warwick Business Information Service, Gower Press, Aldershot

NLUC 1975 *National Land Use Classification* HMSO, London

Newby H 1988 Geographical information systems: introduction. *ESRC Newsletter* 63: 6–8

OPCS 1980 *Classification of Occupations, and Coding Index* HMSO, London

Openshaw S 1987 Spatial units and locational referencing. In DoE *Handling*

Geographic Information: Report of the Committee of Enquiry Chaired by Lord Chorley HMSO, London pp 162–71

Openshaw S 1990 Spatial referencing for the user in the 1990s. In Foster M J, Shand P J (eds) *The Association for Geographic Information Yearbook* Taylor and Francis, London pp 76–84

Openshaw S, Goddard J 1987 Some implications of the commodification of information and the emerging information economy for applied geographical analysis in the United Kingdom. *Environment and Planning A* **19**: 1423–39

Owen T 1988 *Mind your Local Business* Eurofi, Newbury

Pearce N 1990 Taking the risk out of system selection. *Mapping Awareness* **4**(1): 33–7

PPRU 1988 *A Guide to Northern Ireland Statistics* PPRU Occasional Paper 16, Department of Finance and Personnel, Stormont

Rayner, Sir D 1980 *Review of Government Statistical Services: Report to the Prime Minister* CSO, London

Rhind D, Hudson R 1980 *Land Use* Methuen, London

RSS 1990 *Official Statistics: Counting with Confidence. The Report of a Working Party on Official Statistics in the UK* RSS, London

SSF 1989 *Official Statistics: A Statement of Principle for their Collection and Use* SSF, London

Waterhouse R 1989a Anxiety grows over integrity of statistics. *The Independent* 9 Oct

Waterhouse R 1989b Wide support for impartial check on official statistics. *The Independent* 10 Oct

Webster F, Robins K 1989 Plan and control: towards a cultural history of the information society. *Theory and Society* **18**: 323–51

Worrall L 1990 Improving the quality of 'Database UK'. *BURISA Newsletter* **93**: 3–8

Part One

Measuring economic activity and land use

2

Employment

Alan Townsend

2.1 Introduction – why employment?

Most reports on local or regional economies are studded with employment statistics. This is true whether one takes the example of county council plans, district council analyses or consultants' reports on local economic development. Tables and graphs of the area's structure of employment, the way it has been changing, and outline forecasts of future change, are very much the order of the day.

Why should this be the case? After all, employment is not necessarily the most direct measure of economic activity or of land use in an area. The value of output per person employed varies immensely between subsectors of the economy, and so too does the number of hectares of land needed to support one job. On both dimensions, it is salutary to compare, say, a petrochemical works with a multi-storey hotel. In view of such difficulties, the thought must occur fairly early in this chapter that the common use of employment data occurs *faute de mieux* – because there is usually nothing better on economic structures below the regional level of analysis; filling such a vacuum, government instituted an annual Census of Employment (CoE) from 1971; while this became less frequent after 1978, and sampling methods were introduced from 1984 onwards, its geographical versatility and access have been greatly improved as one may see below.

The central role of employment

Added to these pragmatic reasons, however, is the effect of the national reduction of employment from 1979 to 1983 in making unemployment (Ch. 3), and employment recovery, leading issues in many different kinds of locality. Employment was a leading issue in its own right for most of the

1980s in most kinds of places. This was principally because of the economic, social and political costs of large-scale unemployment, including their effects on an area's overall income and expenditure. Government measures for regional 'assisted areas' and inner cities were increasingly supplemented by local authority economic development policies and by schemes to help the unemployed from voluntary groups, the churches and local businessmen. All this required monitoring; and while progress seemed pitifully slow in many areas of the North, the recovery of employment levels in the South from 1986 to 1989 was relatively rapid and raised renewed questions over labour shortages and the threat of development to the environment in the South East. A direct *reversal* of these trends in 1990 to 1991 was equally critical.

The central role of employment in the activity systems of subregions

The situation of congestion in the South East is an adequate reminder of the central role of employment in the whole development process. In any computerized model of the process of subregional town and country planning, it enjoys a central box in the analysis and forecasting of the whole activity system (Ch. 13). The relationships of employment to other variables have, however, changed somewhat and it is essential to review separately the key aspects.

The dynamics of population change

Areas of employment growth generally attract an increase of population, and vice versa. The precise arithmetical relationship was always complex, however, because of the varying size of the dependent (non-working) population of children and pensioners. The 1980s have demonstrated two specific changes; firstly, an increasing lag in the migrational response to employment growth due to the shortage of affordable housing in the South East; and secondly, an increase in the proportion of migration to retirement areas on the coast and in outer rural areas. In the last case, population increase is generating employment, in the service sector, rather than responding to job growth.

The role of employment in transport planning

It is peak-hour work journeys which place the greatest strain on an area's travel facilities, and it is therefore no accident that it is the country's principal office centres which continue to support railway commuting systems. It is a standard part of subregional planning to classify journeys by origin and destination, mode of travel and purpose (Ch. 19). Along with shopping centres, it is the principal zones of work which are found to have the strongest mathematical impact on trip generation. The location of employment is frequently surveyed by local planners; thus one of the principal *ad hoc* sources of local- or establishment-level employment figures is that of local authority surveys of transport.

2.2 Concepts and definitions

The definition of who is at work, and the best means of disaggregating the total workforce, are open to increasing conceptual debate (Department of Employment (DE) January 1986). Not only are there survey problems, from people working at or from home or from unrecorded 'double-jobbing' and work in the black economy, but the growth of a wide variety of part-time jobs makes for an arbitrary element at the margins in the definition of 'employment'. It may be that 'core' jobs in the economy deserve separate recognition from 'peripheral' ones.

Table 2.1 Economic and employment status of the GB population, spring (thousands). (*Source:* DE, April 1991.)

	1984	1990	Change 1984–90
All people aged 16 and over resident in private households	42 675	43 838	+ 1 163
Economic activity rates (%)	62.1	64.0	
Men	75.9	75.5	
Married women	52.0	58.6	
Non-married women	44.8	44.7	
Economically active of which:			
Unemployed	3 094	1 869	− 1 225
Employed, including	23 387	26 168	+ 2 781
Full-time employees, incl.	16 076	17 195	+ 1 119
Men	11 111	11 345	+ 234
Part-time employees, incl.	4 378	5 055	+ 677
Married women	3 260	3 581	+ 321
Non-married women	692	879	+ 187
Full-time self-employed	2 168	2 913	+ 745
Part-time self-employed	450	557	+ 107
On government employment and training schemes	315	447	+ 132
Second jobs as employees	452	726	+ 273
Second jobs as self-employed	251	351	+ 100

Note: The LFS is also extended to Northern Ireland, by the Department of Economic Development in Northern Ireland, in order to provide estimates for the whole UK, as required by the EC. However, the questionnaires for Northern Ireland are slightly different from those used in GB and the source therefore refers to GB only. Note that data are normally provided for Northern Ireland to match those of the main CoE (Table 2.2).

Economic status

Table 2.1 adopts the most precise definition of the GB population which is available at the date of writing. The annual *Labour Force Survey (LFS)* is based on interviews with about 60 000 private households throughout Great Britain. It includes in the 'economically active' all people who are in employment or unemployed, and defines the latter on the International Labour Office (ILO)/Organization for Economic Co-operation and Development (OECD) definition (see Ch. 3); those 'in employment were people aged 16 and over who did some paid work in the reference week (whether employed or self-employed); those who had a job that they were temporarily away from (on holiday, for example); and those on work-related government employment and training programmes' (DE April 1989a: 196). The inclusion of the latter was at first regarded as controversial.

There have been many amendments and adjustments to the series represented in Table 2.1. What is important for present purposes is the *diversification* in the nature of employment nationally. The fall in unemployment, 1984–90, was accompanied by a greater increase in the number employed over these 6 years. This increase was shared only marginally by the traditional field of male full-time employment. Looking at economic activity rates, the proportion of adult males at work fell to the level of three-quarters. By contrast the proportion of married women at work continued to grow further beyond the half-way level.

Employment status

The explanation becomes clearer in light of the types of work which people were undertaking. Part-time work, defined here by the respondent's own assessment, expanded at twice the rate of full-time work, due to its attraction for previously economically inactive married women. Self-employment increased by nearly a third, and there were major increases in government trainees and second job holders. In all (Table 2.1) by 1990 full-time male employees made up only 43.4 per cent of the employed, while part-time employees contributed 19.3 per cent and the self-employed no less than 13.3 per cent. The changes experienced from 1984 to 1990 represented well-established trends; forecasts to the year 2000 suggest further increase in female activity rates (DE April 1989b).

All these data are *published in outline for standard regions*. Most of the data which are available for local areas below the regional level arise from the CoE, which is confined strictly to *employees* (although it may include some double job-holders twice). Before turning to that key source, however, it is important to make a distinction between the CoE, which is undertaken by workplace and industry, and the Census of Population (CoP), 1971, 1981 and 1991, with its coverage of employment by industry and occupation, principally by residence, but also with 10 per cent samples by workplace. It should be noted that results for the whole country for 1981 showed inevitable discrepancies between the two censuses. Not only are there obvious effects from the exclusion of the self-employed from the CoE, but there are further differences arising from self-reporting and sampling in the CoP.

Occupation

The prospect of the 1991 **Census of Population** revives the potential of occupational analysis for economic and other purposes. 'Occupations' comprise headings such as 'accountant', 'cleaner' or 'labourer', as opposed to 'industries' such as 'forestry', 'textiles' or 'education'. All industries may include occupations, such as accountant, cleaner or labourer, in proportions which vary not only by industry, but also by place within those industries. Only the CoP cross-tabulates the two variables, and then only for regions and metropolitan areas (Office of Population Censuses and Surveys (OPCS) 1984). No local or regional data for occupations were provided between 1981 and 1991, except for notified vacancies (see Ch. 3). However, taking the 1981 Census, full details for 549 occupations (and intermediate groupings) were made available, cross-tabulated by employment status. These are provided for local wards and districts in OPCS computer tapes which are accessible in the National On-line Manpower Information System (NOMIS, see Ch. 8). The significance of this is, for instance, that the potential of an area for new investment, say an incoming new firm, lies arguably in the area's occupational (rather than industrial) mix.

Social class and socio-economic groups

As an extension of this point, there is now renewed interest in the role of socio-economic mix in influencing the economic dynamics of different areas. There was a period in the 1960s and early 1970s when the location of mobile manufacturing plants appeared to depend on the availability of an undifferentiated mass of semi-skilled but trainable manual workers, recruited into large mass production plants described as 'Fordist' (after the pioneer of mass car production, Henry Ford). In today's era of 'flexible specialization', the onus appears to lie more on the existence of flexible skills for use in smaller workforces, often run by subcontracting firms in the service sector. It has even been suggested that areas of the UK may be divided between those of 'enterprise culture' and 'dependency culture' according to their occupational mix, and certainly that multi-plant firms have effected a 'spatial division of labour' through systematically exploiting the different skills of different areas for the different functions of routine manufacture, skilled craft work, R&D, head offices and routine 'back-' offices (Massey 1984).

These processes affect and reflect the different structures of different places; indeed, in the view of Cooke (1989), working groups in the political economy of different 'localities' interact in different ways. Arguably the best descriptors for the evaluation of an area's bias for instance towards professional or manual resources for work lies in the *Classification of*

Occupations, (OPCS 1980), which is manipulated by the census coders to classify heads of households into the social classes and socio-economic groups. As an example, OPCS and Registrar-General for Scotland (1984) show that in 1981, the Easington district of Co. Durham had 1.0 per cent of its resident households in Social Class I, compared with 7.3 per cent in Edinburgh. (Social class and socio-economic groups are available on NOMIS.) The evolution of classification systems for the 1990s, including the 1991 CoP, is considered in DE (April 1988). By contrast, the more familiar process of industrial classification is concerned not with the level or quality of jobs, but with the end product or service provided.

2.3 The Census of Employment

Any published table represents only a minute fraction of the data that are readily available (subject to the confidentiality restrictions of the Statistics of Trade Act 1947) by application to the DE, Watford or by use of NOMIS. This census provides a continuous time series for the all-sector total of employees in employment from 1971: the principal discontinuity lies in the classification of employees in the replacement of the Standard Industrial Classification (SIC) of 1968 by that of 1980, but others have occurred below the regional level (see below). A data-bridge is provided by the availability of 1981 results on both classifications, which may be compared (Central Statistical Office (CSO) 1979). There is also a useful index to the 1980 classification (CSO 1981).

Origins and frequency

The 1971 CoE replaced figures based on physical counts of individual workers' National Insurance cards. It is a comprehensive survey of employees except that HM Forces and employees in private domestic service are excluded, and the government departments responsible for agriculture provide the figures for that sector (see Ch. 14). Except in agriculture, part-time employees are defined throughout as those undertaking less than 30 hours per week. The distinction between full-time and part-time workers has proved of increasing significance in national and spatial terms (Townsend 1986).

An annual series for June was broken after 1978 when the Rayner review of the Government Statistical Service (GSS) recommended heavy economies. To reduce processing costs and the burden to employers of form-filling, it was proposed both to reduce the frequency of the operation and to conduct it on a sample basis. The census therefore took place after 1978 in only the years 1981, 1984, 1987 and 1989, in September in each case, to be followed by one in 1991 which will provide some comparison with the CoP. Processing was also spread out over a longer period than before, leading to delays in

Table 2.2 Dates of publication of the CoE (employees) in the *Employment Gazette*

		Date published
Initial article: 'New series of annual employment statistics'		Jan. 1973
Results for 1971 and 1972	Great Britain United Kingdom Regions	Aug. 1973 Sept. 1973 Oct. 1973
Results for 1973	Great Britain Regions United Kingdom	May 1974 June 1974 Aug. 1974
Results for 1974	Great Britain United Kingdom and regions	June 1975 July 1975
Results for 1975	Great Britain Regions United Kingdom	July 1976 Aug. 1976 Sept. 1976
Results for 1976	Great Britain United Kingdom and regions	Nov. 1977 Dec. 1977
Results for 1977	Great Britain United Kingdom and regions	Feb. 1980 Mar. 1980
Results for 1978	Great Britain United Kingdom and regions	Feb. 1981 Mar. 1981
Results for 1981	Great Britain, United Kingdom and regions SIC 80 (final)	Dec. 1983* (Supplement)
Results for 1984	Great Britain and regions	Jan. 1987
Article: '1984 Census of Employment; limitations on results'	United Kingdom	Aug. 1987 Sept. 1987
Results for 1987	United Kingdom and regions Great Britain	Oct. 1989 Nov. 1989
Results for 1989	United Kingdom and regions	April 1991

*Data on the SIC 68 were available on application to the DE (Statistics Division).

the publishing dates for regional results (Table 2.2), which seriously affected the published use of the material.

Sampling and coding

The census has always been conducted through a register of *employers*, originating from the Inland Revenue's operation of the PAYE system through local establishments. A full census was conducted in 1981 and sampling procedures were piloted. The 1984 Census introduced the use of an average sample fraction of 1 in 7 of employers understood to have less than 25 employees. However, 'the size indicator in many cases proved to be misleading . . . with the result that the figures for some areas may have had large units missing, or employment estimates inflated because of a large unit appearing in the sample' (DE August 1987: 409). Non-response itself was a small problem, with only 3 per cent of forms not returned in 1984 and 1987, and the DE (October 1989a) report improved sampling accuracy for 1987. Major changes by county, 1981–87, are mapped in Champion and Townsend (1990).

To some extent, comparisons over time at industry level are inevitably subject to some additional problems. When a change of business activity is notified by an employer, the industrial classification is amended accordingly. However, a change in the balance between one activity and another may occur only gradually in his premises: yet, when a census coder correctly classifies a factory of, say, 1000 workers in one heading to another, it appears to the unwary that one industry has lost 1000 employees and another has gained that number. In addition, retrospective revisions include incorrect industry coding in the 1984 Census (as reported by DE October 1989a).

Access and availability

In the past, these data have been available free to authorized (mainly academic and local authority) users by post. As from the 1984 Census, however, they have been subject to the new government rules over charging for data. Thus, while Appendix 2.1, provided by the DE, lists a wide range of data as being available on application, it must be remembered that the bill for an apparently simple data set can run into hundreds of pounds. Access to detailed results requires application to the Watford address for a ministerial notice (£55 + VAT), which sets out conditions for the public disclosure of employment statistics (£27.50 for postgraduates with ESRC awards). This procedure has been required by legislation of 1973 to prevent the release of data which might indicate the number of workers at individual establishments, which remains confidential. It is a prerequisite for full access to the CoE on NOMIS (Ch. 8), which now includes numbers of establishments in each size-band as well as the total of employees in each area.

Geographical areas

The fuller development of computer systems has led to a revolution in the application of CoE statistics. A traditional bone of contention among geographers and planners was the lack of comparability between labour market information based on employment office areas on the one hand and data of the OPCS, based on local administrative areas on the other.

The employment office areas (now normally 'jobcentre' areas) represent functionally designed catchment areas of individual offices and in general are well suited to their purpose. They could be compared with local authority boundary areas only where favourable coincidences existed or with local knowledge. Local authority areas remain generally less suitable for labour market work – present district areas often represented only a crude amalgamation of the smaller pre-1974 districts. However, they do represent areas of policy interest, as in the rapid growth of local authority economic development policies, and the only readily available units for some statistical series with which comparison is needed. Added to these problems over time were the opening and closure of employment offices, a problem met in NOMIS by establishing a set of standardized ('amalgamated') office areas to guarantee time series analysis.

NOMIS provided the first reconciliation of CoP data to employment office areas by allocating wards to the digitized boundaries of office areas. That was before, however, a fundamental change occurred in the DE Group's geographical system of statistics in 1984. The new system adopted the local authority ward as the basic 'building block' of statistics, to which employers' and unemployed claimants' addresses are allocated by reference to postcode data. From this building block, NOMIS is able rapidly to sum data for a wide range of standard units, Parliamentary constituencies, districts, travel-to-work areas (TTWAs), local education authorities (LEAs), DE administrative areas, counties and regions.

As a result of these and other changes in the basic building block there are several discontinuities in census data below the regional level. Up to 1976, each establishment unit which supplied data was given a jobcentre code direct. In 1978 and 1981, the postcode sectors (from units' addresses) were assigned to jobcentres. From 1984, however, the local authority ward was derived from the postcode. To improve continuity, a ward-based set of data for larger 1981 areas was produced, including allowances for non-response and undetected duplication. The DE no longer supports the use of jobcentre-based comparisons using 1981, 1984 and 1987 data. The recommended method is to use aggregates of ward-based data for comparisons from 1981 through to 1989). Even so, the possibility of building block changes must continue to be borne in mind, along with sampling error and non-sampling errors, when interpreting the data. There are marked reservations about the use of the data below regional level both in the DE and among some potential users.

'Travel-to-work areas' enjoy a special significance in this field of study because they provide comparison between areas enjoying a common minimum level of economic self-containment. In concept, a self-contained labour market area is an area so defined that commuting to and from work all occurs within the boundary of the area; it is ideal for providing a measure of the mismatch between the supply and demand for labour in the area. In practice, it is not possible to divide the country into entirely separate areas with no commuting across the boundaries. Travel-to-work areas have been developed as approximations to such areas; they cover the whole country and are used as the basis for the publication of unemployment rates (Ch. 3). Data on the 1981 CoP on residence and workplace together were used by the DE (September 1984) to establish TTWAs with a minimum self-containment level of 75 per cent.

The use of such areas for analysis for employment is generally preferable to that of any other areas. This is not, however, invariably the case. Clearly, TTWAs still vary in their degree of self-containment. Due to the growth of commuting, many are now large and amorphous areas; large TTWAs like those of London or Newcastle upon Tyne clearly need disaggregation for the purpose of understanding the employment opportunities available for deprived or immobile populations, such as many females living in inner cities.

For policy purposes, different government departments identify certain kinds of area for development, and a major feature of the 1980s has been an extension of this practice – 'central government localism' as it has been dubbed by Martin (1989). In 1984 the Department of Trade and Industry (DTI) reidentified 'assisted areas' under regional policy by naming new TTWAs as Development Areas or Intermediate Areas, and the Rural Development Commission named groups of parishes as Rural Development Areas. Both of these sets of areas are available on NOMIS; however, there is no general agreement on the identification of standard inner-city areas. The user of NOMIS is able to define her/his own geographical areas using the standard geographical indices held by the system; different kinds of area can be mixed, except that jobcentre-based indices cannot be combined with ward-based.

User-defined industries

In the same way that modern systems allow for the construction of user-defined areas through the aggregation of building blocks of wards (or larger areas), so too can the modern analyst manipulate four, three, two or one-digit SIC headings to provide summary groups of national, regional or local significance. This is not necessarily to imply that the SIC 1980 is itself defective, but that different groupings of industries provide meaningful statistical statements for different stituations.

Government practice provides several precedents in grouping SIC divisions. Within the manufacturing and service sectors, more narrowly defined aggregations may be constructed because of their possibly salient role in national and local trends. International experience in manufacturing has led to recognition of the important role of 'high technology industry' (often abbreviated to 'hi-tech'). There is general agreement about the inclusion of electronics and aerospace industries, but less agreement over how many parts of chemicals should be included. In UK conditions, however, the remarkable feature is that 'hi-tech' is a potent contributor to job decline (Hall 1987).

Attention to the service sector has at times concentrated on different concepts and groupings. There is now a mounting body of research on the entity of 'producer services' (Marshall *et al.* 1989, and Ch. 17). By subtraction, the balance of the service sector may be seen as the 'consumer services' (Ch. 18). Within this group, however, lie the 'tourism-related industries', as defined by government and estimated at national level in the *Employment Gazette* (Table 8.1 monthly). This group is undoubtedly contributing to the economic growth of some localities. However, it unavoidably includes many local pubs and restaurants which are impossible to distinguish from longer-distance tourism, and whose growth generally reflects that of population as a whole and its leisure spending.

2.4   The range of data available at county scale (and not below)

The reform of local government boundaries in 1974 produced a more rational system of subregional units than hitherto. This provides a convenient level for government to publish a certain number of economic statistics. For instance, while all parts of the CoE (above) are available in unpublished form at the level of counties of England and Wales and regions of Scotland, this is the lowest level at which outline data are systematically published, in *Regional Trends* (annual, Table 2.5). (The exception is that a variety of local authority and TTWA data are also published in various Scottish and Welsh Office publications.) The data of Table 2.5 may readily be compared with the succeeding Table (2.6 and 2.7), which provide summary statistics on unemployment, earnings, gross domestic product (GDP) per head (see Ch. 4), expenditure and output in manufacturing, and earnings.

Data on earnings are of greatest importance because they provide the first qualification to the census process of simply counting jobs. They show the degree to which adult male full-time gross weekly earnings exceed those of females in each county, and indicate the geographical range of incomes at this level, in which Greater London average male earnings exceed those of the poorest county by 55 per cent, whereas the South East region exceeds that of the poorest-paid GB regions by only 24 per cent. Further refinement of the data, including hourly earnings before and after taking account of overtime and absence, are published in the *New Earnings Survey (Part E)* by the DE (annual).

The migration response to varying employment and income opportunities is differentiated subregionally. Estimates of total *net* migration by county and *local authority* district are published by OPCS (annual). However, the **National Health Service Central Register** provides a quarterly origin and destination matrix of migration flows by sex and by 19 age-groups, available on NOMIS. The matrix includes all counties (although Greater London and ex-metropolitan counties are disaggregated) and it is therefore possible to compare working-age migration with the employment and unemployment trends of all counties. The **Census of Production** publishes a limited part of its results at the county level (Business Statistics Office, annual); these provide data for the number of manufacturing establishments by employment size bands and by broad groups of industries. Comparison of different years makes a valuable contribution in assessing the increased relative importance of smaller plants in different areas. On request, the *Engineering Industries Training Board* is able to make available the total employment of member firms by county for different years; these data are valuable for being disaggregated not only by subsector and sex, but also by broad levels of skill.

2.5 The range of data available at regional scale (and not below)

It is at the regional scale that economic analysis comes into its own and that the analysis of output, costs, profitability and investment (Ch. 4) is essentially most feasible. A principal input to such calculations is the publication of quarterly regional employment statistics. These were revised on a comprehensive basis (*Employment Gazette*, Historical Supplement No. 2, November 1989) to provide a continuous series for each region from June 1971; this includes self-employed and work-related government training programmes so as to represent the 'civilian workforce in employment'. Valuable as such a series may be, it is important to appreciate that researchers used its equivalent very little when the CoE was conducted and reported on an annual basis, and that the series still depends very heavily on the census, conducted as it now is on a 2- to 3-year frequency.

The central element or bench-mark remains that of 'employees in employment', whose definitive total for any region is that provided by the census. As the September 1989 Census was published in April 1991, this meant that the September 1987 Census results remained the basis of estimation of the quarterly regional time series for nearly 4 years before their replacement in the May 1991 edition of the *Employment Gazette*. From past experience, this replacement has revealed large cumulative errors, which must be borne in mind at all times and are illustrated in Table 2.3. In aggregate, the quarterly series estimated a GB reduction of 431 000 employees from 1981 to 1984, when the census revealed a somewhat larger drop of 463 000, which was concentrated particularly in north-west England. It was considered, however, that the results of successive censuses had shown that the quarterly estimates underestimated the number of employees. Therefore allowances for undercounting were calculated using the results for the whole economy from the *LFS*.

In the event, however, the 1987 Census results (Table 2.3) showed that the scale of adjustment was too large in Great Britain as a whole and in the majority of regions: employees in employment had not been growing as fast as had been estimated through the *LFS*. 'This is the result of a number of factors – including sampling and other errors in both series', and differences in coverage; methods of estimating the employee series were being further

Table 2.3 Changes in employees in employment in Great Britain: comparison between census results and previous estimates, by region (thousands). (*Source:* Employment Gazette (various).)

	1981–84 (Sept.) Census	Previous Estimate	1984–87 (Sept.) Census	Previous Estimate	1987–89 (Sept.) Census	Previous Estimate
South East	− 26	− 3	+ 182	+ 344	+ 198	+ 441
East Anglia	+ 36	+ 24	+ 22	+ 100	+ 62	+ 40
South West	+ 7	− 1	+ 76	+ 54	+ 126	+ 109
East Midlands	− 10	− 41	+ 51	+ 96	+ 66	+ 79
'South'	+ 7	− 21	+ 331	+ 594	+ 452	+ 669
West Midlands	− 52	− 95	+ 8	+ 102	+ 113	+ 91
Yorkshire and Humberside	− 69	− 82	+ 9	+ 50	+ 121	+ 36
North West	− 158	− 82	+ 49	− 8	+ 87	+ 115
North	− 59	− 79	+ 14	+ 42	+ 42	+ 37
Wales	− 51	− 26	+ 38	− 6	+ 62	+ 64
Scotland	− 80	− 49	− 23	− 10	+ 87	+ 96
'North'	− 469	− 413	+ 95	+ 170	+ 512	+ 439
Great Britain	− 463	− 431	+ 425	+ 766	+ 963	+1112

examined in the light of these results (DE October 1989b: 566–7). What emerges from these considerations is that it is dangerous to rely in any great *detail* on current estimates beyond the date of the last CoE, and in particular on any precise comparison of regional trends in those periods, published in the *Employment Gazette* quarterly (Table 1.5). What store should one have set on the last column of Table 2.3, in the light of the previous ones? For instance, further data from the same source, for the last 12 months at the time of writing, show marginal *gains of manufacturing employment* in these regions, but it would seem unwise to herald a 'reversal of regional fortunes' prior to the 1991 Census results.

No data have been published regularly for regions other than the employee estimates. These are disaggregated by sex and by 12 principal industries. The increasingly crucial additional item, the estimates of self-employed, are normally published separately, and are insufficiently accurate to allow for cross-tabulation by industry and region. These data have been published in the *Employment Gazette* between February and July in each year from 1983 to 1991 (covering the previous year), with the addition of the series to March 1989 in the *Employment Gazette*, Historical Supplement No. 2, November 1989. Estimates from 1971 are based on the 1971 and 1981 CoPs and results of the Labour Force Surveys carried out from 1973 onwards. Provisional estimates for dates after the last *LFS* are carried out by projecting forward the average rate of growth between the 1981 CoP and the last *LFS* (DE April 1989a) but the assumption in the current recession is of no change.

2.6 Conclusions

A strong feature of the British system of employment data is that the scope for retrieving and publishing data is now maximized on a wide variety of geographical and sectoral systems of classification. The data are available in internationally comparable form in the publications of the OECD (annual) and ILO (annual), although it is sometimes apparent from gaps for the UK that the CoE was made less frequent than the equivalent surveys of foreign governments by UK government decisions of the 1980s. There remains also the problem that the UK pattern of change is disproportionately affected by the growth of part-time jobs and self-employment. Table 2.1 shows how the growth of the self-employed (full-time and part-time) stood among the largest elements of job growth, 1984–90; yet the reader will have just seen how the element of self-employment is measured only by a sampled household survey in between the dates of the CoP. Perhaps this is the most serious problem faced by any review of British employment statistics. There is debatable point in an accurate CoE if some of the main trend movements are occurring in the separate sector of self-employment.

The alternative scenario, of building up resources for larger samples in the *LFS*, and economizing further with the accuracy of the CoE, would only add to disillusionment with the latter. It is sometimes thought that private-sector databases, for instance of British Telecom business subscribers (Ch. 8), might fill the gap. However, it is not likely that such bases would provide a complete coverage of all sectors, or reveal the names of establishments (Ch. 16).

Nothing said in this chapter refers to the *names* of employers. The Factory Inspectorate in some regions has been able to divulge levels of employment which are part of their operational records (see Ch. 16). The impact of sampling in the CoE has already had a serious impact on the potential for local monitoring of changes in employment by local authorities, especially among smaller named establishments. It should be stressed that published directories provide neither comprehensive nor accurate statements of employment for an area's leading factories. There are relatively few ways in which research workers can overcome the limitations of the sources discussed. The main one is that of tight definition of the objectives of research, economical research design, and hard work in obtaining data direct from businesses (Ch. 11).

Appendix 2.1   Data sets available from the CoE.
(*Source:* As described by DE, Watford.)

---

1. Employees in employment: Summary analysis of employees disaggregated by sex, by full- and part-time status, and by industry according to the SIC. Availability:

   1.1 Current ward-based SIC 1980 series.
   Years: 1981, 1984, 1987 and 1989. Reference date: September.
   Areas based on local authority (LA) wards:

   > Ward (i), constituency (ii), travel-to-work area (TTWA), (ii), LA district, county, standard economic region, Great Britain.

   (i) Ward totals only for September 1981. Full industrial breakdown for later years.
   (ii) Based on 'frozen' wards, ie, in England and Wales, groups of wards as defined in 1981, in Scotland, groups of 1984 wards. Data for all other areas are available by both 'frozen' and 'current' ward, ie, ward boundaries as at census reference date

   1.2 Historical jobcentre-based SIC 1968* series.
   Years: 1971–78 and 1981. Reference dates: September 1981, June earlier years.
   Areas: Jobcentre, JC TTWA, county, standard economic region, GB

2. Size analysis of census data units, according to the numbers of their employees, by industry. Availability:

   2.1 Current SIC 1980 series.
   Years: 1981, 1984, 1987 and 1989. Reference date: September
   Areas: For 1981, Great Britain and standard regions only; later years, also TTWA, LA district and county

   2.2 Historical (discontinued) SIC 1968 series.
   Years: 1976–78 and 1981. Reference dates: September 1981, June earlier years.
   Areas: Great Britain and standard regions only

---

*Available also on SIC 1980 for 1981, and for 1984, 1987 on NOMIS.

References

Business Statistics Office annual *Size Analyses of United Kingdom Businesses* HMSO, London

Champion A.G., Townsend AR 1990 *Contemporary Britain* Arnold, London

Cooke P 1989 Locality, economic restructuring and world development. In Cooke P (ed.) *Localities* Unwin Hyman, London pp 1–44

CSO 1979 *Standard Industrial Classification Revised 1980, Reconciliation with Standard Industrial Classification, 1968* CSO, London

CSO 1981 *Indexes to the Standard Industrial Classification Revised 1980* HMSO, London

DE annual *New Earnings Survey, Part E* HMSO, London

DE September 1984 *Employment Gazette* Occasional Supplement No. 3, Revised travel-to-work areas

DE January 1986 Classification of economic activity. *Employment Gazette* **94**: 21–7

DE January 1987 1984 Census of Employment and revised employment estimates *Employment Gazette* **95**: 31–53

DE August 1987 1984 Census of Employment. *Employment Gazette* **95**: 407–9

DE April 1988 Standard Occupational Classification – a proposed classification for the 1990s. *Employment Gazette* **96**: 214–21

DE April 1989a Labour Force Survey – preliminary results. *Employment Gazette* **97**: 182–96

DE April 1989b Labour force outlook to the year 2000. *Employment Gazette* **97**: 159–72

DE October 1989a 1987 Census of Employment, results for the United Kingdom. *Employment Gazette* **97**: 540–58

DE October 1989b Revised employment estimates – incorporating 1987 Census of Employment results. *Employment Gazette* **97**: 560–6

DE April 1991 1990 Labour Force Survey preliminary results. *Employment Gazette* **99**: 175–96

Hall P 1987 The anatomy of job creation: nations, regions and cities in the 1960s and 1970s. *Regional Studies* **21**: 95–106

ILO annual *Bulletin of Labour Statistics* International Labour Organization, Geneva

Marshall J N *et al.* 1989 *Uneven Development in the Service Economy: Understanding the Location and Role of Producer Services* Oxford University Press, Oxford

Martin R 1989 The new economics and politics of regional restructuring: the British experience. In Albrechts L, Moulaert F, Roberts P, Swyngedouw E (eds) *Regional Planning at the Crossroads: Economic Perspectives* Jessica Kingsley, London pp 27–52

Massey D 1984 *Spatial Divisions of Labour* Macmillan, London

OECD annual *Labour Force Statistics* OECD, Paris

OPCS annual *Mid Year Population Estimates for England and Wales* OPCS, London

OPCS 1980 *Classification of Occupations, and Coding Index* HMSO, London

OPCS 1984 *Census, 1981, Economic Activity, Great Britain* HMSO, London

OPCS and Registrar-General, Scotland 1984 *Census, 1981, Key Statistics for Local Authorities, Great Britain* HMSO, London

Townsend A R 1986 Spatial aspects of the growth of part-time employment in Britain. *Regional Studies* **20**: 313–30

Note: In addition to the particular reports and articles listed above from the DE, the *Employment Gazette* carries monthly, quarterly and annual statistics in its 'Labour market data' on coloured centre pages. The Training, Enterprise and Education Division summarizes a lot of data in graphs and shorter notes in the *Labour Market Quarterly Report*, available from the Training Enterprise and Education Division, LMS, Room W815, Moorfoot, Sheffield S1 4PQ. The Training Enterprise and Education Division Intelligence Units produce regional labour market reports periodically from respective regional offices.

3

Unemployment, vacancies and redundancies

Anne Green

3.1 Introduction

This chapter covers three topics central to labour market studies: unemployment, vacancies and redundancies. Each of these topics is considered in turn. Unemployment statistics are outlined in section 3.2, with particular attention paid to sources of unemployment data; the precise definition of unemployment encompassed by each (and hence the comparability of data from different sources); the scope of current data series and sources – in terms of the range of counts available and their disaggregation; commonly used derived statistics; and an assessment of the key advantages and disadvantages of unemployment data for local and regional research. In sections 3.3 and 3.4 vacancy statistics and redundancy statistics, respectively, are considered. Again, the range of sources, their definitions and scope, and useful derived statistics are outlined. The sections are concluded with assessments of the applicability of such data for regional and local studies. The key points emerging from sections 3.2–3.4 are summarized in section 3.5 and general lessons to be borne in mind when using unemployment, vacancy and redundancy data for local and regional research are rehearsed.

3.2 Unemployment

Sources

Statistics on unemployment can be obtained in two main ways: first, from surveys of the labour force, and second, from administrative records. There are several regular surveys of the labour force yielding unemployment statistics, but the two most important are the *Labour Force Survey* (*LFS*) and the **Census of Population** (**CoP**). Data on unemployment are also recorded in the *General Household Survey* (*GHS*). At the local scale, surveys of the labour force, and sometimes specifically of the unemployed, are undertaken increasingly as part of comprehensive local labour market assessments and skills audits (Haughton and Peck 1988), as exemplified by the London labour market study (Meadows *et al.* 1988) and the West Midlands labour market study (Cooper 1989), and skills audits of Nottingham (Galt and La Court 1989) and Coventry (Elias and Owen 1989). However, the main source of statistics on unemployment at the local and regional scale in Great Britain are **Department of Employment (DE) administrative records**. The Department of Economic Development publishes similar records for Northern Ireland.

Definition of unemployment

The measurement of unemployment is both a conceptual problem and a matter of political debate, since unemployment is not unambiguously defined (Thatcher 1976). As noted by Hasluck (1989: 13), the dividing lines between the totally inactive, not currently active, the semi-unemployed, the unemployed, the under-employed and the fully employed are not clear, and unemployment may be defined alternatively as a *state* (not having a paid job), or an *activity* (looking for a job, wanting a job or needing a job). The level of unemployment in a particular area depends on the definition of unemployment adopted.

The sources identified above adopt different approaches to the definitional problem, and hence the coverage of the count varies according to the data source used. Therefore, extreme caution is necessary when assessing statistics from non-comparable sources.

Unemployment data derived from surveys of the labour force (such as the CoP and *LFS*) are based on the individual respondent's self-description of unemployment status during the reference week in question. In the *LFS* additional data are collected on whether an individual is actively seeking work and available for work, thus enabling calculation of the restrictive definition of unemployment ('aged over 14, without work, actively seeking work and immediately available for work') used by the Statistical Office of the European Communities for international comparative purposes (Commission of the European Communities 1989). By contrast, the DE unemployment count represents a static analysis of the situation on the count day, in accordance with the regulations relating to receipt of benefits at the time in question. Hence, in spring 1988 the international measure of unemployment from the *LFS* was 2.37 million in Great Britain, compared with an average DE unemployment count of 2.41 million over the same period. It is estimated that 790 000 (33%) of claimants measured by the DE count were not unemployed by the international definition, compared with 750 000 individuals in the *LFS* unemployed according to the international measure but not claiming benefits (DE 1989a).

A further indication of the complexity of the definitional problem, and

Table 3.1 Changes to the UK official unemployment count. October 1979–October 1989 (adapted from Taylor 1989:1–4)

No.	Date	Description of change	Estimated effect on monthly count
1.	Oct. 1979	Fortnightly payment of benefits	+ 20 000
2.	Oct. 1987	Compensating downward adjustment to published totals	− 20 000
3.	Feb. 1981	Register effect of special employment and training measures	− 370 000 (− 495 000 Jan. 1986)
4.	July–Oct 1981	Compensation for effect of industrial action emergency procedures	− 20 000
5.	July 1981	Benefit changes for unemployed men aged 60 and over	− 30 000 by May 1982
6.	July 1982	Taxation of unemployment benefit	not known
7.	Oct. 1982	Change in definition and compilation of count – from registrant to benefit claimant basis	− 170 000 to − 190 000
8.	Oct. 1982	Discontinuation of count of those seeking part-time work	(− 52 200 in Sept. 1982)
9.	April 1983	Removal of signing-on requirement for men aged 60 and over	−107 400 by June 1983
10.	June 1983	Change in long-term supplementary benefit entitlements for men aged 60 and over	− 54 400 by Aug. 1983
11.	June 1983	Removal of unemployed school-leavers from June, July and August counts	(−100 000 to − 200 000)
12.	Oct. 1984	Change in Community Programme eligibility rules	− 29 000 by Jan. 1986
13.	July 1985	Reconciliation of N. Ireland data to computer records	− 5 000
14.	July 1985	Payment of unemployment benefit in arrears	Not known
15.	March 1986	Introduction of 2-week delay in publication of count	Av. − 50 000
16.	June 1986	Use of larger denominator in unemployment rate	Av. − 1.4% points
17.	Oct. 1986	Change in unemployment benefit entitlements	− 24 000 after 1 year
18.	Oct. 1986	Extension of voluntary unemployment deduction to unemployment benefit	− 2 000 to − 9 000
19.	June–Oct. 1986	Restart and tighter available for work test (further tightened in 1988)	−200 000 to − 300 000 in 1988
20.	April 1988	Extension of voluntary unemployment deduction to unemployment benefit	− 12 000
21.	April 1988	Change in Income Support regulations	Not known
22.	June 1988	Use of larger denominator in unemployment rate calculation	− 0.1% point
23.	Sept. 1988	Removal of 16–17-year-olds from count	− 90 000 to − 120 000
24.	Oct. 1988	Amendment to contribution conditions in respect of short-term benefits	− 38 000
25.	Oct. 1988	Lowering of age limit for abatement of unemployment benefit to occupational pensioners	− 30 000

Table 3.1 Cont.

No.	Date	Description of change	Estimated effect on monthly count
26.	April 1989	Amendment to Redundant Mineworkers Pension scheme	− 26 000
27.	Oct. 1989	1989 Social Security Act: claimants must be able to prove that they are actively seeking work	} − 50 000
28.	Oct. 1989	1989 Social Security Act: restrictions on level of remuneration as a reason for turning down work	
29.	Oct. 1989	Tightening of requalification conditions for unemployment benefit	− 350

implications for the size of the unemployment count, are revealed by a comparison of the number of benefit claimants (as included in the DE unemployment count), with benefit recipients, the self-defined unemployed, and the self-defined unemployed plus persons not in paid jobs and actively seeking work, from the Coventry skills audit (Elias and Owen 1989: 62). In the DE count, 15 000 benefit claimants aged between 20 and 60 were identified compared with 14 700 benefit recipients, an unemployment count of 13 600 on a self-defined basis, and 17 400 self-defined unemployed plus persons not in paid jobs and actively seeking work. Moreover, these differencs are not uniform across the population: there were 6900 unemployed females in the category of self-defined unemployed plus persons not in paid jobs and actively seeking work, on the one hand, and yet only 4600 female benefit claimants; whereas for males there was much less discrepancy between the two categories: 10 400 benefit claimants and 10 500 self-defined unemployed plus persons not in paid jobs and actively seeking work.

Care is not only needed when comparing unemployment counts derived from different sources, but also when comparing counts from the same source at different points in time. This is particularly true of the DE count, which has been the object of numerous changes in coverage over recent years. The Unemployment Unit (UU) (Taylor 1989) identifies 29 changes to the official unemployment count in the period October 1979 to October 1989 (as detailed in Table 3.1). Many of the changes relate to changes in the payment of, or eligibility for, benefits, but there have also been other, quantitatively more important, changes in the coverage of the count. Key changes worthy of particular attention are:

1. The removal of approximately 0.5 million from the count with the spread of special employment and training measures;
2. The change in the *definition* and *compilation* of monthly figures in October 1982 from a clerical count of people registered for work at jobcentres and careers offices to a computer count (known as the Joint Unemployment Vacancies and Operating System (JUVOS));
3. Changes in 1983 in signing on procedures and entitlement to supplementary benefits for men aged over 60;

4. A change in March 1986 involving the introduction of a 2-week delay in the publication of the monthly unemployment count;
5. A series of measures tightening the 'available for work test'.

As a result of these changes there are numerous 'breaks' in the time series, although attempts have been made to *adjust* for the effects of such changes at the national and regional scales. The DE is involved in adjusting unemployment series retrospectively after significant changes in the count, in order to make the series consistent with current definitions. In order to construct the new series for any past period, estimates have to be made of the numbers of people claiming unemployment benefit in that period who would be excluded or included under current arrangements (DE 1985). Periodic revisions to the adjusted series are documented in the *Employment Gazette* (e.g. DE 1988), with time series published in special supplements (e.g. DE 1989b). By contrast, the UU publishes an index which adjusts the current series to the pre-October 1982 basis. Figure 3.1 shows that the DE count was 162 000 lower than the UU index in December 1982, and 696 000 lower in December 1988; while Table 3.2 shows differentials between the DE and UU unemployment counts and rates at the regional scale in November 1988. It has not been possible to identify the precise impact of all the changes in the coverage of the count at the local scale; but due to variations in the distribution of subgroups of the population over space it is likely that impacts will have been spatially uneven – a factor to be borne in mind in comparative studies (for an example, see Ch. 1).

In addition to the adjustments made in respect of changes in the coverage of the count, *seasonal adjustments* are made by the DE in order to provide

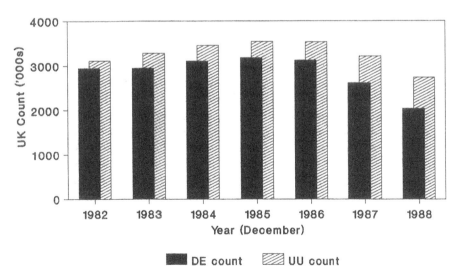

Fig. 3.1 Unemployment: DE and UU comparison (UU count on a pre-October 1982 basis). (*Source:* Based on data in UU 1989.)

Table 3.2 Comparison of November 1988 DE unemployment count and UU unemployment index at the regional scale. (*Source:* UU 1989: 2.)

Region	DE count		UU index*	
	freq.	%	freq.	%
South East	421 700	4.5	560 000	6.0
East Anglia	41 200	4.1	55 400	5.4
South West	113 100	5.4	154 600	7.4
West Midlands	198 300	7.5	266 000	10.0
East Midlands	126 000	6.5	169 100	8.6
Yorkshire and Humberside	203 000	8.5	272 800	11.3
North West	292 800	9.7	393 900	12.9
North	160 100	11.0	217 100	14.7
Wales	113 100	9.5	155 200	12.8
Scotland	260 300	10.5	348 100	13.7
Northern Ireland	108 900	15.7	146 100	20.8
Great Britain	1 930 200	7.0	2 589 000	9.3
United Kingdom	2 039 100	7.2	2 735 100	9.6

*The UU index measures the unemployment count on the pre-October 1982 basis.

a consistent assessment of the underlying trend in unemployment. These seasonal adjustments are made using the 'X 11' programme developed by the US Bureau of the Census (a method which is now used in most industrial countries for seasonally adjusting unemployment figures) (DE 1988).

Scope of data series

The *LFS* is a harmonized and synchronized sample survey used to determine the level and structure of employment and unemployment in the European Community (EC). From 1973 to 1983 the survey was biennial, but since 1983 it has been conducted on an annual basis (Thomson 1989). The survey covers approximately 60 000 households in the UK and includes data on demographic structure, economic activity status, educational qualifications, training, employment – by industry and occupation (for both the current year and one year ago), working time, and the search for work. It is thus possible to calculate a wide range of statistics relating to the unemployed in their economic, social and demographic context, on a basis comparable with those for other EC members. In the case of the UK, the data may be disaggregated to the scale of standard regions and metropolitan counties. Hence, while valuable for studies at the regional scale, the *LFS* can only provide contextual information for local studies. Nevertheless, the *LFS* possesses the key advantage of providing annual data on a consistent basis, and increasing reliance is placed upon the *LFS* as a source of labour market data.

The CoP is the most comprehensive survey undertaken in the UK. The chief drawback of the CoP for local and regional research is that it is conducted on a decennial basis. Nevertheless, it provides an important bench-mark for assessing the coverage of the more regular estimates of unemployment. Moreover, the CoP is an immensely rich data source providing a wide range of information by gender, marital status, place of birth, household type, tenure, current/last occupation and industry, socio-economic group, social class and so on, alongside which data on unemployment may be compared. The data are available for micro-areas (enumeration districts (EDs) and wards at the maximum scale of spatial disaggregation in 1981), and by relating unemployment and labour supply statistics with a range of other indicators, it is possible to derive measures for local areas unavailable from other sources.

DE unemployment counts (as published for selected areas in the DE *Employment Gazette* and available for a wider range of areas through the National On-Line Management Information System (NOMIS), described in more detail in Ch. 8) include not only monthly unemployment stock counts (of persons claiming benefit who are registered at unemployment benefit offices (UBOs) as actively seeking work), but also counts of flows into and out of unemployment. These counts are disaggregated by gender, and in addition it is possible to distinguish school and further training and education leavers (according to age-group), married women, and students on vacation, from 'wholly unemployed claimants'. The frequency of the counts, and their availability with a lag of approximately one month, are among the criteria in making DE unemployment counts a key source of data for local and regional research. Additional disaggregations of current unemployment spells into 14 age bands and 16 duration bands (thus enabling identification of the subgroups such as the long-term unemployed) are available on a quarterly basis. However, information of interest to local and regional researchers on last industry and occupation has not been collected since October 1982.

In the case of computerized records (the vast majority), each unemployed claimant is spatially referenced by residential postcode, and the postcodes are then aggregated to postcode sectors or 1981 'frozen' wards (except in Scotland where 1984 wards are used) for reporting purposes. (This contrasts with the situation prior to the introduction of JUVOS when the unemployed were coded to the offices at which they registered.) The postcode sectors and wards may be aggregated into a range of official (such as travel-to-work areas (TTWAs)) and user-defined areal units.

Derived statistics

The most commonly used derived unemployment statistic is the *unemployment rate*. This measures the incidence of unemployment, and is frequently used in local planning and the formulation of economic development policies. Hasluck (1989: 16) defines the unemployment rate as:

$$U\% = U/L \times 100 = U/(E+U) \times 100$$

where U represents the number unemployed, L the labour force and E the employment count.

At the national and regional scales, the *official* unemployment rate (using the DE unemployment count) is calculated by expressing the number of unemployed benefit claimants as a percentage of the working population (i.e. employees in employment plus self-employed plus unemployed). For subregional areas the official unemployment rate is calculated as the number of unemployed benefit claimants plus the number of employees in employment for June of the previous year. The key source of data for the employees in employment element of the calculation is the Census of Employment (CoE) (Ch. 2). Due to the fact that employees are spatially referenced to their workplace in the CoE, while the unemployed are referenced on a residence basis (as outlined above), it is important that the local areas for which unemployment rates are calculated are relatively self-contained in commuting terms (i.e. that the majority of jobs are filled by local residents and the majority of residents have their workplaces within the area). Otherwise areas of out-commuting will display inflated unemployment rates, while areas of in-commuting will exhibit deflated unemployment rates (as explained by Green and Coombes 1985). In the *Employment Gazette*, unemployment rates are published only for relatively self-contained areas. For this reason no unemployment rate is published for Surrey.

At the micro-scale (e.g. ward level) it is possible to derive a form of 'residence based' unemployment rate by expressing the area unemployment total as a proportion of the population of working age. Use of population of working age as the denominator in the unemployment rate calculations results in a smaller percentage rate than the official rate. For example, the 1983 UK unemployment rate calculated using population of working age as the denominator is 8.8 per cent, but using working population as the denominator the rate is 11.7 per cent, and using employees in employment as the denominator the rate is 12.9 per cent (UU 1988: 3). It is essential to bear in mind, however, that an unemployment rate calculated on this basis is not comparable with official unemployment rates at subregional, regional and national levels.

Comparative change in unemployment rates is often of particular interest from a policy perspective. Discontinuities in the DE unemployment series mean that measurement of temporal change is difficult. Aside from this, however, there is the question of whether change should be measured in either 'percentage' or 'percentage point' terms (as debated by Gillespie and Owen 1981, 1982 and Crouch 1982). If an area had an unemployment rate of 5 per cent in 1979 and of 10 per cent in 1981, the 'percentage' change in the unemployment over the period would be 100 per cent, but the 'percentage point' change would be 5 per cent (i.e. 10% (in 1981) − 5% (in 1979) = 5%). As indicated in Table 3.3, rankings of regions differ according to the measure of change used, with regions displaying a relatively high incidence of unemployment (for example, the North and Wales) tending to be ranked higher on the percentage point change in unemployment rate measure than

Table 3.3 Comparison of measures of unemployment change, August 1979 –
December 1980 (adapted from Gillespie and Owen 1981:190)

Region	Percentage change	Rank	Percentage point change in unemployment rate	Rank
West Midlands	102	1	5.0	1
East Midlands	91	2	3.8	6
Yorkshire and Humberside	80	3	4.1	5
East Anglia	75	4	3.0	8
South East	74	5	2.6	10
North West	66	6 =	4.2	3 =
Wales	66	6 =	4.6	2
North	54	8	4.2	3 =
South West	52	9	2.8	9
Scotland	48	10	3.6	7

on the percentage change measure, while the converse applies for regions
characterized by a lower incidence of unemployment (such as the South East
and East Anglia).

As the proportion of long-term unemployed rose during the 1980s, so
increasing emphasis was placed on particular problems faced by those
unemployed for extended durations. Two areas displaying the same unemployment rate may exhibit different unemployment duration structures which
call for different policies. Various derived statistics relating to the *duration of
unemployment* have been used by local labour market analysts – including the
median duration of completed and uncompleted unemployment spells and
experience-weighted spell measures. Green (1985) has used such derived
measures to reveal that longer duration spells are apparent in northern than
in southern Britain, in metropolitan than in rural areas, and in manufacturing-
dominated than in service-dominated local labour market areas (LLMAs).
Similar statistics may be calculated at the intra-urban scale: Fig. 3.2 shows
spatial variations at the ward scale in the median duration of uncompleted
unemployment spells in Birmingham. The key disadvantage of these duration
measures based on DE data is that no account is taken of recurrent unemployment spells, which have been shown in work-history/cohort studies to be
a crucial feature on the rise of long-term unemployment, and are important
from a policy perspective (White 1983).

Alongside the increasing emphasis on duration of unemployment, growing
attention is paid to the study of unemployment as a *dynamic* phenomenon.
Hence, increasing attention is paid to the analysis of flow, rather than stock
statistics. Green (1986) has cross-tabulated derived statistics on the *likelihood
of becoming unemployed* (on-flows over a specified period as a percentage of
employees in employment) and the *likelihood of ceasing to be unemployed*
(off-flows over a specified period as a percentage of the average number
of unemployed) to identify categories of LLMAs with different types of

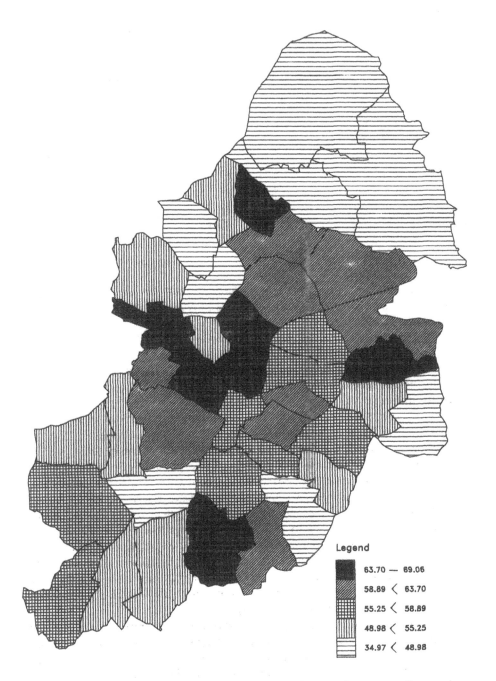

Fig. 3.2 Median duration (weeks) of uncompleted unemployment spells – males, January 1989: Metropolitan District of Birmingham. (*Source:* DE statistics (NOMIS).)

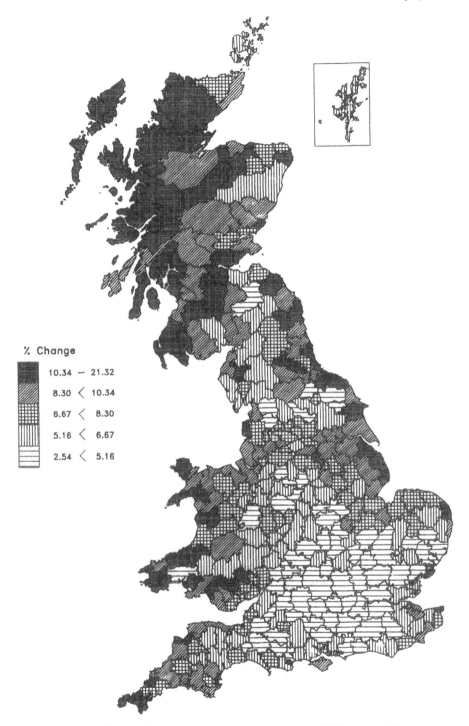

% Change

10.34 — 21.32

8.30 < 10.34

6.67 < 8.30

5.16 < 6.67

2.54 < 5.16

Fig. 3.3 Likelihood of becoming unemployed, October 1988–April 1989: travel-to-work areas of Great Britain. (*Source:* DE statistics (NOMIS).)

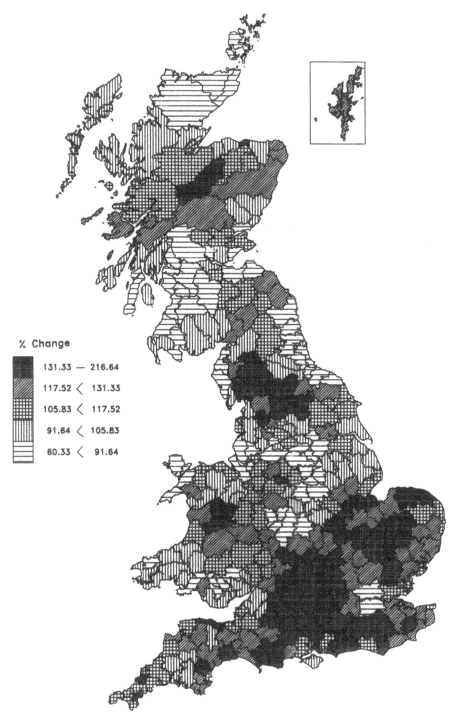

Fig. 3.4 Likelihood of ceasing to be unemployed, October 1988–April 1989: travel-to-work areas of Great Britain. (*Source:* DE statistics (NOMIS).)

unemployment problems, according to different combinations of on-flow and off-flow rates. Figures 3.3 and 3.4, showing the likelihood of becoming unemployed and the likelihood of ceasing to be unemployed, respectively, in the winter of 1988/89, illustrate problems of seasonal unemployment in some coastal areas, and the buoyancy of local labour markets throughout much of southern England.

Assessment

Unemployment statistics display three key *advantages* for local and regional researchers: first, DE statistics are up to date; second, they are published on a regular basis; and third, they are available for small areas. For these reasons, unemployment is often used as a 'proxy' variable for a variety of purposes: for example, as an indicator of the state of the economy, of labour reserves and of social distress (Elton 1985).

The main *disadvantages* of unemployment statistics for local and regional researchers include problems of non-comparability between sources (due to different definitions of unemployment adopted), lack of consistent local time series due to discontinuities in data (as a result of frequent changes in the scope and coverage of counts), and the omission/partial coverage of some subgroups of interest from the official count.

3.3 Vacancies

Sources

There is only one readily available source of data on vacancies in the UK: the official statistics compiled by the DE for Great Britain and similar statistics for Northern Ireland compiled by the Department of Economic Development.

Individual and household survey data are often used to complement the DE vacancy data. For example, both the *LFS* and the *GHS* may be used in association with DE data. Neither of these sources actually provide vacancy data, but they record information on methods of job search.

Employer surveys are becoming an increasingly important source of information on vacancies. For example, the London labour market study (Meadows *et al.* 1988) included a sample survey of private and public sector employers, stratified by size and sector, in order to determine the number and nature of job vacancies in the labour market. In other surveys the main focus is on the identification of skill shortages rather than vacancies *per se*. The Confederation of British Industry (CBI)/Training Agency (TA) Joint Survey of Skill Shortages is a regular monthly business trends survey of 2500 manufacturing companies which seeks to identify shortage occupations, whether shortages are getting worse, and what firms are doing about shortages. Results are reported annually in the *TA/CBI Skills Survey* *in Manufacturing Industry,* with disaggregations by industry and size. The **Computer-Assisted Local Labour Market Information System (CALLMI)** (an ongoing survey of employers undertaken on behalf of the TA) and successor local employer surveys collect information on recruitment difficulties. It is possible to 'gross up' the survey data to derive local, regional and national scale estimates (Owen and Green 1989). If sufficient resources are available it is also possible to conduct a job census for a particular case study area, covering all standard locations of job advertisements (see Marsh 1988).

Definitions

Since jobs are much more intangible things than people, it is theoretically more difficult to identify vacancies than to identify the unemployed. A vacancy may be defined as an unfilled job opening for which a firm is trying to recruit a worker. In some case studies attempts have been made to apply explicit restrictive criteria to the definition of vacancies. For example, for a job census in Chesterfield, Marsh (1988) defined a job opening as a vacancy only if it was available in the reference week (i.e. at the time the census was conducted), it was available to individuals outside the organization and if an employer was making an active effort to fill it.

DE data on vacancies is confined to job openings notified to jobcentres and careers offices. (Although careers offices cater specifically for young people about to enter, or who have recently entered, the labour market, the vacancies notified include some for adults (and vice versa vacancies notified to jobcentres include some for school-leavers) – hence the two data series should not be aggregated together.) Vacancy data at the national and regional scale published in the *Employment Gazette* are seasonally adjusted in the same manner as the unemployment series. All vacancies filled without resort to notifying jobcentres or associated agencies are not recorded in the official DE statistics.

The partial nature of the DE definition of vacancies is a source of difficulty for regional and local researchers, and has been the subject of considerable debate. In 1987 the DE commissioned a research project to determine the number and nature of job vacancies in Great Britain and the jobcentres' share of those vacancies. The survey suggested that in a typical month (in 1987/1988) there were over 700 000 unfilled vacancies in the economy. This confirmed previous estimates that the level of vacancies recorded at jobcentres represents around one-third of all those available (Smith 1988).

However, it is likely that this proportion will vary by local area (partly because of differing occupational mixes of LLMAs but also because in some areas there is a greater tradition of making use of jobcentres than in others), by occupation and by stage of the economic cycle. Considering the spatial dimension first, Marsh (1988) found that 55 per cent of vacancies available in Chesterfield in July 1988 were notified to the jobcentre or careers service – suggesting that in this particular instance it may be more appropriate to multiply the official vacancy figure by two, rather than three (as conventionally used), to derive an estimate of the number of vacancies available. Turning to occupational differentials, vacancies may be handled by private employment agencies, the local and national press, specialist journals and by word of mouth, as well as by jobcentres. Secretarial and accountancy vacancies, for example, are often handled by private agencies. However, the share of a particular occupational market accounted for by such agencies may vary from area to area. Professional and managerial positions, for which applicants are sought from outside the local area, are often advertised in the national press or in professional journals, or are handled through private employment agencies. Jobcentres tend to cater disproportionately for vacancies demanding manual and lower-level skills. This is confirmed by evidence on main recruitment methods used by London employers reported in Meadows *et al.* (1988: 20). Most manual vacancies and catering/retail vacancies were reported to jobcentres, with newspaper advertising used only in a minority of cases. Half of all clerical vacancies were notified to private recruitment agencies and approximately half were filled by this method.

Scope of data series

A monthly basic count of vacancies and placings is available on NOMIS for jobcentres and their aggregates. Notified vacancies, unfilled vacancies, vacancies filled and placings are distinguished. An employer may withdraw a vacancy before it has been filled, and hence unfilled vacancies (i.e. all notified vacancies remaining at the offices on the day of the monthly count) are the most commonly used indicator for studies involving vacancies. Data are also available on the time taken to fill vacancies (under 5 days, 5 days to 6 months, and over 6 months), vacancy flows and vacancies cancelled. Until August 1988 similar counts were available for vacancies and placings under the Community Programme, but with the introduction of Employment Training in September 1988 there were no longer any Community Programme vacancies, and Employment Training places are regarded as training opportunities determined according to the individual needs of unemployed people, rather than as vacancies. In addition, from the end of 1985, two series of vacancy counts disaggregated by occupation (31 occupational groups) and by industry (38 industrial categories from the 1980 Standard Industrial Classification (SIC)) have been available. Since the introduction of equal opportunities legislation, no distinction is made by gender.

Vacancies notified only to careers offices are excluded from NOMIS, but national and regional scale data are published in the *Employment Gazette*. Information at the local scale is available from press notices and from TA employment intelligence units.

As is the case for unemployment, so with vacancies increasing attention is devoted to *flow* rather than *stock* counts. The stock of vacancies at any one point in time (such as the count date) will be unduly clogged up with less popular jobs that take much longer to fill (often because of low pay and poor working conditions). Hence, the stock count will not give a representative picture of the flow of vacancies advertised and filled in a fixed period of time.

Derived statistics

Vacancy data are often used as a valuable complement to unemployment data, in that they represent a measure of demand for labour. The flow of vacancies is often used as an indicator of economic health – but without supporting case-study evidence it is not possible to distinguish 'new' jobs from those becoming available because the previous incumbent left.

As for unemployment, so for vacancies, it is possible to measure the median duration of vacancies. This enables identification of the ease of filling particular types of vacancy by area.

Among the most widely used derived statistics involving vacancy data is the unemployment/vacancy (*U/V*) relation. This represents the number of unemployed per vacancy (it is common for the vacancy count to be multiplied by three to compensate for official estimates that vacancies notified to job-centres represent only one-third of the vacancies in the economy). Analyses of medium-term temporal trends in the *U/V* relationship at the national scale have revealed a tendency over time towards a greater number of unemployed people at any given level of vacancies. In part, this has been attributed to the rise in the number of long-term unemployed – who are often inefficient in their job search and unattractive to employers (Layard 1986). At the local and regional scale, the *U/V* relationship has been used traditionally to identify different types of unemployment – frictional, structural and demand-deficient (Hasluck 1987: 108–11), but the discontinuation of coding of unemployment by occupation and industry in 1982 has limited the scope for such analyses.

Assessment

Whereas unemployment statistics are used to measure the supply of labour, vacancy information is used as an indicator of labour demand. Vacancy statistics published on a regular basis by the DE provide a useful bench-mark. On the positive side, they have the advantage of being available for relatively small areas, they are published frequently and are relatively up to date. Nevertheless, they are of limited value because they are *partial* in nature – covering only those vacancies notified to jobcentres. However, there are no regular reliable data on vacancies advertised or filled via other channels. Increasingly, it is likely that greater use will be made of employer surveys for vacancy statistics.

3.4 Redundancies

Sources

Data on redundancies are available from the DE. A short entry relating to redundancy appears each month in the *Employment Gazette*, and more detailed tabulations remain unpublished. In 1989 complete coverage of all redundancies was available for the first time from the *LFS*. In addition, press reports of redundancies have provided a valuable source of information for researchers.

Definitions

There are two ways in which a permanent job may disappear: first, the occupant can leave and not be replaced – this is *natural wastage*; and second, the occupant can be made *redundant*. The concept of redundancy was established in the 1965 Redundancy Payments Act, which provided for the payment of minimum compensation by the employer to those employees dismissed for reason of redundancy. The distinction between natural wastage and redundancy is not so clear-cut in reality as it might first appear: retirement is one of the chief sources of natural wastage, and redundancy and retirement can often go together: redundancy can be a form of forced early retirement (Harris 1987).

Scope of data series

Three series of redundancy statistics have been collected by the DE as a by-product of the administrative arrangements relating to redundancies. These are outlined below.

The first is *advance notifications of redundancies*: under the 1975 Employment Protection Act an employer is required to give advance notice of one month for redundancies involving 10 or more employees, and of 3 months for those involving 100 or more employees. These advance notifications enable steps to be taken to help the employees involved to find new jobs or training, or to avert the redundancies. The series provides an early indication of redundancy trends.

The second data series is *confirmed redundancies*: the TA follows up prior notifications of redundancies nearer the time they are expected to occur, and records those 'confirmed as due to occur'. Confirmed redundancies represent a subset of prior notifications, since efforts to avert redundancies already notified are sometimes successful and hence not all redundancies notified take place. In the period from 1977 to 1982 approximately one-quarter of prior notifications were withdrawn by employers, and confirmed redundancies accounted for only 38 per cent of advance notifications (the remaining proportion of the difference between advance notifications and confirmed redundancies is accounted for by withdrawals not notified to the DE) (DE 1983). For labour market studies, confirmed redundancies are regarded as a more reliable indicator than prior notifications, and hence this second data series has been used more widely than the first. However, there is evidence that this data series has become more unreliable over time. Information has always been sought on a voluntary basis and there have been instances where employers have not replied. Changes to the Redundancy Fund have also affected the series: from August 1986 rebates from the Redundancy Fund became limited to firms with fewer than 10 employees, and then were abolished entirely in January 1990, so that employers no longer have a financial incentive to notify redundancies. Indeed, the future of redundancy series collected as a by-product of administrative arrangements is under review (Bird 1990).

The third related data series is *redundancy payments*: these relate to the numbers receiving redundancy payments under the 1978 Employment Protection (Consolidation) Act and earlier legislation. Hence, this series relates to redundancies which have actually occurred. As indicated above, rebates from the Redundancy Fund have been discontinued, so payment statistics are no longer produced.

All of the three data series are disaggregated by region and industry, but are *restricted* in their *coverage*. No information is collected on advance notifications or confirmed redundancies involving fewer than 10 employees at any one establishment. Hence, details of small firm redundancies are not included in these two series, and relatively small-scale redundancies spread among departments of establishments within a wider company may be missed – even though on a company-wide level redundancies may involve 10 or more workers. The redundancy payments series relates to all redundancy payments made – it is not restricted to redundancies involving 10 or more employees. The fact that between 1978 and 1982 the number of employees receiving redundancy payments was 40 per cent higher than the number of redundancies confirmed as due to occur indicates that many redundancies take place in small batches (DE 1983). Moreover, redundancy payments are not available to all employees – coverage is restricted according to qualifying criteria relating to age, length of time served and hours of work. With the trend towards flexible work practices – including the growth of self-employment and temporary/fixed-term contract workers, the proportion of employees outside the scope of the redundancy legislation has risen in recent years. The fact that payments are often coded to the region in which a firm's head office is located, rather than to the region in which the redundancies occur, offers further scope for variation between the series.

Further discrepancies between the three data series arise because of the effects of *time-lags*. Whereas data on confirmed redundancies relate to the date when the redundancies are actually due to take place, the advance notifications relate to administrative procedures before the date of the redundancies, and the payments relate to procedures after the

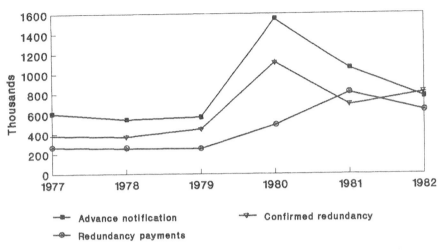

Fig. 3.5 Redundancy statistics: comparison of data series. (*Source:* Based on data in DE 1983.)

redundancies have taken place. Figure 3.5 provides some insight into such time-lag and coverage variations. In the 1989 *LFS* respondents were asked whether they had been made redundant during a specific reference period 3 months preceding the reference week. By contrast with the series discussed above, all redundancies should be covered in the *LFS* estimate. Indeed, in the first quarter of 1989 there were 131 000 redundancies according to the *LFS*, compared with 97 000 and 27 000 redundancies recorded by the advance notifications and confirmed redundancies series, respectively. The *LFS* provides information on individual respondents and hence it is possible to examine redundancies by age, gender, industry, occupation and region. However, since the data are collected in a sample survey the sampling error on some of the more detailed analyses may be quite large and hence results should be interpreted with caution (Bird 1990).

Press reports of redundancies can provide the researcher with an extra valuable source of data for local and regional research. Townsend (1983a) partly based a study on redundancies and closures in big corporations in the period 1976–81 on a 5-year collection of all reports of UK job losses in the *Financial Times*. Although the coverage is not as accurate as the DE data, it has a key advantage over the DE series of providing the employers' names throughout, enabling the construction of a chronology of national corporations' disinvestment strategies (Townsend 1983a: 72). Moreover, Townsend found it possible to spatially reference 90 per cent of reports to the county scale, and to distinguish complete from partial closures, thus providing an indication of whether employment losses through redundancy were permanent or transitory.

Use of redundancy statistics

Redundancy statistics may be used by researchers to trace key events in individual firms and industries within a single region, as an element in a comprehensive sectoral or a regional study, and as a tool in the appraisal and comparison of origins of job loss at the inter-regional or inter-industry scale (Townsend 1983b) and to provide a context for studies which focus on the experience of redundant employees (Mackay and Jones 1989; Daniel 1972). With the availability of LFS data there is also scope for detailed analysis of individual characteristics of those made redundant.

The main use of redundancy statistics has been made in periods of recession. This is because redundancies tend to rise in both absolute and relative terms (as measured by the *redundancy rate*: redundancies per thousand employees) during recession. In such a climate, workers tend to hold on to the jobs they have – hence voluntary separations fall and voluntary flows on to the labour market are diminished. Employers tend not to expand their labour forces or recruit to compensate for natural wastage – hence flows off the labour market are also diminished. In the absence of redundancies, the labour market stagnates during recession. This is confirmed by Jones and Martin (1986) in an analysis which suggested that regional redundancy rates are explicable in terms of variations in the natural turnover rate and the unemployment rate: the higher the unemployment rate, the lower the natural turnover, and the harder it is to contract the labour force through natural wastage, and the more often employers will have to resort to redundancy as a means of slimming down their labour forces. In general, however, redundancies are not a good measure of either changes in levels of employment or unemployment, or the number of people in work.

Using descriptive statistics and shift-share analysis techniques, researchers have highlighted marked regional and industrial differentials in redundancies (DE 1983; Townsend 1982). In general, the south and east of the UK fared better than the north and west in the early 1980s, in relation to their overall employment levels, and service industries suffered fewer redundancies than manufacturing, as may be expected given the shift of employment from manufacturing to services.

Assessment

The definition of redundancy is more clear-cut than that of unemployment or a vacancy, and there are fewer sources of data available to the researcher. However, as is the case for unemployment and vacancies, no single data series provided complete coverage until the advent of redundancy data from the *LFS*, and in using and comparing data from different series it was necessary to bear in mind the relevant restrictions. Redundancy statistics provide a valuable insight into one aspect of job loss. Interest in redundancies tends to vary in accordance with economic conditions, and in general redundancy data tend to be less widely used by local and regional researchers than unemployment or vacancy data.

3.5 Conclusion

A number of key themes emerge from the review of data sources, definitions, scope, and derived statistics relating to unemployment, vacancies and redundancies. For any particular topic, a wide range of data sources are usually available – with no single source providing a total picture. Hence, it is advantageous to make use of various sources, whenever possible. However, it is necessary to be aware of variations in definitions used, and differentials in, and the partial nature of, coverage of the various sources – with consequent implications for different population subgroups/areas. These inconsistencies often lead to problems of non-comparability between data sources at a single point in time and within series over time.

Key advantages of unemployment and vacancy data, and to a lesser extent redundancy data, are frequency of publication, the up to date nature of available data and degree of spatial disaggregation. These positive factors have led to the use of derived unemployment and vacancy measures in a surrogate fashion for other variables/conditions.

In recent years greater use has been made of dynamic, as opposed to stock, measures. This reflects the contention that consideration of flows provides greater insight into the workings of local labour markets, than a limited number of snapshots. Hence the increasing use of unemployment and vacancy flow statistics, alongside more conventional measures of stocks.

Another key trend is for greater use of data from surveys. For labour market statistics at the regional and national level increasing emphasis is placed on LFS data. At the local scale, also, the benefits of surveys in providing a more complete picture, and in complementing data from administrative records, is reflected in the use of skills audits, employer surveys (see Ch. 11) and labour market studies. The advent of Training and Enterprise Councils (TECs) is likely to lend further emphasis to this trend. However, such surveys are often expensive to conduct, and great care is needed in the planning stages to ensure that they yield a representative picture. In the context of growing attention paid to labour market monitoring, unemployment, vacancies and to a lesser extent redundancies, are all recognized as important elements in a composite local picture.

Note: Some of the consequences for information collection and availability of the reorganization of training in the UK are referred to in the Editor's preface (p ix).

References

Bird D 1990 Redundancies in Great Britain. *Employment Gazette* **98**: 450–4

Commission of the European Communities 1989 *Employment in Europe* Office of Official Publications of the European Communities, Luxemburg

Cooper H 1989 *The West Midlands Labour Market* HMSO, London

Crouch C S 1982 Trends in unemployment – a comment. *Area* **14**: 56–9

Daniel W W 1972 *Whatever Happened to the Workers in Woolwich?* PEP, London

DE 1983 Statistics of redundancies and recent trends. *Employment Gazette* **91**: 245–59

DE 1985 Unemployment adjusted for discontinuities and seasonality. *Employment Gazette* **93**: 274–7

DE 1988 Unemployment statistics: revisions to the seasonally adjusted series. *Employment Gazette* **96**: 660–3

DE 1989a Measures of unemployment: claimant count and Labour Force Survey. *Employment Gazette* **97**: 443–51

DE 1989b Unemployment statistics – seasonally adjusted series. *Employment Gazette* **97**: Historical Supplement 1

Elias P, Owen D 1989 *People and Skills in Coventry* Coventry City Council, Coventry

Elton C 1985 Unemployment systems. In England J, Hudson K, Masters R, Powell K, Shortridge J (eds) *Information Systems for Policy Planning in Local Government* Longman, London pp 155–66

Galt V, La Court C 1989 Undertaking a skills audit: the Nottingham experience. *Local Economy* **3**: 263–71

Gillespie A E, Owen D W 1981 Unemployment trends in the current recession. *Area* **13**: 189–96

Gillespie A E, Owen D W 1982 Trends in unemployment – a reply. *Area* **14**: 59–61

Green A E 1985 Unemployment duration in the recession: the local labour market area scale. *Regional Studies* **19**: 111–29

Green A E 1986 The likelihood of becoming and remaining unemployed in Great Britain, 1984. *Institute of British Geographers Transactions New Series*, **11**: 37–56

Green A E, Coombes M G 1985 Local unemployment rates: statistical sensitivities and policy implications. *Regional Studies* **19**: 268–73

Harris C C 1987 *Redundancy and Recession* Basil Blackwell, Oxford

Hasluck C 1987 *Urban Unemployment* Longman, London

Hasluck C 1989 Local labour markets: information needs and issues. *Local Government Studies* **15**(4): 9–19

Haughton G, Peck J 1988 Skills audits – a framework for local economic development. *Local Economy* **3**: 11–19

Jones D R, Martin R L 1986 Voluntary and involuntary turnover in the labour force. *Scottish Journal of Political Economy* **33**: 124–44

Layard R 1986 *How to Beat Unemployment* Oxford University Press, Oxford

Mackay R, Jones D 1989 *Labour Markets in Distress: The Denial of Choice* Avebury, Aldershot

Marsh C 1988 *Job Census in Chesterfield* Report for BBC *Brass Tacks*

Meadows P, Cooper H, Bartholomew R 1988 *The London Labour Market* HMSO, London

Owen D W, Green A E 1989 *Creation of a Prototype National CALLMI Database* Report for the Training Agency, Institute for Employment Research, University of Warwick, Coventry

Smith E 1988 Vacancies and recruitment in Great Britain. *Employment Gazette* **96**: 211–13

Taylor D 1989 *Creative Counting* UU, London

Thatcher A R 1976 Statistics of unemployment in the United Kingdom. In Worswick G D N (ed.) *The Concept and Measurement of Involuntary Unemployment* George Allen & Unwin, London pp 83–94

Thomson I 1989 *The Documentation of the European Communities: A Guide* Mansell, London

Townsend A R 1982 Redundancy and the regions in Great Britain, 1976–1980: analyses of redundancy data. *Environment and Planning A* **14:** 1389–404

Townsend A R 1983a *The Impact of Recession* Croom Helm, London

Townsend A R 1983b The use of redundancy data in industrial geography. In Healey M J (ed.) *Urban and Regional Research: The Changing UK Data Base* Geo Books, Norwich, pp 51–63

UU 1988 *Researching Unemployment Locally* UU, London

UU 1989 Unemployment Unit briefing. In *Unemployment Unit Statistical Supplement January 1989* UU, London pp 1–4

White M 1983 *Long-term Unemployment and Labour Markets* Policy Studies Institute, London

4

Output, costs, profitability and investment

Peter Tyler

4.1 Introduction

This chapter begins by providing a brief indication as to why it is of use to urban and regional scientists to be able to obtain indicators of output, costs, profitability and investment at the local and regional level. The chapter then reviews the information currently available on these key variables in the UK from both government and identified private-sector sources. The chapter concludes by describing how the researcher might make the best of what is available using a methodology which can help to pull together disparate information sources and derive estimates for geographical variations in profits.

4.2 The need for information on geographical variations in output, costs, profitability and investment

The availability of good-quality statistics on geographical variations in output, costs, profitability and investment is essential if some of the most challenging questions concerning the economic development of urban and regional systems are to be addressed. Without evidence on regional output it is not possible to assess by how much regional growth rates may differ and whether they are tending to converge through time.

The available evidence which can be obtained from sources discussed in this chapter points to the existence of substantial differences in economic growth between the regions of the UK throughout the post-war period and with this recognition researchers have sought to understand the causes of this. Research has turned to assess whether there are differences between regions and within regions in the efficiency with which goods and services can be produced and the role of geographical variations in costs and productivity (Hart and MacBean 1961; Luttrell 1962; Fothergill *et al.* 1984).

That some regions have proven more attractive to mobile investment than others has been apparent from published evidence, and the degree to which the attractiveness of an area is associated with relatively lower costs and the potential for correspondingly higher profits (profitability) is clearly of prime research interest.

Perhaps the desire for information on spatial variations in costs and profitability and their contribution to the rationalization of economic activity across space has been at its keenest in understanding the factors behind the pronounced urban–rural shift of manufacturing activity in the UK during the post-war period. At the regional level the impact of regional policy instruments in improving relative cost disadvantages or perhaps attracting companies to high-cost areas and thus potentially impairing national efficiency, has also assumed considerable prominence.

In research addressed to these areas and many others, evidence on geographical variations in output, costs, profitability and investment are of central importance.

4.3 What is available?

General comments

The first observation which has to be made when considering the availability of statistics on output, costs, profitability and investment is that the available published data from government sources are relatively small in relation to other prominent statistics like employment. The geographical coverage of the data is mainly for standard regions or for country (Scotland, England, Wales and Northern Ireland) and the extent and quality of the data have varied significantly through time with, on balance, the quality of data at the spatial level deteriorating as a result of changes in the method of collecting and compiling the statistics. Much of the output information available is for the production industries and there is virtually nothing for the service industries other than that which can be inferred from the **Regional Account** data which will be discussed shortly. Sectoral disaggregation of data is extremely poor and erratic through time.

The availability of information on individual items of costs is extremely variable depending on the cost item considered. There are again substantial difficulties in obtaining data at below regional level, although data are produced in specific cases at county level and the information can usually be obtained from the government department concerned, although with the advent of the Tradeable Information Initiative such information can be quite expensive. The quality and availability of information on any major cost item at the regional or county level have varied considerably through time and this is particularly the case with respect to sectoral coverage at very fine levels of disaggregation.

There are no published statistics on profitability at the regional or subregional level and the researcher interested in these variables is forced to adopt considerable ingenuity in deriving surrogate measures or turn to information provided from private-sector data sources which provide accounting ratio information mainly derived from company accounts. Some information on gross trading profits is available in the Regional Accounts. This is examined later together with a methodology with which to modify Census of Production information in order to provide some tentative indication of regional variations in profits.

Published data on geographical variations in investment are particularly scarce and it is difficult to find much that is both comprehensive in coverage and at the same time capable of providing disaggregation by sector. Information is probably at its best when dealing with countries (Scotland, Wales, Northern Ireland and England). Researchers have often been forced to turn to piecemeal survey-based data (i.e. like the CBI survey of business intentions) which is often not sufficiently disaggregated to be of much direct use in rigorous empirical work.

Major sources of data on output

With respect to output there are three principal sources of information which are provided for the UK or Great Britain as a whole and which have been adapted in specific ways to provide information for specific geographical areas where the emphasis has nearly always been regional or country (England, Scotland, Wales and Northern Ireland). These sources are the Census of Production, the Index of Industrial Production and the Regional Accounts.

The Census of Production

The main source of information on production in the UK is the annual **Census of Production**. The first annual census began in 1970 and before this date information was produced on a quinquennial basis. The current annual census provides information at the national level on working proprietors, employment, wages and salaries, stocks, capital expenditure, cost of industrial and non-industrial services, purchases of materials and fuel, sales, work done and services rendered and indirect taxes. Some of this information is provided at the regional level, although there are limitations which must be taken into account in interpreting what it actually means (which are considered shortly). The last year for which the Census of Production produced any information at below the level of standard region was 1968, when some information was provided for the large British conurbations. The provision of Census of Production data for the standard British conurbations dates from 1951 and it is possible to derive conurbation census data for 1951, 1954, 1958,

1963 and 1968. Researchers interested in these data are referred to Tyler and Rhodes (1985). Nothing has been available in published statistics at below the regional level since this time, although it is possible to acquire special tabulations from the Central Statistical Office (CSO), formerly the Business Statistics Office (BSO), again heavily qualified in terms of what it actually means. In general the level of industrial disaggregation available at the regional level has been good, at least at the level of standard order (1948, 1958, 1968 Standard Industrial Classification (SIC)) and division (1980 SIC) which means that it is possible to undertake some continuous time-series work. A minor problem arose in the 1970 and 1971 censuses. In those 2 years the CSO abandoned the SIC order number presentation in favour of only five industry groups. Fortunately from 1973 onwards the CSO reverted to the full standard order and then division presentation.

Net output is a standard census variable. (The census definition is provided in Appendix 4.1.) It is derived by deducting from gross output the cost of purchases (reduced by the rise, or increased by the fall, during the year of stocks of materials, etc.) and payments for work given out to other establishments. Net output is thus not the same as value added, or gross domestic product (GDP) as measured in the *National Income Blue Book*. The main difference is that 'bought-in services' are included in net output. They are clearly not part of value added or GDP. This is important because bought-in services are a significant and increasing proportion of net output (Tyler and Rhodes 1985). Moreover the proportion which bought-in services takes varies considerably between industries and thus differences in regional industrial structure affect the amount and the growth of bought-in services which will exist in any particular area. Changes in levels and growth of regional net output may be affected by the volatility of the bought-in services components. Therefore census net output may provide a distorted picture of the way in which value added or GDP may be changing at the local level.

Table 4.1 shows more clearly that bought-in services and insurance contributions represent a very large and rapidly growing part of the net

Table 4.1 The components of UK manufacturing net output, 1951–75. (*Source:* Tyler and Rhodes 1985: 333.)

	Census net output (£m.)	Wages and salaries (%)	Bought-in services (%)	Employers' insurance contibutions (%)	Gross profits residual (%)
1951	4 849.1	54.1	8.5	3.3	34.1
1954	6 234.1	55.0	8.5	2.8	33.7
1958	7 853.4	56.7	8.5	3.4	31.4
1963	10 819.7	52.7	8.4	3.6	35.3
1968	15 289.2	50.8	9.7	4.0	35.5
1973	26 600.0	48.2	10.8	4.8	36.2
1975	36 948.0	51.2	12.3	6.3	30.2

Table 4.2 The components of net output in UK manufacturing industries, 1975. (*Source:* Tyler and Rhodes 1985: 333.)

	Net output (%)	Wages and salaries (%)	Net output minus wages and salaries (%)	Payments for bought-in services (%)	Employers' insurance contributions (%)	Gross profits residual (%)
Food, drink, tobacco	100.0	37.6	62.4	14.1	4.7	43.6
Coal, petroleum products	100.0	12.5	87.5	6.2	2.9	78.4
Chemicals	100.0	36.7	63.3	16.3	6.0	41.0
Metal manufacture	100.0	61.5	38.5	12.1	9.3	17.1
Mechanical engineering	100.0	53.2	46.8	11.4	6.3	29.1
Instrument engineering	100.0	60.0	40.0	12.0	7.0	21.0
Electrical engineering	100.0	55.0	45.0	9.7	6.5	28.8
Shipbuilding	100.0	82.0	18.0	6.6	9.7	1.7
Vehicles	100.0	67.3	32.7	10.0	8.0	14.7
Other metals	100.0	54.7	45.3	11.0	6.3	28.0
Textiles	100.0	57.3	42.7	12.2	6.3	24.2
Leather	100.0	51.3	48.7	12.4	5.3	31.0
Clothing	100.0	59.7	40.3	9.8	5.8	24.7
Bricks, pottery, glass, cement	100.0	46.1	53.9	17.1	5.5	31.3
Timber	100.0	52.3	47.7	13.3	5.5	28.9
Paper and printing	100.0	53.2	46.8	14.4	6.5	25.9
Other manufacturing	100.0	51.6	48.4	13.6	6.0	28.8
Total	100.0	51.2	48.8	12.3	6.3	30.2

output minus wages residual. Table 4.2 shows, for 1975, that payments for bought-in services and insurance contributions accounted for 18.6 per cent of net output. Table 4.2 also shows how this proportion varied in the major industry groups. Given that these industry groups are located unevenly through the regions, to draw conclusions about regional efficiency from the net output minus wages residual is misleading. Table 4.1 shows that the proportion of net output accounted for by bought-in services and insurance contributions has grown from 11.8 per cent in 1951 to 18.6 per cent in 1975.

Besides the difficulty of deriving value added information from the Census of Production there is a further major problem with the quality of the data which are available. This is the problem of the *combined return*. Census data are collected at the establishment level. It is quite often the case that activities which are conducted as a single business are carried on at a number of addresses. Where this is so, the CSO asks companies to provide the full range of separate information in respect of each address, whether or not the activities are different. However, as the CSO makes clear in its definitions at the front of all the censuses, for some companies activities carried on at separate physical locations may be integrated to such an extent that they constitute a single establishment. In some cases, a business may involve the assembly, packing, painting, etc. of components from several sites, and the measuring of sales, for example, from an individual site may be rather obscure. A further complication is that many businesses these days only process information at 'establishment' level, not at the individual site level. The 'establishment' is defined to cover the combined activities of the operators of a business at different addresses (termed local units). The consolidated, or combined return, represents the aggregate position for all the plants in the firm. (The only information which multi-plant firms are required to provide on an individual plant-specific basis is employment and capital expenditure (net expenditure on land and existing buildings and all other assets).

To overcome partially the combined returns problem, the CSO have adopted various procedures to allocate the combined return information back to the constituent plants in the regions using the employment data which firms are required to provide on a plant basis. These procedures started in 1970. However, during the 1970s the reallocation procedures changed. Up until 1973, all the output information on the combined returns were reallocated to the regions on a pro rata basis for each industry *en bloc*. Since 1973 the pro rata exercise was completed for the information on each combined return separately. The latter is clearly preferable. One important consideration is that there is a statutory obligation upon the CSO to include separate statements relating to Scotland, Wales and Northern Ireland. The data for these countries should be better than English standard regions as a result.

During the 1970s the CSO adopted what is known as the 80 per cent rule. Under this rule, net output information was allocated by the CSO to the region where individual firms had 80 per cent or more of their activity allocated. All net output which could not be allocated to regions under this rule was left in a single unallocated entry. In the 1980s the CSO have moved to allocating net output from companies with addresses in more than one region on the basis of the employment at each address. (The same allocation principle is used for wages and salaries and gross value added.)

Some indication of the distorting effects which the combined return may have on census variables for geographical areas can be gained by examining the work of Tyler and Rhodes (1985). Table 4.3 presents a table derived from their work which shows the proportion of total net output recorded by the CSO for each region which came from establishments which had manufacturing operations solely or over 80 per cent of their total, in that region. The region least affected by combined processing distortions was that of Northern Ireland – over 90 per cent of its net output reported arose from establishments which had more than 80 per cent of their employment in the region. Scotland and Wales also have a high proportion of such establishments. In the case of East Anglia, where many plants are branch plants of parent companies located in other regions, only 53.7 per cent of total net output attributed to that region by the CSO in 1975 came from establishments with over 80 per cent of their employment in that region. The South East is also badly affected, with a high incidence of multi-region establishments whose headquarters are located in the region but the majority of whose production is in plants located in other UK regions.

The second point which is apparent from Table 4.3 is the extent to which the combined return problem is increasing through time. In the 3 consecutive years chosen, 1973–75, the problem became increasingly serious for the majority of regions. It is to be expected that the severity of the problem

Table 4.3 An indication of the combined return problem.*
(*Source:* Tyler and Rhodes 1985: 335.)

Region	1973	1974	1975
North	67.3	65.6	67.7
Yorkshire and Humberside	65.6	64.3	62.5
East Midlands	77.2	67.4	66.0
East Anglia	60.4	57.6	53.7
South East	59.7	58.8	54.6
South West	61.4	61.5	56.2
West Midlands	64.2	54.8	56.4
North West	63.7	58.6	57.0
Wales	76.9	78.3	64.2
Scotland	79.4	78.1	72.3
Northern Ireland	93.7	94.0	93.2

*Manufacturing net output returned by establishments with more than 80% of their employment in the region as a proportion of total net output in the region.

is likely to have increased during the 1980s although it would need special tabulations from the CSO to confirm this.

Factors like the presence of bought-in services in census net output and the combined return means that considerable care must be taken when seeking to use Census of Production data at the regional level.

Index of Industrial Production

A further source of information on industrial output is the **Index of Industrial Production**. This index is available at the *country* level only but is still a valuable source that has been used rather infrequently by spatial researchers. It is possible to obtain the index with industrial disaggregation to varying degrees for the UK, Scotland, Wales and Northern Ireland in published sources. This spatial disaggregation has proven to be of use when it comes to analysing the impact of regional policies on the depressed regions, since it is possible to use modified techniques of shift-share to allow for the impact of differences in industrial structure.

Further information on the UK index of production (which can often be used to compare movements by sector in the countries of Scotland, Wales and Northern Ireland, can be obtained from the *Annual Abstract of Statistics* (containing yearly indices by broad sector), the *Economic Trends Annual Supplement, Economic Trends* (indices by broad sector), the *Monthly Digest of Statistics, Business Monitor, the United Kingdom National Accounts* and the *Press Notice* issued by the CSO.

The appropriate sources for Scotland are the *Scottish Economic Bulletin* (which has indices of industrial production by market sector and manufacturing industry) and the *Scottish Abstract of Statistics* (which contains a market sector analysis and a breakdown by manufacturing industry). *Welsh Economic Trends* provides seasonally adjusted indices similar to those for Scotland. The *Digest of Welsh Statistics* and *British Business* provide more detail. The *Northern Ireland Annual Abstract of Statistics* is the appropriate source for Northern Ireland and contains indices of output for all industries, broad sectors of industry and for the manufacturing sector by industry headings.

Regional Accounts

Another source of output data at the regional level is provided by the Regional Accounts which appear in *Regional Trends*. The Regional Accounts data provide a considerable body of information which can be used by regional scientists, although it should be realized that considerable manipulation has been required to build the accounts up from disparate sources of data and for some variables the quality of the data is not all it might be. (The sources and methodology used to produce the Regional Accounts are described in CSO 1978.) The Regional Accounts provide estimates by region of GDP analysed by factor incomes and industry groups. Information is also given on personal income, consumers' expenditure and gross domestic fixed capital formation by selected industrial sectors. The time period of the available data extends from the mid 1970s for some variables. Most data are available from the early 1980s, and the information is computed for all eleven standard regions.

Major data sources on costs

A useful starting-point from which to consider the information which is currently available on geographical variations on major items of costs is to refer to the cost structure presented in Fig. 4.1. This refers to the importance of individual items of cost in gross output for all manufacturing in the UK in a recent year (1980). For other sectors, most notably private services, the proportion of individual cost items would obviously vary.

Labour costs

It is clear that labour costs are one of the most important items of cost for which it is important to obtain geographical information. The appropriate source here is the *New Earnings Survey* (*NES*) undertaken by the Department of Employment (DE). The main objective of the *NES* is to provide annual information on the level of earnings of employees in all industries and

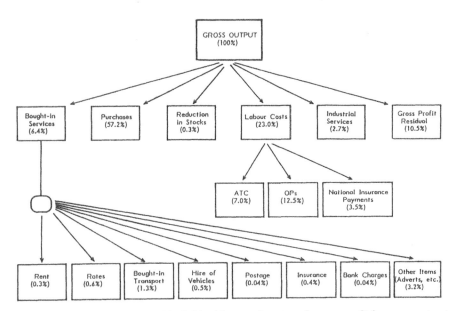

Fig. 4.1 The importance of individual items of cost and gross profit in gross output for all UK. (*Source:* Tyler *et al.* 1988: 21.)

occupations. At the local level earnings data by occupation are published by region and by manual/non-manual occupations for English/Welsh counties and Scottish regions (although further information on occupations by county can be acquired from the DE on request). A similar, but separate, survey is conducted by the Department of Economic Development in Northern Ireland.

A significant advantage of the *NES* data is that they have been produced in a fairly similar form for each year since 1970, and while geographical boundaries have changed it is possible to undertake considerable cross-sectional and time-series analysis. At the level of standard region there is considerable scope for occupationally disaggregated analysis, although small sample sizes can be a problem in specific cases. A description of the methods used to compile the *NES*, and the strengths and limitations of the data source are available in Part A of the *NES* publication produced by the DE and published by HMSO each year (in five parts, of which Part E provides the geographical disaggregation).

A further very useful source of information on labour costs is that produced by the Reward Group entitled *Reward Regional Surveys*. This source contains a range of labour cost and cost-of-living data which is provided down to individual towns in some cases.

Rents

There are no sources of information on geographical variations in commercial property rents published by government departments. However, this gap has been filled to a considerable degree in recent years with the publication of data on a regular basis by some of the major property companies in the City. The information varies by the sector covered and the spatial disaggregation, but examples of the extensive information available can be obtained from Jones Lang Wootton Consulting and Research. In many cases there is little charge. Further information on commercial property data sources is given in Chapter 5.

Rates

The main source of information on geographical variation in rates is the Chartered Institute of Public Finance and Accounting (CIPFA). With the introduction of the uniform business rate the main factor which will cause geographical variations in the rate bill paid by companies will be the rate and value of the properties occupied, and information on the average rateable values applicable to areas will be obtainable from CIPFA (*Financial and General Statistics*). Other relevant CIPFA publications include the *Statistical Information Service* and *Local Government Trends*. Other relevant sources of government statistics include *Financial Statistics*, *Regional Trends*, the *Digest of Welsh Statistics*, the *Scottish Abstract of Statistics* and *Financial, General and Rating Statistics*.

Transport costs

There are no comprehensive data sources which provide information on how the costs of transporting goods or people vary between locations within the UK. The work which has been undertaken to investigate geographical variations in such costs has usually been specific to one region (e.g. Scottish Office 1981; Peida 1984). Existing work has also been based on an *ex-post* assessment of transport costs. That is, the costs of transport to a company are those after the company could have orientated itself geographically in order to minimize the cost of its transport bill. The position for a new company may be quite different. A methodology with which to draw together information from a range of disparate sources is provided in Tyler and Kitson (1987). The approach reported in that paper demonstrates how information can be obtained from a range of public- and private-sector bodies to form an index of likely transport cost variation in Great Britain. A source of useful information to address aspects of the transport question is provided annually by the Department of Transport (DTp) in *Transport Statistics of Great Britain* (see also Ch. 19).

Energy costs

Energy costs can be considered in five main categories; coal/coke, gas, electricity, fuel (fuel oil and gas oil), and a miscellaneous category mainly comprising liquid gas. The evidence suggests (Tyler *et al.* 1988) that for some of these categories of fuel there are considerable geographical variations, the incidence of which is of interest to the regional scientist. Information is available from a variety of sources including the *Digest of United Kingdom Energy Statistics* (mainly at regional level), *Regional Trends*, the *Scottish Abstract of Statistics*, the *Digest of Welsh Statistics*, the *Northern Ireland Annual Abstracts of Statistics* and also information from the major utilities like British Gas (British Gas Corporation *Annual Report and Accounts*), British Coal and the Central Electricity Generating Board (see also Ch. 15).

Insurance costs

Information on geographical variations in insurance and related costs is best obtained from large private-sector insurance groups which will usually provide information for a high level of spatial disaggregation on request at no charge. The information can be obtained for industry and commerce separately.

Major data sources on profits and profitability

Information on profits is particularly hard to come by at the geographical level which is not surprising given the difficulties posed by multiregional and multinational companies allocating value added to specific establishments and the difficulties posed in trying to obtain this information in a statistical rigorous manner through time.

Some information on gross trading profits and surpluses is provided in *Regional Trends* for all sectors. There is further information provided in the *Scottish Abstract of Statistics*, the *Scottish Economic Bulletin*, the *Digest of Welsh Statistics* and the *Northern Ireland Annual Abstract of Statistics* which provide information on the factor income of profits in GDP of those countries.

Statistics for specific sectors is particularly poor. The *Digest of Welsh Statistics* provides some information on the gross trading profits of companies. The absence of sufficiently disaggregated information on profits by sector for regions or local areas makes it particularly difficult to draw any firm conclusions on geographical variations in profits. Attempting to derive estimates of profitability is made even more difficult given the poor quality of capital stock data. It is for these reasons, as well as the difficulty of standardizing for the wide range of factors which can influence the profits and profitability of companies at the local level, that researchers have adopted other approaches than simply appealing to published sources. These other approaches have taken two main forms. The first is to attempt matching pooled sampling of companies and the use of establishment-based information obtained from intensive (and usually expensive) surveying procedures. The work by the Centre for Interfirm Comparison (1977) is an example of this. The other approach has been to develop cost-accounting frameworks at the local level and to derive profits (and, in some cases, profitability) as a residual. These approaches are examined in section 4.4.

Major data sources on investment

The main source of information on investment at the regional level is *Regional Trends*. It contains statistics on gross domestic fixed capital formation by industry categories and by function of government for the UK standard regions. There is not, unfortunately, any information below the standard regions.

Other sources include the *Scottish Economic Bulletin* which provides gross domestic fixed capital formation by industry and the *Northern Ireland Annual Abstract of Statistics* which gives gross domestic fixed capital formation by broad groups of industries.

4.4 Making the best of what is available; an approach which draws together published statistics

When considering cost and profit variables, there are limitations of available published data, and the expense and difficulties provoked in one-off survey-based approaches; some attention has then been given recently to developing a methodology which can bring together information from quite disparate data sources to provide some indication of geographical variations in costs and profits for both manufacturing sectors and particularly private service-based industries for which published information is so obviously deficient.

For those researchers who are interested in identifying how costs and thus profits actually vary for existing establishments across the settlement pattern of the UK, there is no other method of obtaining these data than survey-based approaches combined with company account data. Besides the obvious expense of these approaches there will always remain difficulties in obtaining statistically reliable findings. An alternative approach to assessing how costs and profits vary between existing establishments is to adopt an a priori based accounting framework which allows the synthesis of existing published data from the Census of Production and/or published survey results and a wide range of disparate sources of cost information. A full discussion of this approach can be found in Tyler *et al.* (1988), but a brief insight is possible to conclude this chapter.

A costs-based accounting framework

The first step in establishing a costs-based accounting framework which can be used to bring together disparate data sources and suggest their possible impact on profits (and, with certain assumptions, profitability) is to obtain an indication of the main components of costs for British manufacturing or private service companies.

For manufacturing companies the Census of Production can be used to provide the cost structure for over 100 different sectors of British industry. Evidence on the cost structure of service companies has to be assembled from company accounts or other establishment-based sources.

Figure 4.1 presented earlier in this chapter provides the required information for all manufacturing industry in the UK in 1980 derived from the Census of Production. Each cost component is examined to assess whether there is some possibility of geographical variation. The prime contenders for geographical variations are wages and salaries, rates, rents, energy costs, transport costs and local bought-in services like insurance. Evidence on geographical variations in these costs can be obtained from the sources discussed in section 4.3 and the geographical variation can be indexed relative to the national average. The value of each cost item nationally as given by the Census of Production (or for services cost structures derived from elsewhere) is then taken as a weight and multiplied by the appropriate index of local cost variation relative to the national average for the geographical areas for which spatial cost data can be obtained.

Table 4.4 presents an example of the approach. The table shows the cost structure for the telecommunications industry using national weights obtained from the Census of Production. The principle is illustrated with respect to a hypothetical area A in the country. Column 2 in Table 4.4 provides an

Table 4.4 Generating local industrial cost structures.
(*Source:* Tyler *et al.* 1988: 20.)

	Cost structure for industry nationally (weights) (1)	How units input costs compare in Area A relative to England (2)	Costs for national industry if located in Area A, i.e. (1) × (2) (3)
Purchases (including energy)	1927.6	1.00	1927.6
Stocks used	0.0	1.00	0.0
Cost of industrial services	171.8	1.20	206.1
Rent	32.2	1.30	41.9
Rates	25.1	1.40	35.1
Insurance premiums	22.5	1.20	27.0
Bank charges	2.1	1.00	2.1
Other non-industrial costs	181.4	1.00	181.4
Wages	637.6	1.40	892.6
Salaries	757.3	1.30	984.5
Total costs	3757.6		4298.4

indication of how each item of cost compares in area A relative to the national average. (For items of cost assumed not to vary between areas, the index is set at one.) These indices of relative cost are then multiplied by the national cost weights from column 1. Column 3 gives the resulting cost bill if a firm in the industry was located in area A and paid unit input costs pertaining to area A. The cost index thus provides an index of the geographical variations in costs which would face a firm considering a location in the area used in the analysis and adopting the same factor mix in every area.

The procedure thus assumes coefficients derived from the UK input/output tables (CSO 1988) which represent an average of all companies in the UK. Companies starting up in an area for the first time will have a different cost structure from established companies (the average given by the national input/output tables). Clearly the weights can obviously be varied for the more typical establishment required. This procedure can be used to provide how profits might be expected to vary for a recent year in England for a typical company hypothesized to be locating anywhere in the counties of England and paying the cost per unit of input relevant to the area concerned. The variations in profits so derived are presented in Fig. 4.2.

While results derived in this way are based on average national factor mixes and do not take into account differences in the productivity of inputs between areas (or the factors which affect productivity for existing companies like age of capital, etc.) they do form a starting-point from which accounting results can be 'shadowed to reflect the likely impact of these other factors. As long

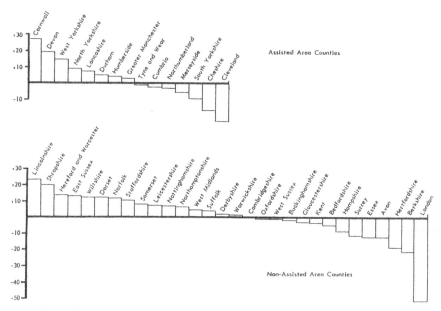

Fig. 4.2 The percentage deviation of gross profits in each county from average gross profits in non-assisted areas (excl. London). (*Source:* Tyler *et al.* 1988: 61.)

as published statistics remain so inadequate in the UK for the key variables which are the focus of this chapter, researchers into urban and regional issues will be forced to adopt such procedures or undertake their own local surveys at considerable expense and possible loss of statistical reliability.

4.5 Summary and conclusion

For those researchers interested in acquiring information on output, some costs and investment for the production industries at the regional level, and within this principally the manufacturing sector, the Census of Production is of value, although the previous sections have drawn attention to some of the grave limitations which must be recognized.

It is possible to adjust Census of Production data to allow for the distortion provoked by inclusion of bought-in services in the net output variable. It is also possible to make tentative estimates of a census-based gross profits residual and the researcher interested in a methodology with which to do this is referred to Tyler and Rhodes (1985) for further information. The problems posed by the combined return will, however, remain and seem, if anything, to be increasing in intensity. Moreover, for geographical areas below that of the standard region, the Census of Production remains of no direct use.

The only comprehensive sources of regional output (GDP) and investment statistics (gross domestic fixed capital formation) remain those described in this chapter and there remains little that can be done to improve the availability of these data by sector (particularly the private services sector) and by local area (below standard region) unless the government can be persuaded to produce more statistics, which looks exceedingly unlikely at present.

In recent years, given the paucity of published information on the important variables described in this chapter, researchers have turned to survey-based approaches to obtain information, often combined with data from company accounts. Examples of this include the work by the Centre for Interfirm Comparison (1977). By necessity most of this survey-based work is partial in terms of both its sectoral and geographical coverage.

This chapter has described the range of data available at present to the researcher interested in obtaining key output, costs, profitability and investment data at the regional and subregional level. Inevitably it will not have covered all the possibilities which exist for piecing together data, and two extremely valuable references are highly recommended for checking both government and non-government sources. The first, concerned with mainly published government material, is published by CSO (1990) *A Guide to Official Statistics*. This is a first-class summary of the data that are available and is a must for any research team. The second, concerned mainly with non-government sources across a very wide range of data, has been produced by the Business Information Service at the University of Warwick (Mort 1990). This volume contains details of virtually all the known private-sector providers of key information and thus the range of data available from sources like *Reward Regional Surveys* and CIPFA referred to in this chapter.

The case for more statistics on output, costs, profitability and investment at the regional and subregional level in the UK is clear. Without such information many of the fundamental questions relating to the causes of interregional disparities in economic growth and development in the UK will remain unanswered, as well as the effect of regional policies designed to equalize growth between regions. Sadly the position has worsened considerably during the 1980s and it can only be hoped that the recent recognition by the government that there is an inadequate statistical base at the national level will filter down to the regional level as well.

Appendix 4.1 Important terms used in Census of Production (as given in *Business Monitor*, CSO PA 1002)

Gross output This is calculated by adjusting the value of *total sales and work done* by the change during the year of *work in progress* and *goods on hand for sale*

Net output This is calculated by deducting from *gross output* the cost of *purchases* of materials for use in production and packaging and fuel and *purchases of goods for merchanting or factoring*, the *cost of industrial services received* and is adjusted for net duties and levies, etc., where applicable. Purchases are adjusted for changes during the year of *stocks of materials, stores and fuel*

Wages and salaries This represents amounts paid during the year to *administrative, technical and clerical employees* and to *operatives*. All overtime payments, bonuses, commission, holiday pay and redundancy payments less any amounts reimbursed for this purpose from government sources are included. No deduction is made for income tax or employees' National Insurance contributions, etc. Payments to *working proprietors*, payments in kind, travelling expenses, lodgings allowances, etc. and *employers' National Insurance contributions, etc.* are excluded

Capital expenditure This represents the value charged to capital account together with any other amounts which rank as capital items for taxation purposes during the year to which each return related. Where expenditure is spread over more than one census year, payments are included in the years in which they were made

---

References

Centre for Interfirm Comparison 1977 *Management Policies and Practices and Business Performance* Centre for Interfirm Comparison, Colchester

CSO 1978 *Regional Accounts* Studies in Official Statistics No 31 HMSO, London

CSO 1988 *Input–Output Tables for the United Kingdom 1984* HMSO, London

CSO 1990 *Guide to Official Statistics* No 5 HMSO, London

Fothergill S, Kitson M, Monk S 1984 Differences in the profitability of the UK manufacturing sector between conurbations and other areas. *Scottish Journal of Political Economy* **31** (1): 72–91

Hart P E, MacBean A I 1961 Regional differences in productivity, profitability and growth. *Scottish Journal of Political Economy* **VIII** (1): 1–11

Luttrell W F 1962 *Factory location and industrial movement* 2 vols National Institute of Economic and Social Research, London

Mort D 1990 *Sources of Unofficial UK Statistics* 2nd edn University of Warwick Business Information Service, Gower Press, Aldershot

Peida 1984 *Transport Costs on Peripheral Regions* Report to the European Commission, Industry Department of Scotland and Department of Economic Development, Northern Ireland

Scottish Office 1981 Transport costs in Scottish manufacturing industries. *Scottish Economic Bulletin* No. 22

Tyler P, Kitson M 1987 Geographical variations in transport costs of manufacturing firms in Great Britain. *Urban Studies* **24**: 61–73

Tyler P, Moore B C, Rhodes J 1988 *Geographical Variations in Costs and Productivity* HMSO, London

Tyler P, Rhodes J 1985 The Census of Production as an indicator of regional differences in productivity and profitability in the United Kingdom. *Regional Studies* **20:** 331–9

5

Floorspace and commercial developments

Martin Perry

5.1 Introduction

The availability of data on commercial floorspace has not responded to
the widening interest in economic development by public authorities. The
political importance of housing and the responsibility of local authorities for
the welfare of their population led to the collection of data on the condition
of the housing stock and levels of investment in housing. In contrast, the
growth in concern for employment has not produced comparable data for
commercial property, although the provision of land and buildings is often
the main component of a local economic strategy. This chapter covers four
types of commercial property: manufacturing, warehousing, retail and office.
Across these sectors the level of information varies, but in most cases the
available information remains more limited than the residential sector.
Following recent cut-backs in the publication of commercial floorspace
data, researchers are now fortunate if they are able to identify how much
property their area contains. Knowledge of more qualitative aspects of the
built stock still depends largely on original survey collection.

The deficiency in the commercial property database exists despite the
collection of much relevant information by the Inland Revenue Valuation
Office. The excessive encouragement of secrecy and expenditure reductions
by the present government prevents the release of information collected in
the normal function of the Valuation Office. As the costs and confidentiality
question are minor, a greater release of information should eventuate, but
until this happens the database remains uneven.

Estate agents and property analysts collect data related to the interests of
their clients and many of these are published and available to researchers.
Such data reflect an interest in property as an investment commodity and
tend, therefore, to concentrate on the financial performance of new leasehold
buildings in the most sought-after locations. The information collected by

private agencies follows (and seeks to influence) investment trends, as shown by a recent increase in the monitoring of the retail sector, particularly in terms of managed shopping centres.

The much larger stock of owner-occupied buildings and rented property in locations not favoured by financial institutions is overlooked in private databases. Of course, this is a large and highly varied market and therefore difficult to sample. The stock of property changes through the addition of new buildings, the demolition of existing and the modification of older buildings. There are different routes through which these changes are accomplished: property developers provide speculative and bespoke premises, and occupants may undertake their own developments. There are different types of developer, including public agencies motivated primarily by employment generation, developers with no long-term interest in the property and developers who are also investors.

The research worker investigating the commercial property market potentially requires a range of information. Among the more important issues are: the condition of the built stock and design attributes relative to changing needs, so that the suitability of buildings and need for renewal can be judged; changes in the stock of floorspace caused by demolitions and changes of use, as well as new additions; changes in the intensity of floorspace use as indicated by vacancy rates, employment densities and output per unit of floorspace. Information on the accommodation of particular activities such as small manufacturing firms, technology-based enterprise and the availability of property suited to new firms is required when developing specific policy measures. For the most part, published data sources are inadequate to accomplish any of these tasks and are most deficient when dealing with individual settlements rather than broad regional trends. The withdrawal of the *Commercial and Industrial Floorspace Statistics (CIFS)* from England (they continue to be published for Wales) means that in most LA areas it is not even known what the total stock of floorspace amounts to or how this has been changing. Improvements and innovations have arisen in the data collected by private organizations and on the particular issues they address, a rich base for analysis now exists. None the less, the overall picture is of a fragmented collection of sources with original data collection often the only option to complete the floorspace component of a local or regional economic strategy.

In this chapter buildings are classified according to the nature of the occupant (current or intended), rather than the character of the property which may be acceptable to several types of activity. This form of classification can be ambiguous where small enterprises are involved due to their propensity to switch activities as the firm develops (Kennett 1985). A further difficulty is the increasing proportion of property supplied for use by warehousing, services and manufacturing. The merging of property types partly reflects the functional coalescence between firms as manufacturing has become more service orientated and warehousing has moved to single-storey buildings in industrial areas. The extent of this coalescence is, however, sometimes overstated. Hennebery (1988) points out that while with modest alteration a factory may be used as a warehouse, a warehouse is less easily converted into a factory. Warehouse and manufacturing floorspace tends to be combined in the data provided by estate agents because investors prefer buildings that may be occupied by the widest range of tenants, while purely industrial buildings are considered less attractive. In the present chapter, as far as possible, a distinction between manufacturing and warehousing space is maintained.

Before discussing the data sources relating to individual sectors, a short outline of the significance of floorspace data in local and regional economic analysis is provided to set the data sources in context. The review of data sources follows, commencing with those that span all types of activity.

5.2 Using floorspace data

Floorspace statistics monitor development activity. It is conventional to view the demand for property as a derived demand, in that buildings are required for the activity they facilitate rather than as an end in themselves. Floorspace data could be used as a surrogate for other measures of economic activity, although in practice this is unlikely due to the unavailability of data and variations in the use made of buildings. Estimating the amount of floorspace in a region from other indicators such as employment or output is similarly unsatisfactory. One of the most important reasons for requiring separate data on floorspace is that the intensity of property use is highly variable. For example, in comparing the employment densities of manufacturing and warehouse property, Tempest (1982) noted how the range of densities within these categories was often larger than the difference between the averages.

Temporal contrasts in floorspace use are also significant. Over time, output and/or employment ratios per unit of floor area may change. For example, there has been a long-term decline in the number of employees per unit of factory floorspace: from 31 per 1000 m^2 in 1964 to 20 per 1000 m^2 in 1985, or around 3 per cent per annum (Fothergill *et al.* 1987). In contrast, the long-term output per unit of floorspace ratio has remained constant, although there have been wide short-term variations. The consequence has been that an increased volume of floorspace has been required to accommodate the growth in output. The advent of microprocessor technology which has facilitated a radical alteration in production methods, and the possibility of divergent patterns between industries, suggests that neither trend can be safely assumed to continue. Such changes in floorspace densities need to be understood when relating space requirements to output and employment projections and determining the levels of service provision to different types of property.

A second reason for measuring floorspace is the importance of commercial property as an investment commodity, rather than being supplied solely in response to demand, for owner occupation. This shift in the motivation for property provision gives rise to the possibility of conflict between owners and occupants. The conflict may arise over the type of provision, its design attributes, management and location. Moreover, the flows of money into property do not necessarily reflect demand, but the condition of financial and investment markets and increasingly investment decisions are made within an international context. The speculative provision of property is also undertaken by public development agencies in an attempt to stimulate economic activity. Rather than property provision reflecting demand, the supply side may be more significant in determining where development occurs. In the late 1970s, for example, the reluctance of private institutions to invest in small-business property was alleged to be frustrating the growth of small enterprise.

The transformation of property provision into an investment commodity should not obscure the modest scale of new provision compared with the existing stock. New provision represents typically just 2 or 3 per cent of the total stock. The suitability and condition of existing buildings are therefore of prime importance in determining floorspace efficiency and a third reason for requiring floorspace data. To take another example from the industrial property market, the unsuitability of a large proportion of the existing factory stock is creating widespread difficulties in British industry (Fothergill *et al.* 1987). These problems arise from the inheritance of old buildings, designed for earlier technologies, and the sustained underinvestment in new property. In the office sector, the increasing use of information technology (IT) is changing the demands for accommodation, requiring the refurbishment or replacement of an increasing proportion of existing provision (Barras 1987).

There are, therefore, three aspects to floorspace in which researchers are potentially interested, namely:

1. *The characteristics of the building and the site.* Here the focus is on the physical and financial attributes of property such as age, layout, rental, size, room for expansion and floorspace/plot ratios. These data assess the operational suitability of buildings and volume of production space.
2. *The ownership characteristics.* The basic distinction is between owner-occupied and leasehold premises, but identifying individual landlords may sometimes be important. Ownership details identify levels of investment by different types of developer/investor which is of interest in studies of urban change and because design and management specifications can vary according to the agency involved.
3. *The characteristics of occupants.* The prime issue is whether a property is occupied and, where appropriate, the causes of vacancy. The characteristics of occupants may be required for the purposes of evaluating the impact of property provision and to assess the accommodation needs of particular activities.

5.3 Multi-sector data sources

Building and site characteristics

The single most comprehensive data source on commercial floorspace is the *CIFS*. Unfortunately, *CIFS* are now available only in respect of Wales. The Department of the Environment (DoE) last published the English floorspace statistics in 1985 after commencing publication in 1964 in *Statistics for Town and Country Planning* (changed to *Commercial and Industrial Floorspace Statistics* in 1974). Data for 1986 had actually been collected but were not published by the DoE. A summary of the English data for 1986 was subsequently compiled by Hillier Parker (1988), providing regional totals by use class and changes in the stock since 1983. In Wales, the data continue to be published annually by the Welsh office in *Commercial and Industrial Floorspace Statistics: Wales*. The Welsh publication provides stock and components of change data by local authority areas. Floorspace data have never been available in Scotland or Northern Ireland.

The procedure for collecting the floorspace data and resulting output are the same for Wales as they were in England. The Valuation Office of the Inland Revenue obtains floorspace measurements as part of its valuation procedure for rating purposes. 'Effective floorspace' is measured, which covers the usable area excluding the space taken by features such as pillars, stairwells and plant rooms. The data are collected by local valuation offices as part of their continuous monitoring of the property market, with returns made to the valuation head office every 3 months. As a consequence, the data have two particularly attractive features: they are regularly updated and available by district, county and national boundaries. Two other features add to the usefulness of these data. First, the data are available for eight separate use classes: central government, local government, shops and restaurants, shops with living accommodation (the residential component is excluded), warehouses, stores and workshops, warehouse open land and industrial; and in each case the number and floorspace of hereditaments (single buildings or part of a building if rated separately) are measured. Second, the annual components of change are recorded in terms of demolitions (whole or partial), new additions, extensions and changes of use.

While the data for England are no longer published, it should be noted that district valuers continue to hold and collect similar data as part of their continuing valuation function. This will continue to be the case under a regime of uniform business rates as property values are still required to determine individual payments. This administrative responsibility also applies to Scotland and Northern Ireland. A much reduced reporting of floorspace additions is made for England and Wales by the Inland Revenue Valuation Office (see below), but in England additional data will be required for any worthwhile analysis. Individual valuation offices may be prepared to release additional data, but this is a matter of negotiation and even where successful is unlikely to yield the detail available in the *CIFS* tables.

As the *CIFS* have facilitated much of our understanding of floorspace changes and given their continuance in Wales, some reference to the use of these data is appropriate. Table 5.1 shows the components of change in the industrial property market during 1974–85. Caution is required in interpreting the figures as some double counting occurs where change over several years is summarized. For example, a new unit completed in one year may be affected by a change of use in the next. Also, the figures for demolitions include only the demolition of whole factories; floorspace lost through partial demolitions is included with 'other reductions', along with changes to other uses. Bearing these points in mind, the data illustrate a number of important features of the industrial property market.

Additions to the property stock arose in three ways. The single most important addition was made by the additions to existing buildings. The expansion of the property stock through entirely new factory buildings was of secondary importance, although they tend to attract more attention in property market literature. Changes from other uses to industrial were of lesser importance. On the other hand, losses of industrial floorspace were dominated by 'other reductions'. As noted, this other category comprises partial demolitions as well as changes of use. As the most likely change of use is to warehousing, a comparison with changes in the warehousing stock provides an estimate of the losses due to use changes and partial demolition. Fothergill *et al.* (1987) estimate that of the 55 million m^2 in other reductions, around 30 million m^2 represent changes of use and the balance partial demolitions. There is, therefore, a much greater rate of demolition of industrial property than the statistics first indicate.

The net balance of additions and reductions during 1974–85 produced a modest reduction in the total stock. An inspection of annual changes indicates that this loss coincided with the post-1979 recession in manufacturing activity. (The recent balance between space in new factories and extensions also shows a contrast with earlier periods.) Consequently when more recent data are considered (Table 5.2), the contrast between the loss of industrial

Table 5.1 Components of change in the stock of industrial floorspace, England and Wales, 1974–85. (*Source:* Fothergill *et al.* 1987.)

	Million m^2	As % of stock in 1974
Additions		
New units	+ 23.9	+ 10.2
Extensions	+ 29.3	+ 12.5
Change of use	+ 13.5	+ 5.8
Reductions		
Complete demolitions	− 14.2	− 6.1
Other	− 55.0	− 23.5
Net change	− 2.5	− 1.1

Table 5.2 Changes in the stock of commercial floorspace, England, 1983–86
(*Source:* Hillier Parker 1988.)

| Sector | Million m² | | Change 1983–86 | |
	1983	1986	Million m²	%
Industrial	228.9	218.6	− 10.3	− 4.3
Warehousing	124.5	135.3	10.8	8.6
Retail	75.1	77.8	3.7	4.9
Commercial offices	46.7	50.4	3.7	7.9

floorspace and growth in other sectors is particularly marked. The trend between 1964 and 1980 saw the stock of manufacturing floorspace grow by around 1 per cent a year. In the office, retail and warehousing sectors, new additions rather than the extensions are the overwhelming source of additional floorspace. The ratio of additional space in new buildings to that in extensions provided during 1983–86 was warehousing 3.4 : 1, retail 3.2 : 1 and offices 5.8 : 1 (manufacturing 1.07 : 1). Other contrasts affecting output and employment ratios exist between use classes. The slower growth in retail floorspace compared with warehousing, for example, is partly the result of more intensive use of retail space through longer trading hours.

Commencing in 1983, the Inland Revenue Valuation Office has produced an annual *Property Market Report*. This report also collects data from district valuers in England and Wales and provides some reporting on property development trends provided by the Northern Ireland Valuation and Lands Office and the Chief Valuer Scotland. For the purposes of the report, valuation officers are required to record a sample of property transactions as an indicator of regional and sector trends. In that part of the report dealing with 'development activity', the data cover new floorspace in representative centres within the valuation districts. Even in the centres that are surveyed, the data are unlikely to be comprehensive. None the less, as the sole remaining source of floorspace data in England, some reference to the data may be required. The data distinguish between new floorspace of schemes completed, under construction and proposed within the 6 months prior to the end of the reporting period. A supplementary bulletin to the main annual report updates the development activity data to provide an annual coverage. Activity is reported for English regions and Wales in three use classes: offices, out-of-town retail and other retail. The partial recording, exclusion of industrial and warehouse floorspace and lack of data on demolitions or change of use make the annual property market report a poor substitute for *CIFS* data. It should be noted, however, that the report is primarily concerned with property transactions and on this issue, while not breaking with its minimalist approach to data reporting, it does offer some advantages (see below).

A different kind of data relevant to floorspace change are the indicators of building activity. The motivation for the collection of these data is to monitor the output of the construction industry as an indicator of national economic performance. Housing is given particular attention in these data series, presumably because of its greater sensitivity to economic change. There is considerably less detail on commercial property. None the less, with the loss of the *CIFS* from England, building activity data are of some value as an indirect measure of new provision and investment in the existing stock.

Part 2 of *Housing and Construction Statistics Great Britain* provides a quarterly series of the value of private-sector new industrial and commercial buildings; public-sector output is enumerated separately, but with no distinction by sector. The value of repair and maintenance (includes extensions) is aggregated for all types of commercial property in the public and private sectors. Published data are broken down by English regions, Wales and Scotland so that for local planning purposes special data requests would be required. The same data are replicated in other official publications including the *Monthly Digest of Statistics for Great Britain, Regional Trends* and the *Scottish Economic Bulletin*, but with no additional detail.

Planning application records are a potential source of information on development; they record the size of the proposal, the initiator of the project and the existing site/property owner. In practice a number of limitations mean that the information needs to be combined with other survey data. Planning applications are generated for proposed new developments and changes of use between use classes. The data are, therefore, more useful for monitoring additions; demolitions generally fall outside planning jurisdiction. Moreover, the range of use changes requiring planning permission has declined with, for example, no application now needed where the change of use is between office, high technology and light industry. The application may not necessarily lead to the proposed development being built, or being built precisely as intended. The information available from planning applications has formed the basis of a development monitoring system operated by the London Research Centre since 1987. The monitoring system combines planning records with other development information to provide a register of all developments over 1000 m^2 or more than 10 dwellings. In this type of databank, planning records become a useful source for updating *CIFS* data and monitoring local change.

The final aspect of the building where data are available relates to rental and property values. There are no official indices of commercial land and property values. The DoE publishes an index of average residential prices, further underlining the bias in official data sources. However, the greater practical difficulties in deriving regional indicators of value and rental trends relating to commercial property should not be overlooked.

Leading estate agents and property analysts produce indicators of both rent and capital value changes. These data result from the growth of institutional involvement in property (see Cadman 1984; Debenham Tewson and Chinnocks 1987) and their demands for information to guide investment decisions. The number and variety of indices produced by estate agents have multiplied in the competition to earn a share of the fee income from advising major investors. Crosby (1988a) divides these indicators into three categories:

1. Whole fund indicators constructed from data supplied by several financial institutions in respect of the overall performance of their property portfolio disaggregated by the type of property (office, retail, manufacturing); for example, the *Morgan Grenfell Laurie/Corporate Intelligence Group Property Index*.
2. Individual property indicators constructed from details of individual properties owned by several institutions amalgamated into a single data series; for example, the Jones Lang Wootton *Property Index* and the Investment Property Databank (IPD).
3. Individual location indices constructed by the valuation of all properties in the centres of selected locations; for example, the *Investors Chronicle/Hillier Parker Rent Index* and Healey and Baker's *Property Rent Indices*.

Discussion of the coverage and relative merits of the various indices can be found in Crosby (1988a) and Moody (1989). In most cases, a summary of the data is available from the firm constructing the index. A number of deficiencies should, however, be borne in mind when using such data. All three types of index rely primarily on estimates of market value derived from valuers' judgements. The small number of actual property transactions and the tendency for rents to increase periodically (usually at 5-year intervals) makes it in practice difficult to derive trends from actual market prices. Moreover, as no two properties are alike, a single index figure can hide significant contrasts and provide limited guidance to the value of individual properties. On the other hand, none of the indices purport to sample the property market as a whole. The focus is on property held in institutional investment portfolios or, in the case of location-based indices, locations of interest to financial institutions. Hillier Parker's *Secondary Rent Index* measures property 'on the border of what institutions would normally buy', but this still leaves a large gap as it concentrates on secondary property in the prime property markets.

None the less, investment property does represent an important slice of the property market – possibly as much as 20 per cent of the value of all property in England and Wales (see Crosby 1988a). Broadly speaking, the data set derived from the largest sample is the most reliable. On the latter criteria, the IPD has established a lead over the other market indicators (see below for further details of the IPD). Analysis of secondary property markets requires original data collection which can be assisted by the regional property market surveys published in the property press (such as the *Estates Times* and the *Estates Gazette*). Crosby (1988b) provides guidance on the construction of such a local property index.

The ease of collating local and secondary market indices would be enhanced if the Valuation Office released the information collected from its recording of property transactions and valuations. At present access to this substantial databank is via the *Property Market Report* which, as noted above, provides a biannual summary derived from Valuation Office records. At present, the potential of the property market data is unrealized because individual offices report only a sample of up to three transactions from which regional summaries are compiled. The small sample frustrates the analysis of temporal or cross-sectional trends from what otherwise would be a valued source of financial data. This abbreviated reporting results from the treatment of all transaction data as confidential, notwithstanding the Valuation Office's own desire to see greater reporting (see Ch. 9).

Reflecting the growth of a transnational market for property, several estate agents now publish international comparisons of office occupancy costs. Jones Lang Wootton's *Worldwide Office Occupancy Costs* compares net rents, regular outgoings and one-off costs in 12 international centres including London. Hillier Parker's *International Property Bulletin* covers 44 centres in 15 countries and includes information related to investment and occupancy.

Ownership characteristics

Information about the ownership of buildings is difficult to obtain. The British Land Registry records properties, both freehold and leasehold, that have changed ownership since 1879, but access to this data source has been limited mainly to conveyancing solicitors. Public access will be allowed from 1990 (see Ch. 9 for further details) on an individual property basis, rather than allowing all properties under the same ownership to be identified by one enquiry. Thus while the Land Registry will facilitate local area studies, measuring aggregate investment flows over a wider area will be tedious.

Assembling information on the location of individual properties within the portfolios of investing institutions is difficult. Property unit trusts identify all their individual holdings in the company prospectus, but these institutions have declined in importance and account for a small fraction of investment funds (Debenham Tewson and Chinnocks 1987). The annual reports of property companies and investing institutions provide little more than an overview of their holdings which, at best, may highlight significant additions to their investment portfolio. Direct requests for data are unlikely to yield significantly more detailed information, such is the secrecy of the property industry. On the other hand, public development agencies such as English Estates are generally co-operative in identifying the location of their property.

The flow of institutional investment funds into property is monitored by the IPD. The IPD contains details of the property holdings and new investment made by most insurance companies, pension funds, property companies and financial institutions in the UK. In March 1989, the total value of this investment was £35bn held by over 120 property funds. The databank comprises a record of all buildings held by the participating funds since 1980 and these currently comprise over 14 000 properties. In terms of that proportion of the property stock held as an investment commodity, the record is impressive but, of course, it remains a small proportion of all floorspace. In 1988, over 58 per cent of the value of property in the databank was represented by offices and almost 80 per cent was in south-east England. For each property, IPD holds information on:

1. *The property*: its location, tenure, use, condition, age, method of acquisition, purchase or development cost and sale price;
2. *The tenancy*: rent review and expiry dates, rent passing, rental value, floor area, basis of rent reviews and repairing clauses;
3. *Financial history from 1980*: capital and rental value, gross and net income, capital receipts and expenditure, development expenditure and levels of occupancy.

The primary purpose of the IPD is to provide participating institutions with information on the financial performance of their portfolio in absolute terms and in relation to that achieved by other institutions. The IPD will also assist investors by analysing trends in particular sectors and locations based on the performance of properties with these specifications that are in the databank. The data are available to outside researchers, except in so far as individual properties or institutions will not be identified. The cost of *ad hoc* research reports and the £1750 annual subscription (1989) to the full list of research reports (known as the *Property Investors' Digest*) will preclude most individual researchers; however, a summary of data can be obtained in the *IPD Annual Review* and Nabarro (1989).

The databank is of interest to urban and regional planners for its ability to relate investment flows and performance to individual locations. For example, regional authorities wishing to attract investment funds can obtain a regional or town profile of investment patterns and performance and this may be compared with other locations. The reliability of this analysis will vary according to the number of comparative investments available for any particular location or property type. Outside the prime investment markets, the smaller number of properties renders the data less reliable. Cross-checking the floor area of the investment portfolio for an individual local authority area with the total stock indicated by the *CIFS* will help set the significance of the data in context. None the less, the IPD represents a significant advance in the monitoring of institutional property investment.

Occupational characteristics

There are no generic sources of data concerned with the occupational character of property. Good information exists in relation to retail floorspace in shopping centres in respect of the store name and specialization. Recent research in London has resulted in the publication of floorspace per worker ratios for different land uses (Barker 1988). The latter information was collected through surveys in several London boroughs and is reported by summaries giving a range for each land use, together with average ratios. Floorspace ratios may also be constructed from the estate records of public development agencies, but these are limited to small industrial property. More generally, where the property is leased, the landlord or managing agent will at least be able (although not necessarily willing) to identify the level of occupation and names of tenants. To identify the operational characteristics of tenants and in respect of owner-occupied property, there is no alternative to conducting a direct survey (Ch. 11).

5.4 Sector-specific data sources

Manufacturing

The additional data available on industrial floorspace relate largely to specific components of the stock. A characteristic of the industrial property market is the importance of a small number of major providers. Public-sector provision of land and buildings for industry has been a long-standing activity in the more depressed regions. Fothergill *et al.* (1987) estimated that in 1984 the Scottish Development Agency accounted for 10–12 per cent of the industrial property stock in Scotland, while the Welsh Development Agency accounted for just under 10 per cent of industrial property in Wales. In the seven counties where English Estates has been most active (Tyne and Wear, Durham, Cleveland, Cumbria, Northumberland, Merseyside and Cornwall) it held 8 per cent of the total stock. When the supply of new factory space is considered, these public agencies accounted for over half the floorspace in new buildings in their respective area during 1983–84. In addition, local authorities were a further significant provider in these regions. Within the private sector, the single largest developer and landlord is Slough Estates. No other private developers match the industrial property holdings of Slough, but within individual locations a few developers account typically for a large proportion of recent provision. Local estate agents or reports in the property press will identify these developers.

The information held by these various landlords varies. Public agencies retain more information on their tenants, in terms of their activity and employment, and are more willing to make their estate records open to research. The annual reports of the various regional development agencies, English Estates and the Development Commission, as well as those from large private-sector developers, provide a summary of development activity and different levels of detail on their tenant profile. A number of studies have now been completed which discuss tenancy profiles of different development agencies, including Slowe (1981), Hillier (1982), JURUE (1983), Coopers and Lybrand Associates (1984) and Perry (1986).

Two types of industrial property have attracted particular investment and policy attention in recent years; these are 'high technology' units and shared-workspace schemes. In relation to these two types of property the UK Science Park Association (UKSPA) and the National Managed Workspace Group (NMWG) are sources of information. While both these associations are discussed here in relation to industrial property, the range of occupants will be wider.

The UKSPA was formed in 1984 to bring together those directly involved with the planning, commissioning, development and day-to-day management of science parks. Of course, there has been much debate on the qualities which differentiate science parks from traditional industrial estates (Hennebery 1987). UKSPA's membership and information base relate to property which is managed actively or at least designed to facilitate: (1) operational links with a higher education institution or major centre of research; (2) the transfer of technology and business skills; (3) the formation and growth of knowledge-based enterprise. The UKSPA recognizes these attributes in a diverse range of estates that are variously labelled science parks, research parks, innovation centres and high-technology centres. It is doubtful that the character of the property or estate management is clearly distinctive from other industrial estates or that the tenant profile is correspondingly unique. These issues can be explored further using the UKSPA database which lists the majority of tenants, their activities, business objectives and employment. At the end of 1988, there were 800 tenants and around 10 500 employees covered by the database.

The NMWG is an organization covering shared-service accommodation. This type of property is aimed at small enterprises (five employees would be a big firm in this context) with modest resources and property requirements satisfied by renting part of a building where on-site office services are available (secretarial, business advice, meeting rooms and computing). This type of property has increased, particularly as a use for old mills, although there is little evidence that the potential functional advantages are the reason for the growth of this type of property provision (Green *et al.* 1989). None the less, a recent study of good practice in managed workspace provision identified around 200 schemes throughout Great Britain (DoE 1987). This list can be updated from information in the NMWG's newsletter. Managed workspace is often provided in the context of the refurbishment of old buildings: Urbed's reused industrial buildings service is an additional source of information for these shared-workspace schemes.

The qualitative assessment of buildings in terms of their suitability for production cannot be determined from secondary data. Work by Fothergill *et al.* (1985, 1987) has established the importance of three variables in determining how far a particular property satisfies an occupant's needs: whether there is production on more than one floor; the proportion of the site covered; and whether there is vacant land adjoining the site. Old, multi-storey buildings on densely developed sites and in densely developed locations tend to be the most inefficient types of property. The significance of these variables is that the suitability of industrial buildings can be assessed partly on the basis of an external survey alone, although a full assessment should also examine the internal design. It should be noted, however, that the period of occupancy also affects the suitability of the building in that an old building may be acceptable to a new occupant. Problems arise over time as the firm seeks to adapt the property to changing needs. A number of studies have been completed in various parts of the country to provide a guide to collecting information on the suitability of buildings and their availability for reuse (West Midlands County Council 1984; Fothergill *et al*. 1985; Ball 1989).

Another indicator associated with building suitability is the level of vacancies. There are two sources of information on the vacancy rate of industrial premises. Statistics on complete vacancies, where the whole factory stands empty, are compiled by King & Co., a leading firm of industrial estate agents, and published in their *Industrial Floorspace Survey*. The survey figures, which cover England and Wales, are derived from several sources – their own records, the records of other estate agents, government development agencies and the property press. The survey does not distinguish between industrial and warehousing property and there are a number of omissions (premises smaller than 465 m², premises not actively marketed, semi-derelict or multi-storey properties requiring refurbishment). Allowing for these omissions, the King & Co. data are a reliable guide to the extent of vacant floorspace (Fothergill *et al*. 1985). The data have become more useful with the introduction of three categories of vacancy: new buildings, buildings over 9300 m² and high-technology buildings.

An alternative estimate of vacant property is provided in Hillier Parker's *Survey of Industrial Voids*. This is confined to rented industrial and warehousing space and is based on a survey of 48 financial institutions, pensions and property companies with investments in this type of property. The usefulness of these data is reduced further by the irregular publication of the survey. On the other hand, the void data provide two interesting indicators of market trends: the average length of time new properties have been on the market and the proportion of vacated premises where a rent is still being paid.

Warehousing

The growth in warehousing floorspace has been greater than any other component of the property market. There are two causes behind the growth of warehousing space. Changes in the organization and ownership of distribution, notably the trend to larger retail units, have encouraged the concentration and relocation of warehousing activity. More importantly, the expansion of warehousing floorspace responds to the volume of retail sales, which has been increasing, while industrial property responds to the volume of production which is more severely affected by recession.

There has been comparatively little research on the warehousing sector. This lack of interest originally resulted from the alleged insignificance of warehousing as a source of employment. Tempest (1982) and McKinnon and Pratt (1984) demonstrate that employment densities in warehousing may not deviate far from those in certain types of manufacturing. A more enduring influence on the comparative lack of research is the vague distinction between warehousing and manufacturing in property and employment databases (McKinnon 1983). In the *CIFS*, for example, the definition of warehouses includes workshops, stores and a miscellany of small service establishments while it omits public-sector warehousing. These overlapping or omitted activities tend to occupy small units, and by excluding property of less than 1000 m² from the *CIFS* category the data more accurately reflect the total warehousing stock (Bone 1987). While problems with the *CIFS* classification can be surmounted, the data sources produced by private organizations increasingly treat industrial and warehousing property as one category.

Retail

The substantial investment that has occurred in retail property over the last decade has led to the enhancement of the database provided by private-sector organizations. Unfortunately, while parts of the retail sector are now well covered, the demise of the Census of Distribution and reduction in the availability of *CIFS* has weakened the overall basis for retail planning (see Ch. 17 for further discussion).

The data derived from *Goad Shopping Centre Plans* are now the single main source of retail floorspace information. These plans were published originally for fire insurance purposes, but are now the main source of floorspace data for retail planning. Goad Plans are published for about 1100 shopping centres in towns and cities in England, Wales, Scotland and Northern Ireland. The population threshold for inclusion has been 7500 but smaller settlements are being added as well as an increasing number of suburban centres in larger cities. The coverage is extending by around 20 new centres a year. The plans are published at a scale of 1 : 1056 and, among other functions, identify public transport facilities, pedestrian and vehicular routes in relation to individual named stores. The naming and identification of stores facilitate the cross-referencing with other data sources such as retail directories which may contain additional data on employment and trading area. The plans for major centres are revised annually and biannually in the case of smaller centres. The plans are well suited to the evaluation of land use change in central areas, but for retailing planning purposes the data extracted from the plans by Sales and Marketing Information Ltd (SAMI) are used more frequently. This commercial organization has abstracted floorspace, store name and activity data from the plans into a computerized database.

The data abstracted from Goad Plans by SAMI is available in two formats. *Shop Count Reports* enumerate the various store types, express this as a percentage of the total number of stores within a centre and provide a comparison with the national average profile. *Ground Floorspace Census Reports* indicate the ground floor area occupied by different store types and express this as a proportion of total retail floor area (of that store type or total activity) and give a comparison to the corresponding national average pattern.

As a tool for evaluating the character of shopping centres and assessing the opportunities for new development, Goad reports are now used widely in retail planning, although they are far from ideal. The plans record the outside perimeter of an outlet at ground level. Floorspace is measured from the map, rather than on site, and to allow for inaccuracies in this process, all figures are rounded to the nearest 93 m^2 (1000 ft^2). The number of retail floors is noted so that the overall total can be derived but, at best, the result approximates only to the gross selling area. The delineation of a shopping centre is determined by the judgement of individual Goad surveyors. The main problem arises in metropolitan areas where there is no clear end to the shopping area and arbitrary boundaries are established. In some cities, central areas may be divided between two or more plans where clear physical boundaries dissect the centre.

A further difficulty is that the focus on shopping centres is increasingly out of line with the growth of superstore and suburban retailing. Out-of-town retail warehouse parks now represent around two-thirds of all new shopping proposals (Brown 1989). Goad has started to incorporate out-of-centre superstores and SAMI has extended this coverage in its own out-of-centre database. The Data Consultancy (formerly the Unit for Retail Planning Information – URPI) also maintains a database covering new forms of retail development. At the end of 1976 there were less than 100 superstores (stores with over 2500 m^2 of selling area); Data Consultancy's 1988 list of UK superstores and hypermarkets (over 5000 m^2 of selling area) covered over 500 stores that were either open or had planning permission. In addition, Data Consultancy maintains a record of managed shopping schemes, including floorspace information and a five-part location classification (town centre, district centre, other traditional centre, new district centre, free-standing) as well as individual addresses for more detailed investigation. Several estate agents also monitor new retail investment, including Hillier Parker in a series of reports including an annual review *British Shopping Developments* and a quarterly survey *Shopping Schemes in the Pipeline*. A database covering all planning applications for food superstores, both successful and unsuccessful, has been compiled by Davies (1987) with a subsequent analysis in Davies and Sparks (1989).

Company annual reports are a further source of information on retail floorspace. Retail organizations tend to include data on their trading floorspace in their annual reports as one indicator of operational performance, unlike industrial and commercial organizations for whom the amount of floorspace occupied is an incidental consideration. The figures in annual reports tend to be limited to a national total but can be combined with turnover figures to provide a broad indicator of the trading performance of different categories of retailer. Data Consultancy collate the information in annual reports as a further component of their retail information service.

Offices

Additional information on office floorspace reflects the focus of investment institutions on the prime investment market of London and south-east England. Jones Lang Wootton maintain a database of the London office market based on information derived from planning department records, contact with the property development industry and their own surveys. The database comprises three components: (1) a record of every new or substantially refurbished office building of over 930 m² completed since 1981; (2) a record of all buildings under construction above the size threshold; (3) development proposals. For each aspect of the database, the developer, owner, present occupiers and letting availability are recorded. A biannual report covering the central London office market and an annual report covering Greater London are published from the database. The pressure for office development in the capital city was also reflected in the efforts of the Greater London Council to establish an office development information system (Barras 1981). The former GLC database has been superseded by that managed by the London Research Centre which operates on a more streamlined basis, but covering all property sectors (see above).

The volume of accommodation available for occupation in central London is monitored by various estate agents. Richard Ellis, for example, publish a quarterly report, *West End Office Market*, which identifies the level and character of available office space. The autumn 1989 report found that of 213 000 m² of floorspace being marketed, 12 per cent was in newly completed buildings and 23 per cent in new buildings awaiting completion. Vacant space was, therefore, concentrated in second-hand properties and was partly explained by the unsuitability of much of this space for modern office users. The same report also records the amount of space purchased in the previous quarter: overseas companies were identified as the principal purchaser (37% of the transactions by value) followed by UK property companies (35%) and UK financial institutions (17%). Leading estate agents monitor office market trends in other major cities, but generally on a less formal basis.

The high cost of office accommodation in London compared with surrounding regions makes the issue of relocation of particular concern to institutional investors and property developers. The extent of office decentralization was monitored by the Location of Offices Bureau until its demise in 1979. Jones Lang Wootton have maintained this database as far as decentralization from London is concerned involving relocations of over 100 jobs. Moves are monitored in terms of their origin and destination, the company involved (name and/or activity) and the motivation for the shift. The findings are published in *Central London Offices Research*, *Greater London Offices Research* and *Decentralisation of Offices from Central London*.

The focus of office-sector data is therefore on the uptake and disposal of space. Trends in the use of office floorspace remain poorly understood, which is a major shortcoming at a time when office property requirements are changing. Service organizations are occupying proportionately larger areas of floorspace, both because of rising standards and because of the increasing use of new technology which creates its own space demands (Barras 1987; Black *et al*. 1986). The shift in office employment and output per unit of floorspace is a more complex calculation compared with the corresponding figure for industrial floorspace. The occupants of office space are not confined to service industries, rather they include primary and manufacturing sector head offices and service divisions, public-sector organizations, as well as 'pure' service firms. Consequently, industry employment and output statistics cannot be compared directly to the *CIFS* record of office floorspace. Without such analyses, however, the understanding of office floorspace trends remains limited.

5.5 Conclusion

The evolution of business in respect of the changing character of its floorspace demands and intensity of floorspace use and the potential for mismatch in the supply and demand of property require continuous monitoring. In contrast, the database on commercial floorspace is partial and fragmented. The withdrawal of *CIFS* from England is in no way compensated by the multiplication of private-sector sources which reflect the interest of developers and investors in the marketability and profitability of a narrow segment of the property market. The ability to monitor institutional property investment has improved, notably through the IPD and the growth of retail databases, but the loss of information on the total stock is a major blow to urban and regional analysts.

The key to improving the database is the Inland Revenue Valuation Office, both in respect of the reinstatement of the English *CIFS* and the release of the information it derives from the recording of property transactions. Relatively minor marginal costs are involved and the Valuation Office is in support of access to its data. The major stumbling block is, of course, the intransigence of a government befixed by public expenditure reductions and wilfully obstructive to the free flow of information. The immediate prospects for a greater release of information are poor, but as the case for publication is strong researchers should continue to press for a change in attitude. The possibility of building comprehensive databanks of property information is one of the advantages flowing from a greater release of information. The Valuation Office has, for example, proposed the co-ordination of Land Registry and local authority records with its own in a national register of property information. Even without the access to Valuation Office records the greater exploitation of planning records offers the basis for local databases. Planning applications are a source of information on the size of new developments, the initiator of schemes and the ownership of land and property. To be useful, however, such data need to be supplemented with information on actual building commencements and losses of floorspace through demolitions and use changes. In this context, the extension of the London Research Centre's development monitoring system is a model for others to adapt.

A further improvement in relation to the presentation of official data would be to refine the classification of building activity statistics. A distinction between types of commercial property and investment in new and refurbishment schemes, so that the series provided the same detail for each type of commercial property as it does for the housing market, would at least provide a financial measure of floorspace additions. Again this is an issue of the manner in which existing information is reported, rather than requiring additional data collection. For the present, local and regional planners and researchers can only be frustrated by the amount of data lying dormant and out of reach.

References

Ball R 1989 Vacant industrial premises and local development: a survey, analysis and policy assessment of the problem in Stoke-on-Trent. *Land Development Studies* **6**: 105–28

Barker R 1988 *Floorspace per Worker Ratios in Commercial Premises in London* London Research Centre, London

Barras R 1981 *An Office Development Information System for London* CES Paper 18, CES Ltd, London

Barras R 1987 Technical change and the urban development cycle. *Urban Studies* **24**: 5–30

Black J T, Roark K S, Schwartz L S 1986 *The Changing Office Workplace* Urban Land Institute, Washington DC

Bone R 1987 Warehousing. *The Planner* **73**(17): 20–4

Brown S 1989 *Retail Warehouse Parks: Their role in the Future of Retailing* Longman/Oxford Institute of Retail Management, Harlow

Cadman D 1984 Property finance in the UK post war period. *Land Development Studies* **1**: 61–82

Coopers and Lybrand Associates 1984 *Impact of English Estates and the Welsh Development Agency on Private Sector Provision of Industrial Property* Report to the DTI, Coopers and Lybrand Associates, London

Crosby N 1988a An analysis of property market indices with emphasis on shop rent change. *Land Development Studies* **5**: 145–77

Crosby N 1988b Shop and rental value change in Nottingham. *Land Development Studies* **5**: 185–205

Davies K 1987 The effects of land use planning and development control on the location of hypermarkets and superstores. Unpublished PhD thesis, University of Wales

Davies K, Sparks L 1989 Planning applications for food superstores. *Land Development Studies* **6**: 147–64

Debenham Tewson and Chinnocks 1987 *Money into Property 1987* DTC, London

DoE Inner Cities Directorate 1987 *Managing Workspace* HMSO, London

Fothergill S, Kitson M, Monk S 1985 *Urban Industrial Change* HMSO, London

Fothergill S, Monk S, Perry M 1987 *Property and Industrial Development* Hutchinson, London

Green H, Boyland K, Strange A 1989 Managed workspace: suitable homes for small business? *The Planner* **75**(19): 25–6

Hennebery J 1987 *British Science Parks and High Technology Developments: Progress and Change, 1983–1986* PAVIC, Sheffield

Hennebery J 1988 Conflict in the industrial property market. *Town Planning Review* **59**(3): 241–62

Hillier J 1982 The role of CoSIRA factories in the provision of employment in rural eastern England. In Moseley M J (ed.) *Power, Planning and People in Rural East Anglia* Centre of East Anglian Studies, University of East Anglia, pp. 177–98

Hillier Parker 1988 *Commercial and Industrial Floorspace Statistics England 1983–1986* Hillier Parker May & Rowden, London

JURUE 1983 *An Evaluation of Development Commission Activities in Selected Areas* JURUE, Birmingham

Kennett S 1985 *The Small Workshops Scheme* HMSO, London

McKinnon A 1983 Development of warehousing in England and Wales. *Geoforum* **14**(4): 389–99

McKinnon A, Pratt A 1984 *Jobs in Store: An Examination of the Employment Potential in Warehousing* Occasional Paper 12, Department of Geography, University of Leicester

Moody M 1989 *Property Information Sourcebook* Estates Gazette, London

Nabarro R 1989 Investment in commercial and industrial property development: some recent trends. *Development and Planning 1989* 65–70

Perry M 1986 *Small Factories and Economic Development* Gower, Aldershot

Slowe P 1981 *The Advance Factory in Regional Development* Gower, Aldershot

Tempest I 1982 Warehousing as an employment source – a study of employment density figures and local authority attitudes. *Planning Outlook* **25**(3): 105–10

West Midlands County Council 1984 *Vacant Industrial Property in the Black Country* WMCC Planning Department, Birmingham

6

Urban land use

Philip Kivell

6.1 Introduction

The need for careful land use planning in such a small and highly urbanized country as the UK is indisputable and for 40 years there has been a comprehensive planning system based mainly upon land use considerations. Despite this, the state of our knowledge of urban land use is still far from satisfactory. Even the most basic facts are contentious. Numerous independent commentators have drawn attention to the shortage and patchiness of urban land use data, and the way in which this handicaps effective land planning and allocation, yet paradoxically the acquisition of land use statistics does not appear to be a top priority for local authorities. Coppock's judgement that 'the collection of adequate data on urban land use and land use changes is always likely to present difficulties' (Coppock 1978: 55) unfortunately remains true.

Although recent years have not seen great improvements in the availability of information, they have seen major changes in both the planning and technical contexts of land use studies. Planning evolved from its domination by development plans and highly detailed land use maps in the 1940s and 1950s to a more generalized system of structure plans in a new local government framework after 1974. By the mid 1980s even the validity of these structure plans was being questioned and a new system of unitary plans was proposed at the end of the decade. Greater attention was also given to a number of specific land use issues, notably those in the inner city and the green belt concerning vacant land and land for houses. During the same period the technical context also changed as traditional ground surveys were supplemented by aerial photography and remote sensing, and as manual analysis and draughting gave way to digital mapping, computerized land management systems and geographical information systems (GIS) (see Chs 8 and 9).

Even with these advances the stark fact remains that the best available comprehensive survey of urban land use in England and Wales is based upon 1969 aerial photography (Department of the Environment (DoE) 1978). It contains only five crude categories (mainly residential, mainly industrial and commercial, educational and/or community, transport and open space) and measures only areas of developed land of 5 ha and above.

6.2 Procedural matters

The collection and analysis of urban land use data present immense problems, not least because as Coleman (1980) observes, land use survey in the UK is not entrusted to a unified professional organization. Instead it is split among a multiplicity of planning authorities with only a modest degree of central control being provided by the DoE. In many cases the purpose of survey, and the definitions and classifications used, vary from one local authority to another and there is no overall consistency of either input or output. In order to understand what is available, and some of the constraints, it is useful to consider a few preliminary matters.

Purpose of survey

Research workers using existing land use data encounter the problem that the nature of the data is heavily influenced by the purpose for which it was originally collected and this inevitably limits its more general utility. Broadly speaking, three uses have governed such exercises. First, a number of studies have been undertaken to measure the overall extent and expansion of urban areas. Such studies are commonly large scale, being national or regional in scope, but rely upon a very coarse subdivision of perhaps no more than half a dozen categories. Second are the inventory type of land use exercises undertaken mainly by local authorities to help them analyse and monitor their planning policies, or as part of a more general property/land management system, or for related needs such as rating purposes. Third are the more specific, subject-based surveys relating to such problems as derelict/vacant land or the availability of housing land. Within the public sector the main guidance on land use statistics comes from the DoE which spells out its requirements to local authorities through periodic circulars. This process gives some degree of comparability among the data, but it remains the general case that land use surveys in the UK have been compiled by different agencies, at different times, using different, often incompatible, techniques.

Defining 'urban'

Defining what is meant by urban presents a number of other initial difficulties. The emphasis upon local authorities as the primary collectors and users of land use information gives the exercise an immediate framework of administrative areas. This is far from ideal, for administrative boundaries rarely coincide with the physical extent of urban growth and large urban agglomerations are commonly divided between a number of separate authorities. Attempts to refine the definition have included the imposition of a population size threshold (Best 1981; Guerin and Mouillart 1983), consideration of continuous developed areas covered by buildings and urban structures (DoE 1978), the use of a residual urban definition from agricultural surveys (Best and Anderson 1984; Deane 1986) and attempts to generalize urban/rural boundaries by statistical techniques (Ward 1983).

Nature of urban land

The particular nature of urban land poses difficulties at two levels, conceptual and practical. At a conceptual level it is important to consider the overall political economy. It is commonly assumed that a rational pattern of land use evolves, mainly by activities competing for sites through the process of supply and demand, yet it is equally clear that the urban land market functions imperfectly. The balance between public and private sectors has shifted in recent years and many external features such as inflation, credit availability, social change and growing affluence have produced additional turbulence in the patterns of urban land use. Old buildings survive alongside new and vacant land persists alongside intensively developed sites. The net result is that a difference exists between the observable land use of a given plot and its potential in a planning context.

At a practical level the main problem is that urban land forms a very dense and small-scale mosaic of development. Questions arise over the choice of the basic unit for survey. This may be decided on grounds of cost, in which case some sort of grid overlay or sampling may be appropriate (Dickenson and Shaw 1982) or there may be a technical constraint as in the case of the level of resolution applicable to remote sensing. For a detailed survey, however, it is desirable to consider individual properties (Coppock 1978) or curtilages including the land attached to buildings (Dickenson and Shaw 1977). Even within individual curtilages there may be several land uses. In such cases it is normally appropriate to record the principal use, i.e. that use upon which all others depend for their existence.

Classification schemes

To allow order or patterns to be recognized, a system of classification is needed. Not surprisingly no ideal system of land use classifications exists and it is unlikely that one could ever be devised. In practice most schemes are not classification (*sensu stricto*), where individual observations are grouped on the basis of similarities, but rather a form of discriminant analysis where each observation is compared with an a priori scheme, and pigeon-holed accordingly. Frequently there is conflict over the number of classes used, with a small number of classes giving ease of allocation but much loss of information and a large number of classes becoming confusing and unwieldy. The ideal requirements for classification schemes have been outlined by Rhind and Hudson (1980) and Hill (1984).

A distinction may initially be made between, on the one hand *land form or cover*, and on the other *land function or activity*. Form or cover is essentially the nature of the elements in the landscape, for example, types of buildings, structures or open spaces, whereas function or activity concerns what the land is actually used for. The distinction is important because it relates to the methods of gathering information. For example, land cover may be discernible from remote sensing imagery but because cover does not give a reliable guide to activity, the latter normally requires a ground survey or documentary evidence.

Most of the large, general purpose classification schemes provide poorly for urban land uses. The *Second Land Utilization Survey* gives only 4 classes (out of 13) to broadly urban uses and more recent schemes designed for use with remote sensing techniques are even less discriminatory. The United States Geological Survey, for example, has only one urban category out of eight first-level groups (Anderson 1976) and the classification proposed for the European CORINE (Co-ordinated Information on the European Environment) project has one category of 'Built-up and related areas' in a group of eight.

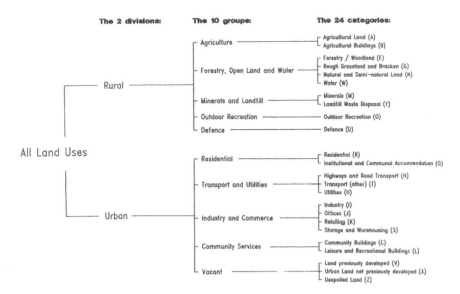

Fig. 6.1 Land use change statistics: classification structure (any recorded land is classified to one category only). (*Source:* DoE 1987: 10.)

An attempt to standardize the individual classification schemes devised by British local authorities in the 1940s and 1950s was not made until the mid 1970s when the National Land Use Classification (NLUC) was promoted (NLUC 1975). Like most others this scheme is hierarchical. It has 15 major orders, 78 groups and 150 subgroups and some compatibility exists with the Standard Industrial Classification (SIC) (Markowski 1982). Dickenson and Shaw (1977) considered applying this scheme in their study of Leeds, but concluded that it had a number of significant shortcomings for use in urban areas. Also in the mid 1970s the DoE attempted to collect statistics on land use changes from local authorities, but that exercise was largely unsuccessful (Dickenson and Shaw 1982; Sellwood 1987). A renewed attempt, on a different basis, was started in 1984 using the classification in Fig. 6.1 (DoE 1986).

In a more detailed sense, and for the application of town planning legislation, all land is deemed to have a use, as defined in the Use Classes Orders of 1972 and 1987 and a comprehensive gazetteer of these has been produced by Godfree (1988).

6.3 Urban land use statistics

Sources

The sources of statistics on urban land use in the UK are many and varied but unfortunately they do not add up to either a coherent or a comprehensive coverage. Valuable summaries of available sources have been made by Coppock (1978), Gebbett (1978) and Best (1981) but there have been significant developments since then in the four spheres discussed below.

Maps and ground surveys

The very comprehensive sets of topographic maps produced by the Ordnance Survey (OS) are a useful basis for measuring the overall extent of urban areas (Fordham 1974) as well as providing more specific land use information. Changes noted by OS surveyors as a part of their regular work on map revision form the bases of attempts by the DoE to monitor land use changes (DoE 1986; 1987; 1988a). Despite the accuracy of the OS maps, some doubts must surround the quality of the land use data thus gathered. In particular there are questions about how systematically the information is gathered and the variability of the time-lags between changes taking place and being recorded.

Detailed land use maps compiled from ground surveys formed the basis of both the *First* and *Second Land Utilization Surveys* and of the many town maps which were required by the post-war planning system. The latter were extensively used by Champion (1974), Best (1981) and others to derive their estimates of the extent of urban land, but as these authors point out, the town maps were subject to many inaccuracies and inconsistencies and by 1970 they ceased to have any contemporary purpose. Land use maps compiled from ground surveys have a number of advantages – they are detailed, accurate and direct records but they are also cumbersome, expensive, difficult to analyse and present only a static picture. For these reasons their use today is mainly restricted to *ad hoc* surveys, for example, of derelict land, or they are used selectively to check and calibrate other methods of gathering information.

Aerial photography and remote sensing

The use of aerial photography for land use survey (see Ch. 11) and urban analysis has been well established since the 1940s and Berlin (1971) has provided a useful bibliography of early studies. From the early 1970s greater attention was being paid to the use of sequential photography to monitor urban change (Dueker and Horton 1971; Hathout 1988) especially in North America, but some British local authorities began to use aerial photographs in conjunction with other data such as that from the census. Improvements have been made possible by advances in photography (Lindgren 1985), but the quality of the land use information remains limited by the nature of urban areas and by the amount and kind of data an aerial survey can provide.

Dickenson and Shaw (1977, 1982) used aerial photography extensively in their study of Leeds and panchromatic photography taken in 1969 at a scale of 1 : 60 000 was used in the DoE's comprehensive study of developed urban land (DoE 1978). More specifically the use of aerial photography in studies of derelict land has been explored by Collins and Bush (1974) and a review of remote sensing in this context will be found in Kivell *et al.* (1989).

New methods of gathering land use information were introduced with the launching of the first Landsat satellite in 1972, but the imagery was, and still is, limited by the poor level of spatial resolution of the sensors and the complexity of the urban scene. Significant advances have been made recently, especially with imagery from the French SPOT satellite which is capable of resolution down to 10 m in the panchromatic mode and 20 m in multispectral mode. (Broadly speaking, panchromatic covers the visible part of the spectrum whereas the multispectral mode is any reflected radiation including the visible and infra-red bands.) The application of remote sensing to studies of urban land use has recently been reviewed by Harrison and Richards (1987) and Whitehouse (1989) who suggest that, as yet, such techniques offer poor accuracy in built-up areas. Conversely, high levels of classification accuracy using SPOT-1 imagery are claimed by Collins and Barnsley (1988), but their classes distinguished only between high-density residential, low-density residential and commercial/residential uses.

Remote sensing (including aerial photography) offers a number of advantages including speed and cost-effectiveness, the ease of time sequence comparisons and the ability to overcome site access problems. Alongside these must be set a number of disadvantages including low levels of resolution, the limited success of land use classification algorithms, the variety of responses given by land under different conditions, the imperfect relationship between land use and land cover and the need for sophisticated equipment to analyse data. Aerial photographs are widely used, but mainly for specific purposes such as individual site evaluation rather than for compiling comprehensive land use inventories. Satellite imagery is currently little used, but as techniques improve its potential will be increasingly realized. A brief list of sources is provided in Appendix 6.1.

Published statistics

A number of government bodies and departments publish statistics on land use, but only a fraction of them relate directly to urban areas. The publication of sequential statistics on agricultural land (see Chs 7 and 13) permits crude residual estimates to be made about the total amount of urban land (Best 1981) and allows the calculation of 5-yearly moving averages of transfers from agricultural to urban uses (Sellwood 1987). These changes may also be monitored more directly for the mid 1980s from the DoE figures of land use change referred to above. An indirect view of the composition of urban land use may be gained from statistics collected for rating purposes by the Valuation Office of the Inland Revenue (Central Statistical Office (CSO) 1990). These provide a breakdown into seven broad categories, but they refer simply to hereditaments and give no indication of relative sizes or areas. In any case, as explained below, the whole rating system is currently being changed.

A limited amount of information is also published on more specific land use activities. Statistics on derelict land are published sporadically (DoE 1974, 1982) and a rolling register of publicly owned vacant land is maintained by the DoE (see below). At one time the DoE also used to publish *Commercial and Industrial Floorspace Statistics (CIFS)* but this series ceased in the mid 1980s (see Ch. 5). In addition there exist a number of other mainstream data sources on employment and population which may be used with mixed success, as Rhind and Hudson (1980) and Champion (1972) have shown, for indirect measurements of urban land.

All of the above sources have shortcomings in terms of the directness of their relationship to urban land use, their level of aggregation, the partial nature of their coverage or the continuity of their data.

Local authority administrative sources

A number of local authority departments collect a wealth of data which can be used in the study of urban land use. Most obviously routine planning statistics relating to allocated uses, development control and planning applications yield much information at local level. These, for example, were the basis of the DoE's unsuccessful attempt to gather annual statistics on land use change in the mid 1970s, and they are used today, rather more successfully, as important inputs to local authority land information and management systems (Ch. 9). Architects' and engineers' departments collect information on building starts, completions and demolitions, and sundry others such as building inspectors' and education departments gather data relating to land and property for their own purposes. Clearly the effective use of such information requires a high degree of co-operation and integration, but it offers the potential of a high-quality, detailed local database which can be updated regularly.

Other local authority sources include rating lists, although relatively little use has been made of these (National Environment Research Council (NERC) 1978) and substantial changes in their organization are under way. Domestic rates were replaced briefly by the Community Charge in 1989. Effectively this levied rates on the individual person, rather than on the individual property, and it produced no information of value for land use studies. In 1991, however, the government announced plans to replace the Community Charge with a new property based charge. Although details are not yet clear, it seems likely that this system will generate some information of value in urban analysis.

In parallel with the community charge there is a system of national non-domestic rates (NNDR), sometimes called the unified business rate, to cover virtually everything except private residential properties. Under the provisions of the Local Government Finance Act (1988), the valuation officer for the charging authority is required to compile and maintain local non-domestic rating lists. These initially date from April 1990 and will be updated every 5 years. In addition a central list, to include government buildings and Crown properties, will be deposited with the DoE. The local non-domestic ratings lists are available for public inspection at the offices of the district authorities. As with the old lists, they do have some value for land use studies. They contain details of the property type or activity (typically in as many as 150 categories), the address and the rateable value (based, as before, upon a notional rental value).

Analysing and manipulating data

Once land use statistics have been gathered they need to be manipulated and analysed. In many respects recent advances in manipulation and analysis have been far greater than those in gathering data.

Traditionally land use data have been treated as a nominal scale variable, amenable to simple tabular and map presentation and only the simplest statistical analysis. The manually coloured map, a mainstay of local authority studies until the early 1970s, fell from favour because it was inflexible, slow and costly to prepare and offered limited scope for analysis (NERC 1978). At first, few alternatives were available, but some authorities began using computer-based techniques such as Synagraphic Mapping (SYMAP) and made progress on digitizing basic map boundaries.

Quite rapidly there began to develop more sophisticated techniques for the management, analysis and presentation of large, spatially referenced databanks in the form of land information systems (LIS) and GIS. These are discussed fully in Chapter 9, but it is important to look at them briefly in the present context.

Land information systems and inventories began in North America, notably in Canada, in the early 1960s. Here and elsewhere they were at first largely confined to rural applications (Gierman and MacDonald 1982; Jones 1986) but they were also used in the analysis of patterns of land use change on the urban fringe.

Relatively limited planning application information systems have been in use in the UK since the mid 1970s (Grimshaw 1985), sometimes standing alone and sometimes linked to property information systems or attempts to monitor broader patterns of land development and potential (Brown 1985). More recently local authorities have increasingly realized the potential of land information systems for specifically urban land use studies.

Humphries (1985) provides a succinct review of land and property information systems and he draws an important distinction between Local Authority Management Information System (LAMIS)-type approaches, in which spatially defined areas are used with full boundary digitizing, and the gazetteer type which depends on a central index based upon postal addresses and/or map references to site centroids. The Tyne and Wear scheme was an important pioneer of the gazetteer type in the mid 1970s (Charlton and Openshaw 1986) but other approaches followed, for example in London through the Central London Land Use and Employment Register (CLUSTER) consortium (Markowski 1982; Home 1984), Warwickshire (Grimshaw 1988), Manchester (Bourke and Davies 1988), Birmingham (Gault and Davis 1988) and Kingston (Weights 1988). As with any technology there remain problems of comparability and compatibility, not least over conventions for spatial referencing of the data.

Major technical problems also remain in the analysis of land cover and land use data provided by remote sensing techniques. Whitehouse (1989) suggests that the traditional spectral classification approaches are inappropriate for the high-resolution data produced by the new generation of satellites and explores new approaches based upon texture and context.

Much of the analytical effort to date has been devoted to the recognition and interpretation of existing patterns, but from a planning viewpoint some indication of developing and future patterns is desirable. Sequential databases, such as the DoE statistics on land use change, allow changes to be monitored retrospectively, but few attempts have been made to predict future patterns. A rare attempt is that by Charlton and Openshaw (1986) who used linear and multiple regression techniques and models borrowed from demography to forecast land use trends, but they conclude that none of these were satisfactory.

Results

A brief examination of the findings of some recent studies will illustrate what is possible using the sources and techniques discussed above. Substantive, although now rather dated, summaries will be found in Coppock (1978), Rhind and Hudson (1980) and Best (1981).

The conclusions of Best and Anderson (1984) are that in 1981 the UK contained 2.05 million ha of urban land (8.5% of the total) and for England and Wales the figures were 1.76 million ha and 11.7 per cent. In the 1960s the loss of farmland to development had averaged 18 750 ha/yr but this fell to 15 000 ha/yr in the first half of the 1970s and to 10 000 ha/yr in the latter part of the decade. Using different definitions, the DoE/Office of Population Censuses and Surveys (OPCS) estimated that 89 per cent of the population of England and Wales were living in urban areas which covered 7.7 per cent of the land area and that the total extent of urban land was 3.31 million ha (DoE 1988b).

The most recent figures for the composition of urban land on a national basis remain those from the 1969 aerial survey (DoE 1978) and these are shown in summary in Table 6.1. A breakdown by local authority district is also available.

An indication of the dynamic processes of land use change is given by the statistics collected by the DoE (1986, 1987, 1988a, 1989a). Broadly these show that in the mid 1980s approximately 24 000 ha of land saw a change of use each year; Table 6.2 summarizes the changes for 1987. In total, 24 660 ha experienced a change of use. On 9375 ha (38% of the total) the changes were from one rural use to another, so these are of little concern for the present chapter. About one-third (34%) of changes were from previously rural to new urban uses, and on 7075 ha (29%), land use changed from one urban activity to another. Net changes in all urban categories were positive, except for the vacant land category which declined by just over 1500 ha. Other figures from the same source show that on average 45 per cent of land developed for residential purposes had been previously developed or was lying vacant in urban areas, a finding which broadly endorses that of Dickenson and Shaw (1982) in Leeds.

Table 6.1 Developed areas 1969 (England and Wales)
(*Source:* DoE 1978: 5.)

	%
Developed area as % of total	9.8
Predominantly residential	60.8
Predominantly industrial/commercial	17.5
Predominantly education/community	1.0
Transport	7.2
Urban open space	13.4

Table 6.2 Changes in land use, England (recorded in 1987) (figures in ha, rounded to nearest 5 ha). (*Source:* DoE 1988a: 7.)

				New use				
				Urban subdivisions				
					Industry			
Previous use	Rural total	Urban total	Resid-ential	Transport	comm-unity	Comm-unity	Vacant	All use
Rural	9 375	7 410	4 160	1 530	750	445	525	16 790
Urban	795	7 075	3 250	935	1 620	345	915	7 870
All uses	10 170	14 485	7 410	2 465	2 370	790	1 450	24 660
Net change	− 6 620	6 620	5 770	1 425	640	340	−1 560	

6.4 Thematic studies

The development of the urban economy in recent years, and the related planning policies, have resulted in the need for specialized, problem-based land use surveys rather than general purpose inventories.

Among the most important of these have been surveys of land which is derelict, vacant or otherwise poorly used. By the early 1970s it became clear that the extent of derelict land was growing alarmingly in urban areas and that it was being created increasingly by the collapse of manufacturing, transport and public utilities rather than by the traditional cause which was mining. In order to monitor the problem, and promote reclamation through derelict land grants, the DoE requested local authorities to undertake detailed ground surveys. Information thus gathered was published in two summary volumes (DoE 1974, 1982) and the results of a further survey in 1988 are awaited. The figures show that 46 per cent of derelict land was found in urban areas in 1982 with the seven major conurbations alone containing one-third of the total.

Closely related to the problem of derelict land, and in some cases overlapping with it, is vacant land. Getting this land back into active use has been an important part of the government's programme of inner-city regeneration and to this end Land Registers were introduced in 1980. Landowners in the public sector are required to reappraise their vacant or underused land with a view either to making use of it themselves or putting it on the register which would signify its availability for development. The register is held by the DoE in the form of a computer database which can be interrogated from terminals at DoE regional offices, together with maps and documentary details of each site. Local councils also hold their own copy which is open for public inspection. At 31 March 1988 a total of 40 000 ha of vacant land was on the register, with 55 per cent of this being in local authority ownership. The government has recently decided that public bodies should themselves additionally publish information about their unused land. A number of studies of vacant land have been undertaken, using local authority and DoE source material (Bruton and Gore 1980; Adams *et al.* 1988) and a comprehensive literature review has been provided by Cameron *et al.* (1988).

Given the importance of derelict and vacant land as a planning issue, it is disappointing to record that official figures suffer from many shortcomings (Chisholm and Kivell 1987), the net result of which is markedly to under-estimate the extent of the problem.

The recent recovery of the urban economy, together with various social changes, has highlighted another major land use theme relating this time to the availability of land for housebuilding. Two issues in particular have been important: the total availability of land together with its regional pattern and the relationship between land on the urban fringe and vacant sites within the city. Residential land surveys, undertaken jointly by planners and builders, were instituted in Manchester in 1979 and were subsequently extended by Circular 9/80 to all English local authorities. The issue has given rise to a number of disputes between planners and developers (McKenzie 1983), and although the government has been encouraging pro-development policies (DoE 1988c), by the end of 1985 fewer than half the counties in England had undertaken joint land availability studies (DoE 1989b). The situation regarding industrial land has been the subject of fewer centralized directives, but most urban authorities maintain records of available sites for their own promotional purposes.

Retailing is dealt with in Chapter 17 but it is worth noting here that it gives rise to particular problems in land use study because of the rapidity of change and the complexity of uses on ground-floor and upper levels. The main sources of information are shopping centre surveys, rating lists and trade directories, but detailed plans and associated listings for over 1000 centres are available commercially from Chas E Goad Ltd.

6.5 Landownership and values

Landownership is important to an understanding of land use and develop-ment, not least because of the vexed relationship between the private and public sectors and because the behaviour of landowners, be they profit maximizers or utility maximizers, profoundly affects the urban development pattern. Most European nations had a register of landownership by the eighteenth century, but not so the UK. In 1925 a Land Register was eventually established in England to ease conveyancing procedures but even today it covers only two-thirds of all land. The registers in Scotland and Northern Ireland are less comprehensive but at least they have been open to the public, a feature which is only now becoming applicable to those in England and Wales as a result of the 1988 Land Registration Act. It is the government's intention that public access to the Land Registry records will be allowed from 1990. The need for a full cadastral survey, giving details of landownership and related matters, has been noted by numerous researchers (for example, Bruton and Gore 1981; Norton-Taylor 1982; Chisholm and Kivell 1987), yet the situation remains unsatisfactory.

Within the public sector information on ownership is slightly more accessible, but even here the statistics are fragmented and have to be gleaned from many disparate sources such as the property information systems and records of local authorities and the piecemeal records of other public bodies. Despite the importance which publicly owned land has in shaping the morphology and planning of major cities, there exists no comprehensive and accurate record of landholdings by such bodies as central government departments, local authorities, nationalized industries and statutory undertakers. Recent attempts by government to make the public sector more efficient and the privatization of a number of utilities and nationalized industries have revealed a surprising degree of ignorance about the extent and status of their landholdings. A number of studies suggest that in large urban areas the majority of land is in fact in public ownership, with local authorities commonly owning more than half of the total. In Manchester, Kivell and McKay (1988) identified 14 significant public-sector landowning bodies which between them accounted for approximately 65 per cent of the city's land.

One further facet of the land use question to be considered is that of land values. Here again the familiar pattern occurs; information is sparse and fragmented, especially in comparison to countries such as Austria, Denmark and Sweden where land-value maps are often used for taxation and other fiscal purposes. In the UK sources of detailed information are handicapped by confidentiality and, at best, data can be obtained only in highly aggregated or small-scale sample form. Some of the sources have been summarized by Howes (1980) but essentially there are only two. District valuers and the Valuation Office of the Inland Revenue regularly produce statistics on site values and capital values. Many of these are reported to the DoE (1988b) which publishes a few of them regularly, for example housing land sales and prices. Additionally private valuers, surveyors and property advisers compile statistics on land values and transactions. Some of this information is published by the firms concerned or through professional journals such as the *Estates Gazette* or *Estates Times*, but usually it represents only a sample of land actually sold.

6.6 Conclusion

In conclusion it is difficult to demur from the overall findings of Coppock (1978) a dozen years ago. The availability of urban land use statistics is still unsatisfactory, coverage is inadequate and patchy and there remain large gaps in our knowledge.

At the local level there exist a number of sources of urban land use information, notably within the records of district and county authorities. These sources, however, vary in the reliability and regularity of their cover, most of them collect information for purposes other than dedicated land use studies and the classification systems which they use are frequently incompatible. At a national level the DoE has made a number of attempts to collect and analyse land use data but these attempts suffer from many shortcomings, notably their restriction to sample studies and their reliance upon extremely crude classifications. At a supranational level some influence is now beginning to be felt through European Community (EC) activity. The most relevant programme, CORINE, has been under way since 1985 with the purpose of providing information on the environment to assist in policy formulation. Information on land use and land cover from remote sensing sources will form part of this programme, but at present this has a lower priority than data relating to topography, soils and biotopes (Briggs and Mounsey 1989).

Clearly the particular nature of urban land poses enormous problems in terms of its smallness of scale, the complexity of activities and the importance of human factors such as ownership and the planning context. The different means of gathering information all have shortcomings; ground surveys are expensive and cumbersome, remote sensing techniques are unproven and documentary evidence is fragmented and discontinuous both in time and space. Rapid advances have been made in managing, mapping and analysing information especially through GIS techniques, but the nature and provenance of the raw land use data have seen few such improvements. In the light of this it is perhaps remarkable that urban planning has functioned as well as it has for over 40 years.

Appendix 6.1

Aerial photography
Within the public sector the main sources of aerial photography in the UK are as follows

Central Register of Air Photography, Ordnance Survey
Central Register of Air Photographs for Wales, Welsh Office
Central Register of Air Photography, Scottish Development Department
Department of the Environment (NI), Ordnance Survey of Northern Ireland

In addition the Air Photo Unit at the DoE provides a restricted service for government departments and some other public organizations.

Local authorities and government-funded bodies such as research council institutions also hold collections of aerial photography for their own use as do a number of universities and polytechnics. Notable among the latter are the Universities of Aston, Bristol, Cambridge, Dundee, Durham, Keele, Reading, Sheffield, Swansea and University College, London.

In the commercial sector there are a number of large air survey companies, for example:

Clyde Surveys Ltd
Hunting Technical Services Ltd
BKS Surveys Limited
JAS Photographic
Cartographic Services Southampton Ltd
Geosurvey International Ltd
Committee for Aerial Photography

Many small companies offering smaller format and oblique photography also exist and these may be found in the Yellow Pages under 'aerial photography'.

Satellite imagery

The main sources are:
Royal Aircraft Establishment
SPOT Image

References

Adams C D, Baum A E, McGregor B D 1988 The availability of land for inner city development: a case study of inner Manchester. *Urban Studies* 25: 62–76

Anderson J R 1976 *A Land Use and Land Cover Classification System for Use with Remote Sensor Data* US Geological Survey Professional Paper 964, Washington DC

Berlin G L 1971 *Application of Aerial Photographs and Remote Sensing Imagery to Urban Research and Study* Exchange Bibliography No 222, Council of Planning Libraries, Monticello, Illinois

Best R H 1981 *Land Use and Living Space* Methuen, London

Best R H, Anderson M 1984 Land use structure and change in Britain 1971–81. *The Planner* 70(11): 21–4

Bourke A, Davies J 1988 The use of GIS in Greater Manchester. *Mapping Awareness* 2(5): 27–30

Briggs D, Mounsey H 1989 Integrating land resource data into a European geographical information system: practicalities and prospects. *Applied Geography* **9**(1): 5–20

Brown I D 1985 Land potential and development monitoring systems. In England J R, Hudson K I, Masters R J, Powell K S, Shortridge J D (eds) *Information Systems for Policy Planning in Local Government* Longman, Harlow, pp 226–37

Bruton M, Gore A 1980 *Vacant Urban Land in South Wales* Department of Town Planning, University of Wales Institute of Science and Technology, Cardiff

Bruton M, Gore A 1981 Vacant urban land. *The Planner* **67**: 34–5

Cameron G C, Monk S, Pearce B J 1988 *Vacant Urban Land: A Literature Review* DoE, London

Champion A G 1972 *Variation in Urban Densities between Towns of England and Wales* Research Paper 1, University of Oxford School of Geography

Champion A G 1974 *An Estimate of the Changing Extent and Distribution of Urban Land in England and Wales 1950–1970* RP 10 Centre for Environmental Studies, London

Charlton M, Openshaw S 1986 Planning and land use in a UK metropolitan county. In Kivell P T and Coppock J T (eds) *Geography, Planning and Policy Making* Geo Books, Norwich pp 113–40

Chisholm M, Kivell P T 1987 *Inner City Wasteland* Hobart Paper 108, Institute of Economic Affairs, London

Coleman A 1980 Land use survey today and tomorrow. In Brown E H (ed.) *Geography Yesterday and Tomorrow* Oxford University Press pp 216–28

Collins M, Barnsley M 1988 *Energy Saving through Landscape Planning: A Study of the Urban Fringe* Property Services Agency, Croydon

Collins W G, Bush P W 1974 The application of aerial photography to surveys of derelict land in the UK. In Barratt E C, Curtis L F (eds) *Environmental Remote Sensing* Arnold, London pp 167–81

Coppock J T 1978 Land use. In Maunder W F (ed.) *Reviews of UK Statistical Sources* vol VIII Pergamon, Oxford pp 1–101

CSO 1990 *Guide to Official Statistics* 6, HMSO, London

Deane G 1986 Statistics measure post-war change. *Town and County Planning* **55**(2): 346–7

Dickenson G C, Shaw M G 1977 *Monitoring Land Use Changes* TP20, Centre for Environmental Studies Planning Research Applications Group, London

Dickenson G C, Shaw M G 1982 Land use in Leeds 1957–76: two decades of change in a British city. *Environment and Planning* **A14**: 343–58

DoE 1974 *Survey of Derelict Land in England* London

DoE 1978 *Developed Areas 1969, a Survey of England and Wales* London

DoE 1982 *Survey of Derelict Land in England* London

DoE 1986 *Land Use Change in England* Statistical Bulletin 86/1, GSS, London

DoE 1987 *Land Use Change in England* Statistical Bulletin 87(7), GSS, London

DoE 1988a *Land Use Change in England* Statistical Bulletin (88)5, GSS, London

DoE 1988b *Urban Land Markets in the UK* HMSO, London

DoE 1988c *Land for Housing* Planning policy guidance 3, London

DoE 1989a *Land Use Change in England* Statistical Bulletin (89)5, GSS, London

DoE 1989b *Land for Housing* Progress report 1988, London

Dueker K J, Horton F E 1971 Urban change detection systems. In *Proceedings of Seventh International Symposium on Remote Sensing of the Environment* University of Michigan, Ann Arbor pp 1523–36

Fordham R C 1974 *Measurements of Urban Land Use* Occasional Paper No 1, Department of Land Economy, University of Cambridge

Gault I, Davis S 1988 The potential for GIS in a large urban authority: the Birmingham City Council GIS Plot. *Mapping Awareness* **2**(5): 38–41

Gebbett L F 1978 Town and country planning. In Maunders W F (ed.) *Reviews of UK Statistical Sources* vol. VIII Pergamon, Oxford pp 103–219

Gierman D M, Macdonald C L 1982 Land use monitoring in the urban-centred regions of Canada and the Canadian land data system. *Computers Environment and Urban Systems* **7**(4): 275–82

Godfree S 1988 *Land Use Gazetteer* Leaf Coppin, Deal

Grimshaw D J 1985 Planning application systems. In England J R, Hudson K I, Masters R J, Powell K S, Shortridge J D (eds) *Information Systems for Policy Planning in Local Government* Longman, Harlow pp 215–25

Grimshaw D J 1988 Monitoring the use of land and property information systems. *International Journal of Information Management* **8**: 188–202

Guerin G, Mouillart M 1983 L'occupation des sols urbanisés. *L'Espace Gégraphique* **2**: 153–7

Harrison A R, Richards T S 1987 *An Evaluation of General Purpose Classification Techniques for the Discrimination of Urban Land Use in SPOT Panchromatic and Multispectral Data* University of Bristol, Remote Sensing Unit, Final report of MPSI Systems Ltd, Bristol

Hathout S 1988 Land use change analysis and prediction of the suburban corridor of Winnipeg, Manitoba. *Journal of Environmental Management* **27**: 325–35

Hill R D 1984 Land use change. *Geoforum* **13**(3): 457–61

Home K 1984 Information systems for development land monitoring. *Cities* **1**(6): 557–63

Howes C K 1980 *Value Maps: Aspects of Land and Property Values* GeoBooks, Norwich

Humphries A M 1985 Property information systems. In England J R, Hudson K I, Master R J, Powell K S, Shortridge J D (eds) *Information systems for Policy Planning in Local Government* Longman, Harlow pp 196–214

Jones G 1986 A summary of different land information systems. In Selman P (ed) *Environmental Conservation and Development* Planning Exchange Occasional Paper 24, University of Glasgow pp 160–6

Kivell P T, McKay I 1988 Public ownership of urban land. *Transactions of the Institute of British Geographers* **13**: 165–78

Kivell P T, Parsons A J, Dawson B R P 1989 Monitoring derelict urban land: a review of problems and potentials of remote sensing techniques. *Land Degradation and Rehabilitation* **1**: 5–21

Lindgren D 1985 *Land Use Planning and Remote Sensing* Martinus Nijhoff, Dordrecht

McKenzie A 1983 Land for housing. *Town and Country Planning* **52**(3): 68

Markowski S 1982 *Land and Building Use: Analysis and Surveys* SSRC Planning Review 4, Edinburgh

NERC 1978 *Land Use Mapping by Local Authorities in Britain* Experimental Cartography Unit, Architectural Press, London

NLUC 1975 *National Land Use Classification* Joint Local Authority, Local Authorities Management Service and Computer Committee, Scottish Development Department and DoE Study Team Report, HMSO, London

Norton-Taylor R 1982 *Whose Land is it Anyway?* Turnstone Press, Wellingborough

Rhind D, Hudson R 1980 *Land Use* Methuen, London

Sellwood R 1987 Statistics of changes in land use; a new series. *Statistical News* **79**: 11–16

Ward R M 1983 Land use mapping techniques for city and regional planning. *Journal of Environmental Management* **17**(4): 325–33

Weights B 1988 Kingston's GIS pilot set for promotion. *Mapping Awareness* **3**(2): 8–11

Whitehouse S 1989 A multistage land use classification of an urban environment using high resolution multispectral satellite data. Unpublished PhD thesis, University of Keele

7

Rural land use

Terry Coppock

7.1 Introduction

It might be expected that governments would need to know how the surface of their territory was used; but there is little evidence that this was so until recent times. It was not until 1984 that there was explicit recognition by central government of the need for reliable information at a national level (Department of the Environment (DoE) 1984) and, despite the introduction in 1947 of statutory land use planning by local authorities, there is still no comprehensive information on rural land use for administrative areas within the country. The purpose of this chapter is to review what is meant by land use and to examine the principal sources of information on rural land. It is complementary in this respect to Chapter 6, in which urban land uses are reviewed, and Chapter 12, where ways of collecting data *de novo* are considered. The division between urban and rural is not, however, clear-cut, especially on the urban fringe; some sources cover both and some essentially urban uses, such as residences, roads and railways, are also important features of the rural environment.

The meaning of land use/land utilization (the terms are used interchangeably) might appear to be self-evident. They seem to have become technical terms quite late and the first official use appears to have been in the Committee on Land Utilization in Rural Areas (Ministry of Works and Planning 1942). In the National Land Use Classification (NLUC), the terms are interpreted as activity/activities undertaken on land (DoE 1975); but while this is easy to state, it is much more difficult in practice, and the activities are often inferred from the appearance of the land. Indeed, what is described as *land use* often refers to *land cover*, especially the vegetative cover of land in rural areas. This is much easier to identify, whether by direct observation or by the interpretation of remotely sensed imagery, but it can often be misleading.

Almost a third of the country is occupied by what the *First Land Utilization Survey* called *heathland* (Stamp 1948), which is a general description of a form of vegetation dominated by low shrubs and coarse grasses. Such land may be used for a variety of purposes. Most is in agricultural use, but some will be used to support grouse or red deer, to be shot as a recreational activity, some for the collection of urban water supplies and some for informal recreation by members of the public. Identifying these activities and, more especially, the land they use, is very difficult. *Land cover* is most likely to be employed where activities occur at a low level of intensity and occupy indeterminate areas.

Such terms as woodland and heathland are thus often used as descriptors of a range of activities, but there is an obvious danger in assuming a constant relationship between cover and use. Woodland may be employed as a surrogate for timber production, but not all woodland is so used; some provides private or public recreation or is kept for its amenity value. While such uses may be the only activity, a variety of uses may take place on the same land, i.e. it is in *multiple use*. In a strict sense, multiple use should be reserved for activities which occur simultaneously on the same ground, as with sheep grazing and water collection, but it is widely used to describe land on which more than one activity takes place, irrespective of whether they are coincident in time or each occupies exactly the same area, as with hunting in winter on land used for crop production in summer. Most records of land use refer only to the dominant activity and not a great deal is known about the land occupied by many of the subordinate activities.

Records of land use will also need to identify the land area where the activity takes place. For reasons of confidentiality, land use records will normally be aggregated, usually for administrative units. These will necessarily contain a number of *impurities* (i.e. other land uses), and may correspond only in a general way with the land on which the activity occurs. Even where individual records are available, they too will often contain impurities which have been ignored in enumeration, for example, patches of open land within woodland. In interpreting land use data collected by others, especially where these have been aggregated, it is important to be aware of the problems of internal variability and imprecise location, so that the information can be correctly interpreted.

Three other aspects of land use merit mention, although lack of space precludes treatment of them here. First, land varies greatly in *quality* and its suitability for different activities, and interpreting rural land use requires some knowledge of land quality, a topic that receives some mention in Chapter 14. Second, and related to land quality, land uses vary greatly in their *intensity*, the level of activity/production per unit area, although most data on land use relate only to area occupied. Third, decisions about land use are made by owners and occupiers, and their characteristics may be an important consideration in interpreting spatial differences in land use and changes in use. Unfortunately, no comprehensive records exist of patterns of rural land *ownership and occupation*, although some not very satisfactory information is available in respect of land used for agriculture and for forestry.

Any system of collecting data on land use will require some classification of uses, which will range from the very simple, as with the *First Land Utilization Survey* (Stamp 1948), to the very complex, as with the Institute of Terrestrial Ecology's (ITE) (1981) *Survey of the Rural Environment*. Because of complexities of rural land uses, which often grade into each other, such classifications are necessarily simplifications. Careful study of definitions and their interpretation is essential so that misleading conclusions are not to be drawn from comparisons of data from different sources or from the same source at different dates.

7.2 Existing sources

Budgetary constraints often require that existing sources be used, because data are expensive to collect. It is, however, important to recognize that such data may have been collected for a variety of purposes, as a housekeeping operation, as an inventory of use at a particular point in time, as a basis for monitoring change or for some reason not directly or explicitly concerned with land use. Identifying the objectives and the way in which these have influenced the classification of uses and method of collection adopted is essential for the proper evaluation of such sources.

No data are collected on an identical basis for all parts of the UK, although the agricultural censuses come closest to this. The emphasis here will be on those data that are available on a consistent basis for the component countries. Other data are available for individual districts, counties and regions and some policy areas, such as National Parks; but no complete record exists of such holdings and they will not be discussed here.

Few of these national sources have been subject to independent evaluation; where they have undergone an internal audit of any kind, this has rarely been published. It is true that the appraisals of official statistics undertaken in the early 1980s, the so-called Rayner reviews, do provide some critiques, although their principal purpose was to eliminate the collection of data not required for the managerial purposes of the agency collecting them. Claims of high accuracy and consistency must be viewed with suspicion, and although some checks can be made by comparing data from different sources and by examining their plausibility, any proper evaluation will require the co-operation of the collecting agency. Public bodies are not keen to draw attention to any limitations of data they collect; they may also be willing to tolerate deficiencies which are irrelevant to the purpose for which the data are collected, even though these may weaken their value as land use records. Weaknesses and strengths can sometimes be inferred and there is a danger of underestimating those sources which have been evaluated simply because their weaknesses have been exposed.

The widespread use of computers to handle numerical data also makes it important to know whether such data are available in machine-readable form. This is increasingly the case with contemporary data, but large quantities of historic data, important for establishing trends or changes, exist only in analogue form and must first be converted by scanning or key punching. In the accounts which follow, attention will also be given to the classifications used, the areas for which data are available, the date(s) or frequency of collection, the method of measurement and resulting accuracies, and the compatability of data over time, together with some general appraisal of their quality and usefulness. Attention will first be given to sources which deal with a wide range of uses; those which provide information on only one category will then be reviewed.

7.3 Ordnance Surveys

The Ordnance Survey (OS), the national mapping agency for Great Britain (there is a separate OS for Northern Ireland) does not collect land use data as an end in itself. Its duty is to record the topography of the country, i.e. both its natural features and those man-made features that are of a permanent or semi-permanent kind. Information recorded on the basic-scale maps (at 1 : 1250 for the main urban areas, 1 : 2500 for other lowland areas and 1 : 10 000 for the uplands) does, however, permit some statements to be made about land use, particularly in urban areas (Harley 1975). In rural areas, apart from built structures such as farm buildings and residences, land use can be inferred from land cover, since woodland and various categories of rough vegetation are shown by symbols. It is often assumed that land which is free of symbols, the so-called *white land*, is in agricultural use; but while most of it probably is, there is no way of distinguishing that which is truly agricultural from that which is not. Only two categories of agricultural land are specifically identified, land under orchards and under glass; use of the remainder is inferred.

Apart from the difficulty of identifying uses comprehensively and unambiguously, the other main limitation of data on land use derived from OS maps is that they span a wide variety of dates. The OS operates a system of revising its basic-scale maps under which changes are divided into primary, comprising most urban-related developments but also large-scale afforestation, and secondary, the remainder (Tym and Partners 1985; Latham 1989; Sinclair 1985). When sufficient primary change has accumulated to justify a visit, the changes are surveyed, along with any secondary changes that can conveniently be included. Uneconomic primary change, i.e. that which does not justify a visit, and any secondary change are systematically mapped only when maps are revised, at intervals determined by the age of the map and the amount of secondary change that is believed to have occurred, but not exceeding 50 years. Thus, while the picture of urban land use is kept reasonably up to date, rural changes may go unrecorded for very much longer. It is thus impossible to construct a cross-sectional view of rural land use from this source.

An additional weakness is that such information is available only in map form and must be laboriously measured to convert it into numerical form.

This situation will change when the process of converting existing maps into digital form is completed. Priority is being given to urban areas, coverage of which is expected to be complete by 1995; but no target date has been set for rural areas (OS, personal communication, 1989).

In addition to the detailed information shown on the basic-scale maps, some categories of land cover are recorded on the smaller-scale maps derived from them, for example, the 1 : 50 000 series, although the latter are revised more frequently to incorporate major changes and, in respect of woodland and rough land, can provide a reasonable snapshot for the country as a whole. The maps of its predecessor, the 1 : 63 360 series, were in fact used to prepare a small-scale land use map of Great Britain (Bickmore and Shaw 1963), and the Forestry Commission (FC) (1983) has used the woodland plates of the first edition of the 1 : 50 000 series in the 1979–82 census of woodland.

Since 1983, an attempt has been made to derive statistics on changes in land use from the basic-scale maps, although the emphasis is on urban change (Tym and Partners 1985; see also Chs 6 and 12). This initiative, the Monitoring Land Use Change project, is an adaptation of the process of mapping primary change and uses the same surveyors to identify additional categories of change and to estimate the extent and date of such changes and the land uses they replace. Twenty-four classes of change are identified, and these are amalgamated into 10 groups which in turn form 2 divisions, rural and urban (see Fig. 6.1). This grouping is somewhat arbitrary, with minerals, landfill, outdoor recreation and defence classified as rural, and residential and transport allocated wholly to urban. The remaining rural groups, viz. agriculture, forestry, open land and water, are chiefly descriptions of land cover and assume that white land is all agricultural. Rural changes also present greater difficulties than urban, partly because of the much greater average time-lag before they are surveyed and partly because several of the categories grade into each other. The resulting information is of changes *recorded* in that year, some of which may have occurred much earlier, and it will take time to build up reliable estimates of the changes occurring within any one year. Summary statistics are published each year for the main groups and information published for Wales is similar (Welsh Office 1989). Those for Scotland differ in that forestry, rough grassland and bracken and other uncultivated land are separately distinguished (Scottish Office 1989).

7.4 Land utilization surveys

The only complete cross-sectional views of a substantial part of the country have been provided by direct observation in the field in the two *Land Utilization Surveys*. The first, for Great Britain, was organized by Dudley Stamp (1948) and was undertaken between 1931 and 1941, although most of the fieldwork was undertaken between 1931 and 1934. Each parcel of land on the OS 1 : 10 560 maps was classified in one of five rural and two urban categories of use, viz. forest and woodland, grassland, arable land, heathland, moorland

and rough pasture, houses with gardens, and land agriculturally unproductive. The work was undertaken by volunteers, mainly secondary-school pupils and university students, who were not necessarily expert in either map reading or the interpretation of land uses. The information was checked mainly by matching of edges of adjacent sheets and by traverses, and was transferred by eye to outline maps of the OS at the 1 : 63 360 scale. The areas under the different categories were then measured at that scale by means of transparent squared paper.

Most of the 1 : 63 360 maps were published (the manuscript versions of those that were not are held in the map collection of the Royal Scottish Geographical Society) and estimates made of land in each category. Original maps were often returned to surveyors with a photostat copy being retained, but many of the originals and the photostats were destroyed by fire at the London School of Economics where the remainder are still held. A map of Great Britain was published by the OS at a scale of 1 : 625 000, but this too was compiled only as a very generalized statement. A similar survey was organized in Northern Ireland by Hill (1948) between 1938 and 1946.

No independent evaluation of this survey has been made, although it is to be expected that errors of both location and interpretation will have occurred, especially when field sheets were surveyed by younger pupils. Errors may also have occurred in transferring data from the field sheets and in making measurements. The spread of dates also presents a problem, although the maps can generally be taken as representing the situation in the early 1930s. The categories, too, are also more descriptive of land cover than of land use.

The *Second Land Utilization Survey* was organized by Coleman (1961; Coleman and Maggs 1964) and was intended to be compatible with, but more detailed than, the First. In this survey, too, use was recorded in the field by volunteer observers who classified the use of each parcel on the 1 : 10 560 maps into one of 64 categories. These can be grouped into 13 major categories, again with a rural bias and representing a mix of land uses and land cover. The survey began in 1960 and was completed in England in 1969 and in Wales in 1970. The rate of progress was rather slower than in the first survey, but 60 per cent of England and Wales had nevertheless been completed by 1964. In the late 1960s the survey was extended to Scotland, where coverage was largely restricted to the central lowlands, and the maps, which are deposited in the National Library for Scotland, have never been analysed. The field sheets for England and Wales are held in the Department of Geography at King's College London and some derived maps have been published on OS base maps at a scale of 1 : 25 000 on which land parcels are shown, thereby facilitating data transfer. Unlike the first survey, no county memoirs or national overview have been published, although extensive measurements, using point sampling, have been made to provide, *inter alia*, national estimates (e.g., A. Coleman, personal communication, 1977). Complete resurveys of 3 counties and parts of 19 others (Coleman 1985) were undertaken in the 1970s and 1980s, using a more complex classification, with

25 major categories, 55 minor categories and 270 subtypes (Coleman and Shaw 1980).

According to the organizer, a variety of tests, including joint exercises with the OS and the FC, has demonstrated that the survey was over 99 per cent accurate, although no details of this claim have been published (Coleman 1985). One observer, who completed the mapping of *c.* 260 km^2 in late 1966, found wide variations in accuracy in those parts of the sheets which had already been completed, an observation whose general validity was supported by the views of the organizers of the Welsh part of the survey (Fordham 1974). He believed that its accuracy, if ascertainable, would not be high. Whatever the truth, experience with surveys undertaken by undergraduates (Coppock 1954) and by skilled surveyors in the Monitoring Land Use Change project (Tym and Partners 1984; Gould Consultants 1988) suggests an assessment intermediate between these two extremes. In any case, the maps are a valuable potential source, identifying categories for which no official data of any kind are available and providing a base from which estimates of subsequent changes at a county level can be made.

7.5 Institute of Terrestrial Ecology

One other study that provides a national overview is discussed in more detail in Chapter 12, the ITE's (1985) *Survey of the Rural Environment.* Sample surveys, capable of providing reliable estimates of land use in Great Britain in 1977/78 and 1984, have been undertaken, using field surveys to record on maps of the 1 : 10 000 scale land uses in each of 256 sample 1 km squares of the National Grid. This survey was primarily rural and has been undertaken by specialized staff whose work was checked independently. The complex classification has a total of 473 categories, mainly of types of cover and of landscape features. The data are held by the ITE and any limitations are likely to arise from the small size of the sample and from sampling error rather than from errors of interpretation. An updated survey is being undertaken in 1990.

7.6 Aerial photography

For the most part aerial photography is a potential rather than an actual source and then mainly for parts of the country rather than for the whole (see Appendix 6.1). It can, nevertheless, provide information on a wide range of land cover for both rural and urban areas. Only two major projects involving the interpretation of land use from air photographs has been undertaken at a national level, that of the developed areas survey of England and Wales using air photography in 1969 (see Ch. 6) and that on landscape change in England and Wales using a variety of coverages. There are, however, several near-national coverages and one study of Great Britain and one of Scotland are in progress.

Most of the country was photographed by the Royal Air Force in the late 1940s at a scale of 1 : 28 000, although the quality of the resulting photography is poor by modern standards; however, it is now being used for a number of studies of change in the rural environment (see Ch. 12). Coverages for England and Wales photographed in 1969 at a scale of 1 : 60 000 and for England, Wales and part of Scotland for 1980–81 at a scale of 1 : 50 000 were also taken by the Royal Air Force. The latter coverage, originally a requirement of the Ministry of Defence, was photographed with different cameras; the Ministry's interest lapsed in 1980 and the remaining 55 per cent of England and Wales had a low priority, while only 40 per cent of Scotland was covered before flying was abandoned. A complete survey of Scotland was taken in 1987–88 by commercial companies on contract to the Scottish Development Department (SDD).

Partial coverages are also available for large parts of Great Britain, notably those undertaken on behalf of the OS as a basis for map revision (OS, personal communication, 1989); coverages of Scotland since 1969 are largely complementary, much of the Highlands and Islands being covered at a scale of 1 : 35 000 or smaller, and southern Scotland at scales of between 1 : 15 000 and 1 : 35 000 (Coppock and Kirby 1987). Almost half of Scotland is covered by photography taken since 1969 at scales of 1 : 15 000 and larger, although this comprises many small projects. In England and Wales, air photography has become increasingly popular among local authorities, numbers of which have commissioned complete surveys; according to the British Air Survey Association, coverages of about 40 per cent of counties were taken between 1975 and 1984, with a peak between 1970 and 1974 when 24 counties were taken (Kirby 1985).

Air photographs for national and subnational areas have been interpreted for land use purposes. Sample air photographs have been identified by the FC and by Huntings Surveys and Consultants on behalf of the DoE and the Countryside Commission, in respect of forestry and of landscape change, and other samples are being interpreted by the Nature Conservancy Council (NCC) to monitor changes in the rural environment (see Ch. 12). Complete air photography of Scotland in 1987–88 is being interpreted by the Macaulay Land Use Research Institute to provide information on land cover/use, a task due for completion in 1991. Examples of subnational coverages include a study of land use in the Welsh uplands (Parry and Sinclair 1984), of changes in moorland in seven National Parks (Parry 1982) and of part of the Scottish lowlands (Langdale-Brown 1980). Examples of interpretation at county level include those of Hertfordshire in 1974 and 1981 (Champion and Markowski 1985) and of West Sussex (Lukehurst 1985).

7.7 Satellite imagery

Satellite imagery represents, even more than air photography, a potential source requiring interpretation, with only one national study, of England

and Wales, having been undertaken and then on a research and development (R&D) basis (Huntings Surveys 1986). Since 1972 satellites of the Landsat series have been transiting the UK at intervals of 18 days (reduced with Landsat 4 and 5 to 16), and the first SPOT satellite was launched in 1986, with a repeat cycle of 26 days, although this has not yet been used for any published study of land use. The first three Landsat satellites carried a multispectral scanning system (MSS) capable of identifying some types of rural land cover with a nominal resolution of 80 m, and the fourth and fifth satellites also carried a thematic mapper (TM), with greater spectral sensitivity and with a resolution of 30 m. MSS data have been used by Hubbard and Wright (1982) to generate a map of land use in mainland Scotland and for various other experimental studies in Scotland (see e.g. Stove 1983). TM data have been used experimentally in the DoE/Countryside Commission study of landscape change. Data for all such satellite coverages are held at the National Remote Sensing Centre at Farnborough.

Both air photography and satellite imagery are further discussed in Chapter 12 as providing alternative methods of obtaining data on land use/cover.

7.8 Agricultural land

The remaining sections in this chapter are concerned with individual categories, rather than with a range of land uses. The two most important sources are those on agricultural and silvicultural use, although they are different in character, that on agricultural land being the product of an annual census which covers a high proportion of the total agricultural area, whereas the information on forestry is derived mainly from periodic censuses which are increasingly drawn from samples of such land (see also Ch. 14). Agriculture is in fact the only use for which near-contemporaneous data are regularly provided.

Agricultural censuses have been collected annually in June since 1866, through a postal questionnaire sent to known occupiers of agricultural land. The censuses are, however, intended primarily as guides to agricultural production and the agricultural departments (the Ministry of Agriculture, Fisheries and Food (MAFF), the Welsh Office Agricultural Department, the Department of Agriculture and Fisheries for Scotland (DAFS) and the Ministry of Agriculture for Northern Ireland) seem to have no need for detailed statistics of land use. The method used has remained essentially unchanged, although there have been variations in the minimum size of holding for which a return is sought and in the categories of information requested. Since 1968 eligible holdings below a prescribed area and standard labour requirements have been excluded, although they are periodically the subject of special censuses (see Ch. 14). In Scotland, where such holdings may comprise substantial areas of rough grazing, there was a fall of over 926 000 ha in the extent of rough grazing when such minor holdings were withdrawn from the census in 1969 (DAFS 1970).

The comprehensiveness of the census is affected by the accuracy of the agricultural departments' records of occupiers of agricultural land and by rates of refusal to answer the questionnaires. Securing a complete enumeration is difficult, because there are no cadastral records, especially as there are some 90 000 changes of area each year in England and Wales alone. Additionally, interpretation of what is agricultural is itself a matter of debate, especially on the urban fringe.

Not all occupiers known to the agricultural departments make a return, despite a legal obligation to do so, and estimates have to be made in lieu by reference to previous returns (if any) and national trends. The proportion of such estimates in England and Wales has fluctuated in recent years in the range 11–16 per cent, and is higher in Wales than in England. It is lowest on the largest holdings and highest on mixed and horticultural holdings. In Scotland, the proportion of estimates is well under 10 per cent (DAFS, personal communication, 1989).

The censuses provide annual statistics of the area of land used for agriculture, broken down into categories, of which three are of principal interest here, viz. the total areas under tillage (land in crops other than grass), grassland and rough grazing. Of these, the area under tillage is probably the most accurately known because it is important to farmers and occurs in well-defined parcels whose areas are shown on OS maps. Improved grass also tends to occur in well-defined parcels, but there are problems over permanent grass in semi-agricultural use, for example, as land supporting horses kept for recreational use. Rough grazing is probably the least reliable category of the three since its extent is of little interest to farmers who may not know it accurately; the area given may often be a residual figure, obtained by deducting the area in crops and grass from the total area of the holding. The censuses now include the area under woodland and other land used for agricultural purposes, but both are minor features whose extent is unlikely to be very accurately reported. Some rough grazing in England and Wales is common land, for which separate estimates have been made at infrequent intervals, although their accuracy is unknown; occupiers are required to record only the area of rough grazing in their sole occupation.

Practice in respect of official checks on the census differs in England and Wales and in Scotland. In the former there are now no field checks; the census branch relies on plausibility tests and on comparisons with returns for the previous year, and attempts to clarify major discrepancies with the occupier. In Scotland, census staff refer queries to local staff of DAFS, who know each area well.

Individual returns from occupiers are confidential and are consolidated into parish summaries and then into county/regional and national totals. Only the last two of these are published in annual volumes of agricultural statistics (Ch. 14), but parish summaries are available in printed or machine-readable form from the agricultural departments. In England and Wales, summaries for the previous 10 years can be obtained from the Agricultural Census Branch and the Welsh Office respectively, those for earlier years being deposited in the Public Record Office. For Scotland, copies of all summaries are held in the Scottish Record Office.

The parish thus constitutes the basic unit for most investigations of agricultural use, although it is far from ideal. Not only do parishes vary greatly in size and hence in the level of aggregation, but the land to which the summary refers may not all lie within the boundary of the civil parish (Coppock 1960). While it is probably true to say that the bulk of land in most summaries probably lies within parish boundaries, some may be as far as 32–48 km away within the same county. Parish summaries also do not include any of the land in common grazings or in minor holdings. A further complication in England and Wales is that, to maintain confidentiality, summaries will not be provided where there are fewer than three holdings (see Ch. 14). Cross-sectional comparisons over time are also complicated by changes in parish boundaries and by the fact that the summaries for different dates may refer to somewhat different areas of agricultural land because of changes in farm size.

So long as these limitations are recognized, the parish summaries provide the most up-to-date statement of the distribution of the main classes of agricultural land throughout Great Britain. They have been used to produce maps in agricultural atlases (e.g. Howell 1925; Coppock 1976) and provide the agricultural component of the ITE's National Land Characteristics Database (NATLAC), in which various attributes of land in Great Britain have been assembled for 10×10 km cells of the National Grid. In general, the more aggregated the data, the more reliable will be the trends they reveal, although changes in the basis of collection, in the completeness of the returns and in the definitions of uses must always be borne in mind.

Until the Monitoring Land Use Change projects was implemented in 1985, the agricultural censuses provided the only source of regular information on exchanges between agriculture and other uses. When occupiers of agricultural land give up or acquire land, they are required to indicate the area involved and its origin or destination. The great majority of these exchanges are within agriculture, but others are to urban development, forestry or other uses. Such data were first made public in evidence to the Scott Committee (Ministry of Works and Planning 1942). Subsequent figures were given in answers to Parliamentary questions and then in the annual volumes of *Agricultural Statistics*; a convenient summary of changes in Great Britain to 1966 is provided by tables in the volume published in 1968 to mark the centenary of agricultural statistics (MAFF/DAFS 1968). MAFF has also issued regional and county 5-year averages for the period 1945–70 (Best 1981) and these have been used by Best and Champion (1970) to examine patterns of changes.

These losses from, and gains by, agriculture are broken down into four categories for England and Wales, viz., urban, industrial and recreational development; government departments; FC and private woodlands; and other adjustments. In Scotland, seven categories of loss are given, viz. roads, housing and industrial developments; recreation; mineral workings;

hydroelectric and water boards; service departments; forestry; and others.

The agricultural departments have always had reservations about the data, which are thought to be incomplete; they may also anticipate changes, in that there may be a time-lag between release from agriculture and adoption of a new use. Nevertheless, Best and associates have made extensive use of this source, particularly the data on transfers for urban purposes, to estimate land use statistics for earlier periods (Best 1981). Comparisons with other sources, such as the *Land Utilization Surveys*, provide some confirmation of the validity of these data in respect of urban uses, although those for transfers to forestry are less consistent. Other sources confirm the plausibility of the county data (Agriculture EDC 1977; Blair 1978).

MAFF has, however, become increasingly concerned about their accuracy, since an increasing proportion of such changes go unrecorded by occupiers, and the Census Branch no longer has the resources to pursue non-respondents. Publication was therefore discontinued in 1980 and, in view of the apparent success of the OS-based project, collection of these data may well cease. The Scottish data continue to be published, not only for the country as a whole but also for the regions, perhaps because DAFS contacts the new occupier to confirm that a transfer has taken place.

Despite these limitations, the agricultural censuses remain an important source for the study of land use and it is for this reason that they have been extensively studied and their weaknesses identified. They have the advantages of being in numerical form, much of it machine-readable, of being the only major source where the information is provided by the user, of describing the situation for the whole country on a single day, and of being available every year on a broadly comparable basis. Their main disadvantages arise from the need to preserve confidentiality, from the administrative procedures used to produce parish summaries, from the limited precision with which the data can be located, and from their dependence on the willingness and ability of large numbers of individual occupiers to provide accurate and consistent answers to the questions asked. Their strengths and weaknesses must be considered alongside those of other sources, notably field survey and remotely sensed imagery, on which they provide a valuable check (Ch. 12).

7.9 Forestry

Apart from the information obtained from the OS maps, which record all woodland as interpreted by its surveyors, and that on woodland on farms, available from the agricultural censuses, information on forestry is derived largely from the records of the FC which has a dual role as forest authority (the agency implementing government policy for private forestry) and forest enterprise (as manager of the public forest estate). The terms *state* and *private* used in this connection are rather misleading because the former refers only to the woodland owned/leased by the FC and the latter includes land owned by other public bodies.

Two categories of information exist on land used for forestry: records kept by the FC about its own estate and, in less detail, those categories of private woodland which are grant-aided; and periodic censuses of woodland, although the last of these, undertaken in 1979–82, was confined to *other woodland* (i.e. that which is not grant-aided). These records are, however, limited to area, composition and production; no information is kept on the extent to which woodland is used for other purposes.

The FC keeps detailed records of its own plantings and fellings, by forest districts; such data are then aggregated by conservancies (which, like districts, do not correspond with local authority areas) and by countries to provide totals for Great Britain as a whole. The FC also keeps records of those woodlands which are grant-aided under the Dedicated and Approved Woodland schemes, particularly about planting and replanting, but much less is known about other woodland. The FC does, however, keep records of virtually all fellings, since a felling licence is required except where few trees are involved. District officers also make estimates of any private planting which is known to them (thought to represent about a fifth of all such planting). Published information appears in the FC's annual reports which provide some details of areas planted and restocked; these are useful for monitoring trends, but are of little value for any detailed spatial analysis.

The FC also undertakes periodic censuses, at intervals of approximately 15 years. The last complete census was undertaken by FC staff between 1947 and 1949 as a field survey of all woodland over 2 ha. Detailed reports were published giving county totals in both map and tabular form (FC 1952). A separate sample survey of 1.6 km strips, 40 m wide and chosen at random, was conducted of woodland under 2 ha and a separate report issued (FC 1953). Records of all these surveys are held in the Public Record Office (for England and Wales) and the Scottish Record Office.

The next census, undertaken between 1965 and 1967, was of private woodland of 0.4 ha or more, and was confined to a sample of 15 per cent of 1 km squares of the National Grid, chosen at random within each marketing region. The woods in these squares were then surveyed and measured by FC staff in the same manner as in 1947–49. Estimates of woods under 0.4 ha were made for a smaller sample of 1 km squares. The results are regarded as reliable only for conservancies and countries. Data for the FC's woods were derived from its own records and are not strictly comparable. A report on this census was also published (FC 1968).

The most recent census, undertaken in 1978–82, was also a sample survey, this time of *other private woodland*, data for the FC's own woodlands and for grant-aided private woodlands being provided from the FC's records. The minimum size of woodland and the sampling strategy were both different from those adopted in 1965–67, although the later census is thought to be more reliable. For this census, the woodland plates of the OS's 1 : 50 000 series maps were used to identify these *other* woodlands; these were then digitized, measured and stratified into six size groups, and a sample drawn from each group and compared with the most recent air photography. A revised estimate of area was then made and a subsample of woods visited to determine composition. A report giving data for countries and conservancies has been published (FC 1983); like those for previous censuses, it also gives a detailed account of the methods used.

The data on forest land have probably been more accurately measured than those for any other major source, although data from recent censuses are subject to sampling error. The censuses have been carefully planned and observations and measurements are made by the FC's professional staff. On the other hand, the methodology has changed in each census, the intervals between them are long and there are problems of comparison. The FC's own records are probably highly accurate, but those for grant-aided and for other private woodland are progressively less so. The FC annual reports provide data for monitoring trends in its own woodland and, less certainly, in the other categories. With FC permission, data could be reassembled for other areas, but the task would be difficult since most data are held in district offices and, in respect of grant-aided woodland, on individual files. These characteristics limit the usefulness of forest data, but like the Agricultural Census, they provide a valuable check on the plausibility of other estimates of the area under forestry.

7.10 Other uses

Apart from records of derelict land, which are discussed in Chapter 6, no other category is the subject of regular or periodic enumeration. Recreation is partly covered by the OS's large-scale maps, some information on open space is available on the maps of the *Second Land Utilization Survey* and the Countryside Commission holds records of the extent of land in country parks and picnic areas. Although a great deal of miscellaneous information is also held by the Countryside Commission and the NCC (for example, the Countryside Commission holds data on the extent of land protected for conservation), it is debatable how far amenity can be regarded as a use. Land used for military training, field sports and water gathering are three major categories for which data are not readily available. Nor is there any satisfactory information on the extent to which land is in multiple use.

7.11 Overview

The sources of data on rural land use reviewed here comprise various categories of use and of cover. Their plausibility and the extent to which they are aggregated vary widely, although published figures may be supplemented by the issue of unpublished data for smaller areas. Statistical data are all in aggregated form, with levels of disaggregation controlled by considerations of both confidentiality and convenience. Relatively few historic records are yet available in machine-readable form. Very few of the data have been evaluated and changes in the staff responsible for their collection may mean that those now in post know little of their antecedents. Some evaluations have been undertaken by academic researchers who have compared data from a variety of sources or investigated the mechanisms by which data are compiled. While it is important that the need for such data be kept under review, there appears to be a tendency in central government to take a very short-term view of such questions, as in the Rayner reviews. Such periodic reviews are now to be a regular feature of government statistics, although the Tradeable Information Initiative may have an opposite effect if it reveals that the data are of value to others who may be willing to pay for them or to contribute to the cost of providing them. Electronic data handling increasingly offers the opportunity to assemble data in the form required by users, as the OS has recognized in its digital programme; but such an approach requires sound data, either in discrete form or at the lowest possible level of aggregation.

Note: Changes to the availability of the agricultural parish summaries are referred to in the Editor's preface (p ix).

References

This chapter draws heavily on interviews with officials of the DoE, the FC, MAFF and the Scottish Office, and on two consultancies in which the author was involved, one with Roger Tym and Partners, a feasibility study for a *National Land Use Stock Survey*, for the DoE, the other for the SDD on approaches to monitoring changes in the Scottish landscape. It also draws on work undertaken as part of the project, sponsored jointly by the Royal Statistical Society and the Social Science Research Council, to review statistical sources in the UK and on 40 years of experience in using many of these sources.

Agriculture EDC 1977 *Agriculture into the 1980s: Land Use* National Economic Development Office, London
Best R H 1981 *Land Use and Living space* Methuen, London
Best R H, Champion A G 1970 Regional conversions of agricultural land to urban use in England and Wales 1945–67. *Transactions of the Institute of British Geographers* **49**: 15–32
Bickmore D P, Shaw M A 1963 *Atlas of Great Britain and Northern Ireland* Clarendon Press, Oxford
Blair A M 1978 Spatial effects of urban influences on agriculture in Essex 1960–74. Unpublished PhD thesis, University of London
Champion A G, Markowski S 1985 Land use information held by local authorities. In Roger Tym and Partners (eds) National land use stock survey, appendices, unpublished report to the DoE, London pp 86–98
Coleman A 1961 The second land use survey: progress and prospects. *Geographical Journal* **127**: 168–76
Coleman A 1985 Lessons from comprehensive national ground survey. In Roger Tym and Partners (eds) National land use stock survey, appendices, unpublished report to the DoE, London pp 121–31

Coleman A, Maggs K R A 1964 *Land Use Survey Handbook, Second Land Use Survey* Isle of Thanet Geographical Association

Coleman A, Shaw J C 1980 *Field Mapping Manual, Second Land Utilisation Survey* King's College, London

Coppock J T 1954 Land-use changes in the Chilterns 1931–1951. *Transactions of the Institute of British Geographers* **20**: 113–40

Coppock J T 1960 The parish as a geographical/statistical unit. *Tijdschrift voor Economische en Sociale Geografie* **5**: 22–5

Coppock J T 1976 *An Agricultural Atlas of Scotland* John Donald, Edinburgh

Coppock J T 1978 Land use. In Maunder W F (ed.) *Reviews of United Kingdom Statistical Sources* vol VIII Pergamon, Oxford pp 3–101

Coppock J T, Kirby R P 1987 *Review of Approaches and Sources for Monitoring Change in the Landscape of Scotland* SDD, Edinburgh

DAFS 1970 *Agricultural Statistics Scotland, 1969* HMSO, Edinburgh

DoE 1972 *General Information System for Planning* HMSO, London

DoE 1975 *National Land Use Classification System* HMSO, London

DoE 1978 *Developed Areas 1969, a Survey of England and Wales*, DoE, London

DoE 1984 Research specifications for a feasibility study to develop a land use stock survey. Press release, August

DoE 1989 *Land Use Change in England* Statistical Bulletin 89(5), GSS, London

FC 1952 *Census of Woodlands 1947–49* Census Report No 1, HMSO, London

FC 1953 *Hedgerow and Park Timber and Woods under Five Acres 1951* Census Report No 2, HMSO, London

FC 1968 *Census of Woodlands 1965–76* HMSO, London

FC 1983 *Census of Woodlands and Trees 1979–80* FC, Edinburgh

Fordham R C 1974 *Measurement of Urban Land Use* Occasional Paper No 1, Department of Land Economy, University of Cambridge

Gould L Consultants 1988 Land use change statistics research project. Unpublished report to the DoE, London

Harley J B 1975 *Ordnance Survey Maps: A Descriptive Manual* OS, Southampton

Hill D A 1948 *The Land of Ulster 1. The Belfast Region* HMSO, Belfast

Howell J Pryse 1925 *An Agricultural Atlas of England and Wales* OS, Southampton

Hubbard N K, Wright R 1982 A semi-automated approach to land classification of Scotland from LANDSAT. In *Remote Sensing and the Atmosphere* Proceedings of the Annual Technical Conference of the Remote Sensing Society, Reading, Remote Sensing Society pp 212–21

Huntings Surveys and Consultants 1986 *Monitoring Landscape Change*, 10 Vols Report to the Department of the Environment, Borehamwood

Institute of Terrestrial Ecology 1981 Unpublished Survey sheets, Merlewood Research Station

Institute of Terrestrial Ecology 1985 Personal communication

Kirby R P 1985 Remote sensing: aerial photography. In Roger Tym and Partners (eds) National land use stock survey, appendices, unpublished report to the DoE, London pp 145–67

Langdale-Brown I 1980 Lowland agricultural habitats (Scotland): air-photo analysis of change. Unpublished report to the NCC, Edinburgh

Latham J S 1989 Remote sensing: satellite imagery. In Roger Tym and Partners (eds) National land use stock survey, appendices, unpublished report to the DoE, London pp 168–84

Lukehurst C T 1985 The Sussex land use inventory – lessons for a national land use survey. In Roger Tym and Partners (eds) National land use stock survey, appendices, unpublished report to the DoE, London pp 98–106

MAFF/DAFS 1968 *A Century of Agricultural Statistics: Great Britain 1866–1966* HMSO, London

Ministry of Works and Planning 1942 *Report of the Committee on Land Use Utilization in Rural Areas* Cmnd 6378, HMSO, London

Parry M L 1982 *Surveys of Moorland and Roughland Change* Report of the Moorland Change Project, School of Geography, University of Birmingham

Parry M L, Sinclair G 1984 *The Mid-Wales Upland Study* Countryside Commission, Cheltenham

Scottish Development Department 1989 *Land Use Changes Scotland 1987 & 1988* Statistical Bulletin No 1 (E) 1989, Scottish Office, Edinburgh

SDD 1989 *Land-use Change in Scotland, 1987 and 1988* Statistical Bulletin No 1(F), Scottish Office, Edinburgh

Sinclair G 1985 Broad cost estimates for land use stock survey. In Roger Tym and Partners (eds) National land use stock survey, appendices, unpublished report to the DoE, London pp 132–44

Stamp L D 1948 *The Land of Britain: Its Use and Misuse* Longman, London

Stove G C 1983 The current use of remote-sensing data in peat, land-cover and crop inventories in Scotland. *Philosophical Transactions A RS London* **309** (1508): 359–70

Tym R and Partners 1985 Monitoring land use changes. Unpublished report to the DoE, London

Welsh Office 1989 Personal Communication

Part Two

Monitoring economic activity and land use

8

Manpower information systems

Michael Blakemore and Alan Townsend

Local and regional studies are a very obvious field for the application of geographical information systems (GIS). Whenever a national statistic is disaggregated into its component geographical areas, with their myriad potential different groups of land uses, buildings, employers and house-holds, it invariably involves the application of computer technologies for the storage, retrieval and analysis of data. The necessity of computer assistance is increased when incorporating updates and time-series analysis. As will appear in this and succeeding chapters, most of the fields of data measurement of Part A have produced equivalent information systems. However, progress varies according to the importance of the topic for central government, local government and the private sector. Systems developed, say, within one county planning department may use different standards, technology and definitions from those of another.

8.1 Manpower information systems

One field of national, regional and local importance is that of *manpower*, comprising data pertaining to the supply of human resources (female and male) to the labour market. Employers need up-to-date and accurate infor-mation on available workers and the conditions on which they can employ them. The new Training and Enterprise Councils (TECs, or LECs – Local Enterprise Companies in Scotland) most importantly need to have both detailed local information, and access to statistics such as those provided monthly, quarterly and at longer intervals by the Department of Employment (DE) (Chs 2 and 3), and to any available contextual information germane to their local labour markets. In this way the local labour market can be monitored and compared and contrasted with regional and national trends. Central and local government continue to need accurate information on the economically active workforce, employment, vacancies and unemployment. Traditionally, such *information* has employed small armies of clerks in the statistical sections of government offices. However, these also employed a wide range of conventional paper files on, for instance, the expansion plans of firms in the local economy, employment trends of industrial estates, the circumstances of redundancies and closures, and, somewhat separately, information on training arrangements and labour law. These could be accompanied by bibliographical reference material in the shape of local and national company directories. The speed of change, and range of demands, of contemporary labour markets make it imperative that planners, strategists, employers and others have rapid and effective access to integrated information. Anything less than that puts them at a considerable disadvantage in an increasingly competitive economy.

All these various types of information (see Benjamin 1989), hard, soft and bibliographical, have been progressively transformed into computerized databases. Some of these are very small and hardly merit the description of information *systems*. The requirements of a '*system*' are the subject of section 8.2. A normal requirement would be that the data are networked on-line to a variety of remote terminal locations, or else that the host can provide a rapid postal response to enquiries. These criteria are met by the National On-line Manpower Information System (NOMIS), provided by the University of Durham for the Employment Department Group in the UK. This has increasingly expanded its data domains from that department's statistics to embrace related fields including data on population and migration. It is the subject of the case study with which this chapter is concluded. It has counterparts in the Netherlands and Denmark, and is intending to expand to include European (EC) labour market data. It has, however, no data on named employers or in the form of text, for reasons of confidentiality which are beyond the remit of this chapter.

There are complementary policy and research tools. For instance, the EC's databases include MISEP; this contains *documents* published within the Mutual Information System on Employment Policies programme, which encourages exchange of information on employment and unemployment policies in EC member states. It has always been necessary for the researcher on local and regional studies to ally company data with that of government employment statistics in monitoring local trends in economic activity. Company data are now provided for by no less than by 161 commercial databases in the EC area (Abrahams 1989; Jennings 1989). These are mainly separate national systems, but it is striking that no less than 52 related to the UK (nearly twice the number of the 'runner-up', France, with 27).

These, potentially, are replacing the company reference volumes used by the local researcher. They are given a place in this chapter for that reason, and because they point the way in which GIS in this country may gradually be privatized. Government policy sees the area as one for private-sector competition (Department of Trade and Industry (DTI) 1989a). Government policy begs a number of questions about the aims of information systems, and about the consistency of data over time; these are discussed in sections 8.3 and 8.4 respectively.

8.2 Requirements of an information system

There are a great many databases in the UK which purport to be manpower information systems. At its crudest any set of labour market data which is available to a group of researchers may be an information system. Indeed, even a box file of press cuttings is a form of information system, but its use will normally be idiosyncratic. No matter how many information systems, there are three basic gatherers of the data; central government, who need it for national and regional policy issues; local government, whose use is more geographically restricted; and the private sector, who see a potential profit in gathering and selling information. While they do collect some detailed survey data, the academic sector is mostly a consumer of existing information for purposes of research.

There are a series of criteria which, in running a system, must be met for evaluation in this chapter. The criteria are not a standard set used for information systems in general, but reflect a concern that information availability is a crucial determinant for local and regional research. To be of general and widespread use for research a manpower information system must:

1. *Be readily available to the widest possible research community.* This rules out systems generated in-house for parochial use, or whose availability is restricted for commercial or security reasons.
2. *Be consistent and comprehensive.* Consistency requires suitable documentation on the nature, precise framework of origin and limitations of the data holdings. Comprehensive means that there must be more than a single data domain. The geographical spread is a more debatable point. In general there is an inverse relationship between comprehensiveness and geographical area covered. Local information systems, with domains collected by local government specialists, will better reflect parochial variations. But that then leads to a problem of how to combine a plethora of local sources to give a consistent coverage for a larger area. The Rural Areas Data Base at the ESRC Data Archive (abandoned late 1989) was a case in point where a multitude of rural information was gathered, leading to severe problems of coverage and consistency.
3. *Not practise a policy of delaying the availability* of information to give prior advantage to one particular group.
4. *Have a stable existence*, to allow the user community to build, and utilize, skills of usage. Many academic information systems have transient life spans, oriented towards particular projects. Few are built with a goal of giving general access.
5. *Be responsive to the research community's needs,* and not just have a goal of financial profitability.

6. *Do more than just make data available*. A database is more than a collection of data items. An information system allows users to analyse data, and to customize output for further use in their own systems.

7. *Be on-line, or provide a rapid mail or courier response* to users. It should also be user-friendly. At present that often means attention being paid to expert system techniques (Andrews 1989; Cookson 1989), or innovative methods of delivery such as 'Executive Information Systems' (Abrahams 1989).

In the context of these prerequisites, 'in-house' information systems are ruled out of consideration, no matter what their intellectual attractions may be. Similarly, there may be information systems which have only very limited access or areal coverage, but these tend to be targeted to specific localities and themes, and as such are very difficult to evaluate. In many ways this latter type of information system may proliferate as the government approves more TECs. Many of these will gather their own local employer databases, and already there is concern as to how up to 100 locally collected data sources may be resolved to give an overall national and regional picture.

In this chapter, therefore, the Regional Data System of the DTI, which attempts to provide a database of manufacturing establishments by name, is not considered; this is for the most part not available outside the DTI because of the confidentiality of data secured from firms under the Statistics of Trade Act 1947 (Nunn 1983). Four key industrial data sources were also investigated by Foley (1983), who assessed their relative merits in identifying electrical engineering establishments in West Yorkshire. Recruitment agencies and professional 'head-hunters' often have highly detailed skill information systems, as do human resource development agencies. Not only are such information systems regarded as proprietary, but they are also debarred from wider utilization by the Data Protection Act of 1984. Concerns over invasion of privacy are growing. Brindle (1989) notes the contentious decision to include a question on ethnic origin in the 1991 Census; Evans (1989a) addresses concerns over the possible use of 'community charge' records to update population estimates; Hughes (1990) discusses one individual's experiences with a credit-reference information system; and Evans (1989b) highlights the growing dangers of unauthorized access to information on computer systems.

Quite simply, unless researchers can use them they are of very limited use. Also, relatively little attention will be paid to the 'new' dissemination technologies (see *Information World Review* for monthly reports on this). There is considerable interest in information being disseminated on a particular kind of compact disc, CD–ROMs, and many publishers now provide databases with large static data series (typical of these is the Supermap system marketed by Chadwyck–Healey which carries the Australian and USA Census data). However, this technology (at least in late 1989) was still unable to deliver very large databases (about 600 megabytes a disc), and the technical problems of update had yet to be fully addressed. Of more interest to integrated information are higher-speed communications networks. The growth of VANS (value added networks), using high-speed data transfer and local integration on workstations, seems a more attractive route for the moment (Osborne 1989).

One of the key dilemmas for information systems is setting goals. Should it be a highly specialized, and extremely consistent/accurate local one? But if so, how are these specialized information systems to be linked for wider research? Perhaps a national market for very specialized data will develop – for example one very much needed information source is regular tabulation by area, sector and size bands of employment establishments – not available from the Census of Employment (CoE). The 1947 Statistics of Trade Act forbids this. The advent of the British Telecom Business Database, and the Market Location Database, seems to be a private-sector answer to a government data limitation. But how much more does private-sector information cost and does it meet exacting quality controls? Business Database has a size indicator, but it has been updated only when an enquiry is triggered by an event such as a change in phone number. So while it may be feasible to provide size by type outputs, there is no definitive time stamp. Market Location includes an indicator of employment, but for a more limited sectoral coverage (see also Ch. 16).

While this is not the place to enter an extensive discussion on UK government data policy it should be said that the Rayner Committee (HMSO 1981, and for a discussion of the proposals Hoinville and Smith 1982) formalized the view that data collected by the government are for the government's use, and must be sold at real cost to the public, and also that the government should not collect data which could better be collected by the private sector (a new rule over 'unfair competition'). As a consequence the UK information market is becoming increasingly privatized, whereas the USA still has a strong government presence (contrast, for example, the products, costing and usage profiles of the UK Office of Population of Censuses and Surveys (OPCS), and the DTI, with the massive, and very cheaply available, data outpourings of the USA Department of Commerce). However, as the USA Census was being undertaken early in 1990 there was a growing civil liberties concern, relating to the availability of the TIGER files. These are structural digital map files at a scale of 1:100 000 that include transportation routes, rivers, administrative geographies, and road names with address ranges. A wonderful source for commerce, but maybe a threat for individuals. There even were worries that the viability of future censuses was under doubt, as it is already in several European countries. The availability of information produced by the government must, therefore, be viewed not just in terms of access, cost and integrity (as has been the case with concerns in 1989 by the Royal Statistical Society (RSS) about the integrity of government statistics), but also in the context of vital issues of custodianship of the privacy of individuals and groups.

In the public sector, the ESRC Data Archive at the University of Essex has a long history of providing data to the academic community, and has had considerable success in gaining research access to manpower information.

For example, county tapes of monthly unemployment data by postcode areas are available from this source. Readers are referred to their data-holding descriptions, and to their on-line enquiry and bulletin board which allows any academic on the UK joint academic network (JANET) to browse through their directories.

8.3 The range of UK databases

A basic feeling for the range of UK databases can be found in the BRITLINE Directory of On-line Databases (Ince 1988). This is regarded as a most comprehensive and informative directory. Readers are referred to this for database details, which for manpower information may be examined in respect of local authority and community information (mainly textual information on training, development schemes and other local level issues and events) and in respect of company databases.

This latter sector is dominated by the 'big players' in the public information market, with company databases and information systems that are oriented primarily towards credit referencing. Many local authorities have their own employer databases, oriented towards local purchasing and mail-shots; for example, in 1990 Lancashire County Council marketed a company database of over 7000 businesses with size and classification included. Cleveland County Council have their own employer database called 'Linkline' with over 2700 companies. Spicer (1988) provides an overview of the Tyne and Wear Business Information System, arguing that better and more detailed databases would significantly help the success of new investment into an area. Owen (1988) also reviews databases available for local economic studies (see also Ch. 16). Information on the more localized scale is increasing, but is subject to the previous concerns of consistency, and doubts as to how it may be aggregated to regional and national levels. A 'local' database approach with national coverage is typified by the Training Agency's CALLMI (Computer-Assisted Local Labour Market Intelligence) System. It was set up in the autumn of 1986 to provide four key areas of information covering nationally collected statistical information, locally captured employer information, unstructured textual commentary and information on training provision. The database was developed primarily to meet local area office needs (these offices are now superseded by the TECS), though it was deemed desirable to collect information that could be aggregated to a national picture. As it stands, CALLMI is not publically available, since it holds confidential information on individual employers. There have been calls for regional and national level abstractions, though as yet this has not occurred (see also Ch. 3). CALLMI indicates yet again the very real problems in a bottom-up approach to building a national database, since there must be very severe quality control guidelines for all staff involved in all aspects of raw data collection. At the other end, with a top-down approach there will always be the situation whereby local needs may not be met effectively. Herein lies the fundamental dilemma for all information systems.

At the national level of coverage there are many full commercial systems:

Dialog. Over 200 databases ranging from bibliographic (e.g. the GEOBASE version of *Geographical Abstracts*), through full text to statistical.

Dun and Bradstreet. Information on companies by address.

Financial Times. Databases ranging from free text to securities information.

Infocheck. Detailed analyses of over 200 000 limited companies, and basic information on over 1 million.

Key British Enterprises. Information on companies by address.

Kompass. Information on companies by address and detailed product classification.

McCarthy Information Services. Their on-line service disseminates full text of newspapers and journals for the UK and Europe.

Who owns Whom. Classifies individual subsidiary companies by name of (successive) parent ownership.

Such databases are both textual and statistical in their nature. For the academic researcher they may often seem cost-prohibitive – up to £80–£90 per hour connect charge, plus print costs; some will charge in tens of thousands of pounds for a subscription service. They are updated constantly and are available on-line.

8.4 Temporal consistency of standards

For academic standard research most of the databases described in section 8.3 have some fundamental flaws. The constant update makes it very difficult to carry out analysis over time. Company databases of this sort have a 'now' value given the credit rating objectives. Previous states are not so important. Temporal consistency is a prerequisite for analysis through time, yet in a dynamic economy definitions seldom stay the same. The most often quoted example of this in the UK has been the multiple redefinition of the unemployment series (Ch. 3). In 1989 even the RSS saw fit to enquire into what some have seen as the degradation in the quality of UK government statistics (see Waterhouse 1989a, b; Neuberger 1989), and a well-publicized meeting took place between the RSS and the Director of the Government Statistical Service (GSS). The presence of the Tradeable Information Initiative (DTI 1986; 1990) has not really impacted significantly on the wider availability of data, largely due to Treasury cost rules. There has been little incentive for departments to disseminate data when normal 'gross running costs' rules mean that any profits go to Treasury. This issue was highlighted by the Chorley Committee (Department of the Environment (DoE) 1987). Furthermore, the government's recent attitude to information technology (IT) (DTI 1989a) increasingly seems tied to a dogma of private-sector competition.

An extreme example of temporal inconsistency in manpower information is the decennial CoP. In 1971 it was made available in computer form on a grid-square basis. In 1981 the base for England and Wales was the enumeration district (ED), yet for Scotland it was the postcode sector (so much for 'national' information policy) – this fundamental difference was cited by the Chorley Committee (DoE 1987, 1988), which regarded postcodes as being a very important base geographic unit, yet with the 1991 Census the difference will remain according to the White Paper (HMSO 1989). The tardy decision-taking process for the UK census contrasts badly with the 10-year-ahead decision cycle for the 1990 US Census, and the readiness of complete digital mapping at 1:24 000. Not only will the geographies be different, and the actual units themselves markedly different because of population changes (given the need for EDs of fairly constant population), but also the data items will be different. In 1981 a 'change file' was produced at aggregate geographical levels for a limited number of data items that were the same between 1971 and 1981 (see Rhind 1983 for an in-depth discussion of the 1981 Census, and Openshaw 1987 – in the Chorley Report – for a statement on the basic spatial unit problem; see also Ch. 1).

In all areas of manpower information, and indeed in socio-economic information in general, there is the constant conflict between research needs and immediate policy needs. Government and the private sector put more emphasis on the immediacy of needs, on the snapshots in time, than they do on ensuring cohesive time series. This has a particular impact on space–time modelling where a major assumption is often a stable information series. It should be said, however, that such academic considerations are seldom in the priority list of most information gatherers. Pragmatic use of frozen information comes from the private sector, and in the UK in particular from CACI Ltd. They have two products which are based on official demographic statistics. The first are local population updates – at ward level and including working age population – which are modelled on the 1981 Census and the Registrar-General's district level yearly estimates. The second are small area population projections to the year 2000. Such extrapolation can be a profitable way of filling an information void where national statistics do not fully service the market-place requirements. Being commercially available data, of course, means that access for academic research is limited unless the finance can be raised to purchase them. Brooks (1989) cites lack of cash and unwillingness to pay for data as limitations on the increased use of on-line information services in the UK.

The most temporally volatile company statistical databases are the real-time equities systems – Reuters, SEAQ, Knight-Ridder are typical of these. In one Knight-Ridder advertisement the company gave an example where they had the updated GNP figure 30 seconds ahead of Dow Jones, Reuters and Telerate, the result being that a client made a quick trade of stocks with a profit of tens of thousands of dollars. It is here that companies thrive and fail, and the presence of such systems world-wide, linked in with the computer-based automatic trading programs used by major international stockbrokers, is a major influence on the growth and death of employers. The effects of 'black October' in 1988 are still being felt in the employment structures of the UK. It should be stated, however, that these huge information systems are not always profitable (Oram 1989), though recent reports cite Dun and Bradstreet as experiencing a 12 per cent increase in income for 3 months ending December 1989, and McCarthy Information have grown 152 per cent in the 5-year period to 1989, increasing turnover from £626 000 to £1 580 000. Clearly there are future prospects for profitable information markets, and these are being perceived by the major communications corporations, such as Maxwell Communications Corporation.

Other commercially available databases cover domains such as the property market, trade, commerce and economics – the EUROSTAT REGIO and CRONOS statistical databanks being typical of these for the EC, and the Central Statistical Office (CSO) macroeconomic databank in the UK, along with NOMIS; lastly urban and rural affairs are covered on a regional basis by the London Research Centre's URBALINE. For more details on all of these refer to the BRITLINE directory (Ince 1988).

8.5 NOMIS – a case study

NOMIS is a specific case study of a UK manpower information system that is available to all user groups. It should be stressed that the author's involvement in developing NOMIS is one key reason for its inclusion, though it is arguable that there is no comparable on-line system in the UK which addresses specifically local labour market demands. The development of the system has required the evaluation of factors such as market demand, information quality and consistency, charging and access, research facilities, and technological developments in computer networking and information dissemination.

The system has been in existence since 1978, though operationally since 1982 and fully commercially since 1986. It now contains over 19 billion data

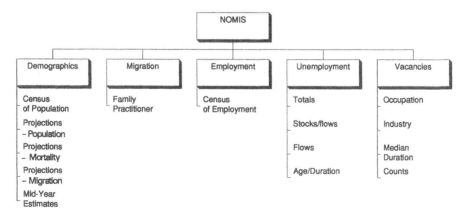

Fig. 8.1 NOMIS data sets–live domains

Table 8.1 Data availability on NOMIS

Data files	Periodicity	Start date	End date	11 Regions	18 Regions and ex-metropolitan counties	27 Rural development areas	Training and Enterprise Councils	66 Counties (and Scottish regions)	26 Assisted areas	133 Local education authorities	115 NHS migration area	214 District health authorities	280 Functional regions	334 1984 Travel to work areas	380 1978 Travel to work area	485 Local authority districts	650 Parliamentary constituencies	852 'Standardized' employment offices areas	916 Employment office areas	8900 Postcode sectors	10756 1984, 1987 wards	11085 1981 wards	13000 1981 enumeration districts
1. Census of Employment, SIC 1980		1981			*		*	*	*	*			*	*	*	*			*			
2. (by sex, full-time and part-time)		1984/1987/1989		*	*		*	*	*	*			*	*	*	*	*		*			
3. Census of Employment, SIC 1968	Annual	1971	1978					*	*				*	*	*			*	*	*		
4. (by sex, full-time and part-time)			1981					*	*				*	*	*			*	*	*		
5. Labour Force Survey	Annual	1988?																			*	
6. Employees in employment	Q	1971																				
7. Census of Population, small area statistics			1981	*	*			*	*			*	*	*	*	*		*	*	*		
8. occupation statistics, 10% sample			1981	*				*	*			*	*	*	*	*		*	*	*	*	
9. occupation by industry 10% sample			1981												*						*	*
10. Population estimates (England and Wales)	Annual	1981						*						*		*	*		*			
11. NHS migration data (by age and sex)	Q	1984									*				*		*	*		*		
12. OPCS projections of population and migration	Annual	1981	2011		*	*		*	*			*	*	*							
13. Vacancies and placings by industry and occupation	Q	8/1978		*		*	*		*	*			*	*	*						
14. Vacancies and placings by median duration	Q	3/1986		*		*	*						*	*	*						
15. Unemployment, claimants by 9 published categories	M	8/1985		*		*	*	*	*	*			*	*	*						
16. Unemployment, claimants by 9 published categories	M	6/1983		*	*	*	*	*	*	*			*	*	*			*			
17. Unemployment, claimants by age and duration	Q	9/1985		*		*	*	*	*	*			*	*	*						
18. Unemployment, claimants by age and duration	Q	7/1983		*	*	*	*	*	*	*			*	*	*						
19. Unemployment, inflow and outflow by age and duration	Q	9/1985		*		*	*		*	*			*	*	*						
20. Unemployment, inflow and outflow by age and duration	Q	6/1983				*	*		*	*			*	*	*			*			
21. Unemployment register by 7 published categories	M	7/1978	6/1983				*	*			*			*	*	*					
22. Unemployment register by 7 published categories	Annual	6/1972	6/1978				*	*			*			*	*	*					

Note: Additional series disaggregate the unemployment register by ethnic origin, and by last industry and occupation until 1981–82, when these data ceased to be coded.

items (and growing at about 2 billion a year) on the following manpower domains (Fig. 8.1), which are articulated in more detail in Table 8.1:

1. *Employment* – dominantly the CoE which has been enumerated only every 3 years from 1978 to 1987, 1989 being the most recently available on-line, and the 1991 Census statistics due for publication early 1993. Geographical disaggregation (Fig. 8.2) is down to electoral ward and postcode sector and thematic disaggregation to activity headings of the 1980 Standard Industrial Classification (SIC).
2. *Demographic and migration content.* The 1981 CoP (small area statistics, special workplace statistics and others). Population estimates and projections. Quarterly analysis of migration using the National Health Service Central Register (NHSCR).
3. *Job vacancies.* These are those registered by the Employment Service jobcentres, and account in general for between 30 and 35 per cent of all job vacancies in the UK (see Ch. 3). Monthly totals, and quarterly analyses by occupation and industry, as well as a complex tabulation of median durations of vacancies down to individual KOs (key occupations).
4. *Unemployment.* Monthly counts down to ward, and postcode sector, and 'seasonally adjusted' at regional level. Flows on and off the register and a large quarterly analysis by age and duration.

Among current developments are the addition of the annual *Labour Force Survey* (*LFS*) and quarterly estimates of employees in employment by region. It is expected that NOMIS aims to be a host for the full range of 1991 CoP, results, from ED level upwards. While the range and periodicity of data

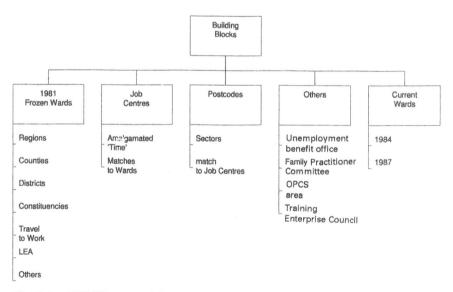

Fig. 8.2 NOMIS geographies

seem considerable, NOMIS is by no means an all-encompassing manpower information system. It does not hold individual level data either for people (Data Protection Act would forbid this), or for individual employers (the Statistics of Trade Act 1947 forbid the government from releasing such statistics, and in the commercial arena individual statistics are too profitable to be let out of the holding company's control). Domains which have been targeted as highly desirable additions by the Employment Department Group (who finance NOMIS), are education, health and social security, and EC statistics on manpower. For over 3 years NOMIS has had the prospect of adding government data on school and further education. Normally a period of one year is expected to elapse between initial enquiries to possible data providers, and on-line release. NOMIS does not hold data sets that have restricted spatial coverage, and is largely a disseminator of official UK government manpower information.

In the light of current government policy on tradeable information (DTI 1986, 1990) NOMIS has been influential in developing charging regimes that address all sectors of users. This policy has been developed in association with the Employment Department Group, who wish to ensure proper commercial return from fully commercial customers, yet wish to stimulate research in the educational sector. There is consequently a three-tier structure of charges. Research students only have to pay basic computer charges to the University

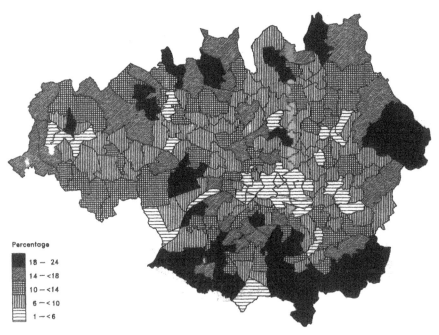

Fig. 8.3   Socio-economic groups 4 and 5 in electoral wards of Greater Manchester as proportion of all SEGs, 1981 CoP. (*Source:* OPCS statistics (NOMIS).)

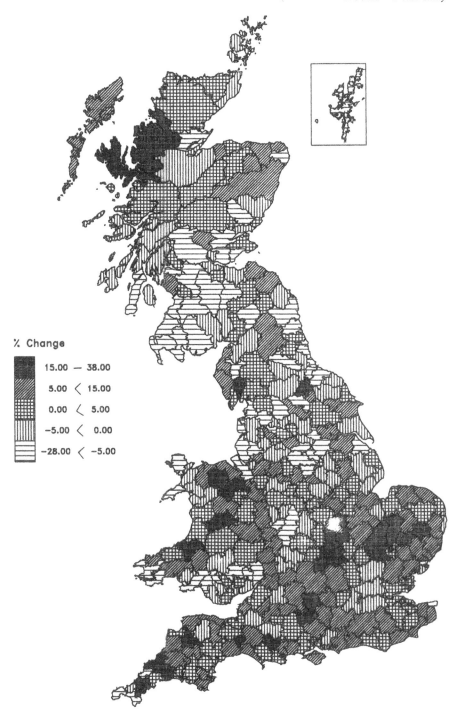

% Change

15.00 — 38.00
5.00 < 15.00
0.00 < 5.00
−5.00 < 0.00
−28.00 < −5.00

Fig. 8.4 Employment change in travel-to-work areas of Great Britain, 1981–87 (0.17% national decline 21 306 418 to 21 270 792). (*Source:* DE statistics (NOMIS).)

of Durham. Local authorities, and other academic users, pay a government surcharge for the costs of developing the system. Only fully commercial users pay additionally, for the data themselves. Costs of government data will be under considerable scrutiny with the run-up to the 1991 CoP release.

NOMIS has all of the facilities that would be expected of an on-line manpower information system. It has high-speed networked access (via the Joint Academic Network and British Telecom Packet Switching Service (PSS) and conventional dial-up), user support and training facilities, and a fluent command language that allows users to generate unique geographical requirements in real time (Ch. 2). The immediacy of NOMIS is such that data series may be processed at fine geographical resolution and classification, using analyses such as ranking, percentages, averaging, location quotients, chi-square tests and shift-share.

Complex statistical facilities are not provided. For these, and indeed for computer mapping, NOMIS can be regarded as a sophisticated 'data engine'. It provides unified access to a wide variety of different national data sources; digitized boundary files provide mapping at all administrative levels (Figs 8.3 and 8.4); analytical techniques allow the combination of data and comparative analysis. All these in real time and on-line. Increasingly on-line information systems are fulfilling the function of the integration, validation and timely supply of information. Analysis is moving down to the level of desk-top workstations where data are downloaded from a mainframe host. NOMIS now supports interfaces into mapping systems such as PCGIMMS and PCMAPICS, graphical packages such as Harvard Graphics, and it assists the supply of raw data into statistical packages, and comma separated value (CSV) for spreadsheets. In this fashion users can be assured that the latest information is available, with the knowledge that it is possible to transfer subsets down on to workstations. Of course this introduces a new dimension of data security. It is a fact of life that anything that passes on to the screen of a PC can be 'captured'. That requires careful attention to the ways in which users can extract data, especially where some data suppliers wish to restrict access to certain user groups. These are technical issues largely beyond the scope of this chapter, but they are germane in the context of UK information policy. Copyright, and the expectation of full commercial data-charging guidelines, mean that potential data providers may increasingly be cautious about dissemination on-line unless integrity and security can be maintained. This in turn will impinge on the available research information.

Note: Some of the consequences for information collection and availability of the reorganization of training in the UK are referred to in the Editor's preface (p ix).

References

Abrahams P 1989 At the frontiers of corporate data. *Financial Times* 4 Jan
Andrews B 1989 The growing role of computerised expertise. *Financial Times* 14 Sept

Benjamin B 1989 *Population Statistics: A Review of UK Sources* Gower, London

Brindle D 1989 Ethnic origin question in 1991 Census. *The Guardian* 14 Nov

Brooks M R 1989 Integrated information services; but at what price? In Anon (ed.) *Proceedings, ONLINE Information 89.* Learned Information (Europe) Ltd, Oxford pp 177–82

Cookson C 1989 Why computers need to learn English. *Financial Times* 20 Sept

DoE 1987 *Handling Geographic Information. Report to the Secretary of State for the Environment of the Committee of Enquiry into the Handling of Geographic Information* HMSO, London

DoE 1988 *Government Response to the Report of the Committee of Enquiry into the Handling of Geographic Information* HMSO, London

DTI 1986, 1990 *Government held Tradable Information: Guidelines for Government Departments in Dealing with the Private Sector* DTI, London

DTI 1989a *Information Technology. Government's Response to the First Report of the House of Commons Trade and Industry Committee: 1988–89 Session* HMSO, London

Evans R 1989a Poll tax data will be used to update census details. *Financial Times* 18 Dec

Evans R 1989b The electronic nightmare. *International Management* Sept; 39–43

Foley P 1983 A comparison of four industrial data sources. In Healey M (ed.) *Urban and Regional Industrial Research* Geo Books, Norwich, pp 91–104

HMSO 1981 *Government Statistical Services.* Cmnd 8236, HMSO, London

HMSO 1989 *1991 Census of Population* Cmnd 430, HMSO, London

Hoinville G, Smith T M F 1982 The Rayner review of the Government Statistical Service. *Journal of the Royal Statistical Society Series A* **145**(2): 195–207

Hughes M 1990 Credit and the credibility gap. *The Guardian* 20 Jan

Information World Review A monthly newspaper specializing on developments in information dissemination. Published by Learned Information (Europe) Ltd, Oxford

Ince S (ed.) 1988 *BRITLINE Directory of British Databases* EDI Publications Ltd, Linfield, Surrey

Jennings M 1989 Europe on-line. *International Management* Dec: 52–3

Neuberger J 1989 Truth, whole truth, and statisticians. *Sunday Times* 15 Oct

Nunn S 1983 Information systems in central government: the role and potential of the Regional Data System. In Healey M (ed.) *Urban and Regional Industrial Research: The Changing UK Data Base* Geo Books, Norwich pp 105–10

Openshaw S 1987 Spatial units and locational referencing. In DoE *Handling Geographic Information. Report to the Secretary of State for the Environment of the Committee of Enquiry into the Handling of Geographic Information* HMSO, London pp. 162–71

Oram R 1989 Profits fall warning from Dun and Bradstreet. *Financial Times* 14 Feb

Osborne J 1989 Freight of the Art. *Management Today* Oct: 141–8

Owen T 1988 *Mind your Local Business* Eurofi, Newbury

Rhind D (ed.) 1983 *A Census User's Handbook* Methuen, London

Spicer J 1988 Tyne and Wear Business Information System (COBIS). *BURISA 83*: 15–17

Waterhouse R 1989a Anxiety grows over integrity of statistics. *The Independent* 9 Oct

Waterhouse R 1989b Wide support for impartial check on official statistics. *The Independent* 10 Oct

9

Land and property information systems

Peter Dale

> The Geographic Information System . . . is as significant to spatial analysis as the inventions of the microscope and telescope were to science, the computer to economics, and the printing press to information dissemination. It is the biggest step forward in the handling of geographic information since the invention of the map (Department of the Environment (DoE) 1987: 8).

9.1 Introduction

This chapter highlights a number of issues that have arisen in the development of geographical and land information systems. The field is fast changing and the concepts and practices are developing at such a pace that the life span of most systems is short. Hence no individual system of hardware or software is mentioned. An analysis of those systems that were available in the UK at the end of 1988 was published by Local Authority Management Services and Computing (LAMSAC) under the title of *GIS Product Profiles* (LAMSAC 1988). More up-to-date information on what is available is published in the journal *Mapping Awareness* and in the annual *Year Book of the Association for Geographic Information* (Shand and Foster 1990). Further developments are reported in the *International Journal for Geographical Information Systems*. A more detailed study of the whole field of geographical information systems (GIS) has recently appeared (Maguire *et al.* 1991), hence what follows is more by way of an introduction for those not familiar with the concepts than a definitive review of a fast expanding and developing field.

9.2 Geographical and land information systems – GIS vs LIS

Traditional economic theory deals with three primary resources – capital, labour and land. More recently, information has been recognized as a primary resource and much effort has been directed towards its better management: 'Effective information resource management means treating information like other corporate resources – such as labor, capital, plant and equipment – and integrating technological capabilities with human resources. As advances in information technology bring the business community into a new era, the strategic management of information will be pivotal, and the organization will be affected at every level' (Diebold 1985:33).

Over the last decade or so, developments in information technology (IT) have had a major impact upon the manner in which spatial data can be handled. Two separate developments have been in progress. One of these, digital mapping, allowed existing maps drawn on paper to be converted into digital form; as a result, graphic data could be processed and displayed in different ways. The other, the development of sophisticated database management systems, has allowed large volumes of textual data to be stored, retrieved and processed. The two developments have now been linked into what are variously called geographical or land information systems.

A central government inquiry into the handling of geographic information, chaired by Lord Chorley, defined a *geographical information system* (GIS) as 'a system for capturing, storing, checking, integrating, manipulating, analysing and displaying data which are spatially referenced to the Earth. This is normally considered to involve a spatially referenced computer database and appropriate applications software' (Department of the Environment (DoE) 1987: 132).

This definition of GIS lays emphasis on the technical aspects of handling spatial data. It makes no mention of the human and institutional elements that are an integral part of any total information system. Outside the UK, what have become known as *land information systems* (LIS) began to emerge in the late 1960s. Many definitions of LIS exist, perhaps the clearest being that of the US Bureau of Land Management which defines LIS as 'a combination of human and technical resources, together with a set of organizing procedures, which results in the collection, storage, retrieval, dissemination, and the use of land information in a systematic manner' (US Department of the Interior 1989: i).

The origins of LIS lie in a range of institutional and management problems that needed to be solved, the most significant of which were concerned with record management. GIS technology offered a prospect for the more efficient handling of large volumes of existing land-related data and records. The central issue was to provide an improved service of overall land resource management and to this end, technology was seen to offer a possible solution. In concept, an LIS does not necessarily require the use of computers; the essential element in LIS is land information management for which GIS hardware and software can provide suitable tools.

9.3 Land and property information

In an LIS, the term 'land' encompasses all things physical and abstract that may be associated with the surface of the earth – whether above or below the ground. Thus crops growing on the land, minerals beneath it, buildings constructed upon it and water that may cover it are all part of the land. Similarly all rights, use and value associated with the land can form part of an LIS. From a legal perspective, land includes what the lawyers describe as 'all appurtenances attached thereto'. On the other hand, professional bodies such as the Royal Institution of Chartered Surveyors prefer to differentiate between land and property, since from a management and economic perspective, the space upon which a building stands and the construction and use of the building itself pose different sets of administrative problems. Although LIS is the more widely accepted term, the distinction between land information and property information has some practical value. The distinction between GIS and LIS is, however, becoming less clear; LIS are making greater use of the technology and GIS are becoming more closely involved with institutional problems and the use of the data.

Land and property information systems service, among others, the land market. The size of the potential land market in the UK is unknown. Her Majesty's Land Registry deals only with those properties in England and Wales that have been registered. By 1988, approximately 11.5 million titles had been registered out of an estimated total for the country of 22 million (Smith 1989). At present, private land is only brought on to the register when subject to the transfer of its freehold or when entering into a lease of more than 21 years. Under existing legislation, some privately held properties may never be brought on to the register and hence a full pattern of private landownership in England and Wales will not be known. The extent and value of central and local government held land is also uncertain. In 1989, the Audit Commission (1989: 2) reported that: 'No-one knows how much local government property is worth, but it is probably more than £100 billion, excluding council housing . . . A typical county may have a portfolio worth some £600 million; a typical metropolitan authority may have a (non-housing) portfolio worth £400 million.'

Local authorities are under pressure to manage their land and property in a more commercial way. The present cost of land administration is high. The Audit Commission (1989: 1) has stated that: 'One way or another the cost of holding and running property represents close to a quarter of total local authority rate and grant borne expenditure for one year.'

To manage their property efficiently and effectively the local authorities need good property information systems (Haines-Young and Doornkamp 1989). These must allow for many different users of the system. Estimates have shown that up to 80 per cent of a local authority's records are spatially related (Mahoney 1989). In an analysis of the use of spatial data by Swansea City Council, 24 different sets of large-scale Ordnance Survey (OS) plans were found to be in use. The authors of the study considered that one significant result of their investigation was 'the identification of considerable inter-departmental data usage, and an awareness of the extent to which such data usage might increase if a GIS were introduced. Although the majority of data-holdings were generated and maintained within a single department, use of data-holdings extended across several departments' (Bromley and Coulson 1989: 33). They concluded that

> there is no doubt that the benefits of introducing GIS are considerable. These benefits will multiply dramatically if GIS is viewed as a corporate resource. Decision-making can be based on a far broader information base than is currently possible, and staff working in different departments, because of a wider access to data, can provide a coordinated response to the public (Bromley and Coulson 1989: 35).

Land and property information systems are not, however, solely tools of local authorities. Large quasi-governmental and commercial organizations need to manage property portfolios. Organizations like British Rail and the Water Boards are major landowners whose needs for land and property information systems are similar to those in the local authorities. Likewise, the leading commercial property management companies handle large portfolios on behalf of their clients, some companies spending in excess of £1m. per year on the handling of relevant land information. The gross number of estate agencies in the UK has been declining as increasing numbers are acquired by the leading property companies and finance houses. The effect is to increase the need for computer networks, linking together branch offices so that each can share access to the same database. In spite of the fluctuations in the land market due to national economic factors, the value of property tends to grow as the land that is available for development becomes more scarce. To meet the demand, there is a need for more efficient handling of land-related information. The market for property information management systems in the UK has been estimated at around £1.5b. (Rowley and Gilbert 1989).

Among the many services offered by professional property and estate managers, Kirkwood (1984: 199) found the most common to be:

(a) undertaking surveys and preparing reports;
(b) valuing property for sale and purchase;
(c) advising clients on how and when to dispose of property;
(d) preparing sales particulars;
(e) maintaining registers of properties and applicants;
(f) marketing property;
(g) obtaining finance for clients;
(h) preparing for, and undertaking, auctions;
(i) finding suitable property for clients.

Some of the above activities can be assisted by the use of standard IT packages such as spreadsheets, word processors and basic database management systems. Some are better served by the more open flow of information within the market – such as data on the availability and value of property. Databases that service the needs of surveyors are already available (Dixon 1988). At the national level, two factors have inhibited the flow of property information as a corporate resource. These have been the secrecy of Her Majesty's Land Registry and of the Inland Revenue Valuation Office property and valuation records. In evidence to the Committee of Enquiry into the handling of geographic information, the Valuation Office stated that: 'We believe there is a strong case for co-ordinating development of Land Registry, Valuation Office and local authority systems to provide the basis for a comprehensive national register of property information. Such a register could be organized on regional lines by linking of computerized data sets' (DoE 1987: 190).

With effect from the end of 1990, public access to some Land Registry property records has been allowed, though the transfer and mortgage value of property will remain a secret until further notice. The opening of access to Land Registry records will stimulate greater interest in land information management. Although access to the portfolios of individual property owners has not been allowed (it is possible to find who owns an individual property but not all the properties that an individual owns), evidence from other countries suggests that there will be a great growth in the demand for spatial information, especially when access can be obtained on-line by computer. In Sweden for example, the records of the Central Board for Real Estate can be accessed from a number of computer terminals around the country. The demand for access to property-related information held within the system currently runs at around 35 000 enquiries each day (Andersson 1989).

Relatively few countries have treated their land registry records with the level of secrecy that has been applied in England and Wales. Even in Scotland

Table 9.1 Data for possible inclusion in a parcel-based LIS

1.	Land rights, restrictions and ownership
2.	Land values and tax assessments
3.	Rural and urban land use
4.	Buildings and other constructions
5.	Population and census data
6.	Administrative
7.	Antiquities and cultural
8.	Geological and geophysical
9.	Soils
10.	Vegetation and wildlife
11.	Hydrological
12.	Water and sewerage
13.	Gas, electricity and telephones
14.	Pollution, health, safety and emergency services
15.	Industrial and employment
16.	Transport
17.	Climatic and other environmental data

it has been possible for any member of the public to examine the land registers. In many countries there is a national cadastre with public access to the records. A *cadastre* is a parcel-based information system that notionally at least should contain details about each land parcel in the country, including its ownership, value and related attributes. The cadastral records also contain a survey plan of each parcel showing its size, shape and location. Records similar to those found in a cadastre may be used to establish landownership (as in England and Wales through Her Majesty's Land Registry, or in the German *Grundbuch*) or, in the fiscal cadastre, for documenting land values for land acquisition and land tax assessment. Both ownership and fiscal value may be combined with additional information such as land use in what is called a multi-purpose cadastre. The multi-purpose cadastre is a tool of land management and administration, compiled at a national level. It can be used to provide land and property information to all potential users at an appropriate fee. A list of data types that may be included in such a cadastre is given in Table 9.1.

9.4 GIS/LIS hardware and software

By 1990, there were several proprietary computer systems for property management on the market but few handle the geographical elements that are available in a computerized geographical or land information system (GIS/LIS). In building a comprehensive system there must be data, software to manipulate them, and hardware to support them. The minimum hardware configuration needed is a good-quality personal computer. This should have a colour screen with high-resolution graphics, significant processing capacity, back-up storage on hard disc, a printer for textual output and a plotter for hard-copy graphics. Depending on the tasks, some facility to digitize maps, that is to convert graphic data into digital form, may be needed. This can be achieved either by scanning the map or by pointing a cursor by hand at objects on a map and thence to record their locations electronically.

The software in a GIS/LIS should be able to handle both graphics and text. The data should be stored in a database, the structure of which should provide access independent of the particular application. Anybody should be able to use the data for his or her own individual purpose without being constrained by the way that the data have been stored in the computer or the way in which the computer handles them. The speed and efficiency with which the data are retrieved from the database will largely depend upon the data structure. Such structures are often described as hierarchical, network or relational (Dale and McLaughlin 1988).

Hierarchical structures allow data to be related in a tree-like arrangement. An example is a post office address where the letters are sorted into countries, then regions, then districts, then towns, then streets then individual properties. Network structures allow some cross-connections between entries on files without having to search the entire hierarchy. With a network structure, the computer could, for instance, retrieve records of other properties in the same street but in a different administrative district without having to go back to the root of the tree and then search back down again.

Relational structures store data as if they were in rows and columns. The software can then search for all characteristics found in a given row or column. Thus it might search through a list of all land parcels to find those that have a particular classification of land use and are over 100 m² in size. In this case, land use and area would be treated as separate attributes of the land parcel and each would occupy a separate column in the parcel record.

A variety of relational database management systems are commercially available and a special language, known as SQL (structured query language),

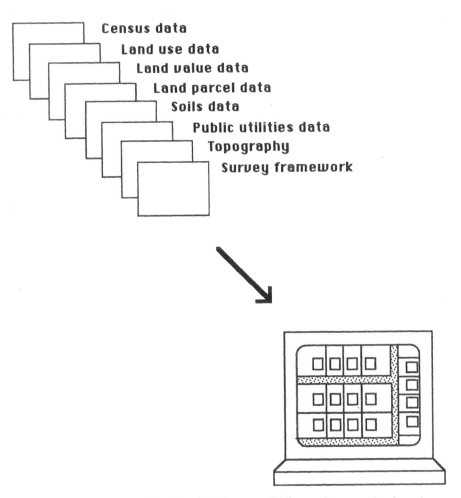

Census data
Land use data
Land value data
Land parcel data
Soils data
Public utilities data
Topography
Survey framework

Fig. 9.1(a) Data layers combined on VDU screen. Different data sets that have been stored on magnetic disc or tape as various layers of information can be combined for display on screen, or else printed out using a plotter

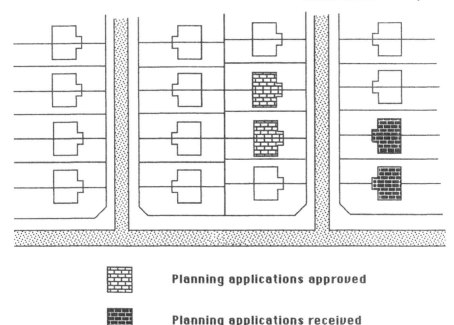

Planning applications approved

Planning applications received

Fig. 9.1(b) Boolean operations on data. Example of immediate map display from planning applications files with large-scale base mapping and selected overlays

has been developed to exploit their potential. The relational model has proved attractive in the development of GIS and LIS since it allows the quick combination and display of overlaid data (Figs 9.1(a) and 1(b)). Relational database structures are inherently simple and can be integrated into distributed networks so that data can be shared easily between different organizations.

The software to handle the attributes of the data will depend on the way that features are defined and incorporated into the database. This is called the *data model*. A *feature*, sometimes called an *entity*, is anything about which data are stored – such as a building or a boundary or a plot of land. An *object* is a collection of entities which, within the data model, comprise a more complex item such as a housing estate formed by a group of buildings or a farm formed by a group of fields and buildings. An *attribute* is a property of an entity or feature, for example indicating that the building is constructed using bricks, that the boundary is an administrative boundary or that a plot of land has a particular category of land use. The data model would normally define specific groups of entities and their attributes, and the relation between the entities.

Within a GIS, the software to handle the graphical elements must allow data and instructions to be entered from a keyboard or by pointing at a menu with a cursor (using, for instance, a 'mouse'). It must be possible to read data in from existing computer files and should process data taken directly from field survey recording devices and from photogrammetric instruments. It must accept data from digitizing tablets that are used when recording the coordinates of points that make up the lines on a map. The system must allow for the data to be edited either to correct mistakes or to update the information by adding new features and deleting old ones. It must allow for data manipulation by extracting features and by overlaying data sets (to display, for example, land use data combined with soils data). It must also allow the output of data either on to screens, to files or to plotting devices that make hard copy.

9.5 Data acquisition, storage and display

Often referred to as data capture, the acquisition of spatial data in computer-readable form is an expensive process especially where large volumes of data are concerned. Various sources of land use data and the problems of surveying and compiling land use records by manual methods are discussed elsewhere (see Chs 6, 7 and 12). The conversion of both graphic and text data into machine-readable form can account for at least 80 per cent of the cost of initiating a spatial information system. Thus although it may be possible to purchase GIS hardware and software to run on a personal computer for under £10 000, the acquisition of data for the system may cost between £50 000 and £100 000.

Textual data may be computerized either by manual entry into the system through a keyboard or by scanning, either of the text itself or of bar codes. Manual entry is a time-consuming and therefore an expensive process unless there is access to cheap labour – for instance, some organizations have used the resources of less developed countries to convert data into digital form. The use of bar codes, for example in stock-taking and at market sales points, is becoming increasingly common. This requires each data item to be coded prior to being recorded. Scanning of conventional typewritten text, although possible, is slower than for bar codes and requires more computer power to process; it can also make mistakes. Although some systems can interpret handwriting, none is yet fully reliable. Similarly, some progress has been made with automatic voice recognition so that data values can be dictated into machines but the systems are not yet foolproof.

Conversion of graphic data into digital form is undertaken either by scanning the image or by digitizing the features on it. Maps consist of points, lines and areas, each representing some object or entity. Each entity must have at least one attribute to define its nature – a line may, for instance, represent the edge of a road, a property boundary or the outline of a building – or all three characteristics. The addition of such attributes in the form of feature codes is costly and time consuming. During the mid 1980s, the OS were using over 160 categories of feature in their large-scale map database. More recently a smaller subset of these has been adopted after discussions with the National Joint Utility Group (NJUG) who represented the public utilities, and with other users (NJUG 1988). This shortened list of attributes does not include land use, the classification of which was addressed in the National Land Use Classification (NLUC) (DoE 1975) and is discussed by Kivell (this volume, Ch. 6)

Graphic data may be held within the computer either in raster or vector format. *Vector* data consist of lists of pairs of coordinate values, interspersed with attribute codes, each coordinate pair representing a point that may be an object in its own right or the turning or end-point of a line. With *raster* data, the area to be recorded is divided up into small rectangles known as pixels, and an attribute value is allocated to each pixel. The records consist of a series of pixel values that are normally obtained by scanning. Examples of each approach are given in Figs 9.2(a)(b) while their relative advantages and disadvantages are shown in Table 9.2.

Vector data may be gathered by field measurements, from photogrammetric surveys or by measurement from existing maps. The OS is currently converting all its large-scale map sheets into vector form by digitizing. Vector digitizing produces a set of coordinates to represent the start, turning and end-points of each line. The map can be redrawn from these coordinates either on a screen, or on paper or film using an electromechanical or electrostatic plotter. With good-quality equipment, it is possible to redraw a map to the same geometric quality as the original.

Raster data can be acquired very rapidly by electronic scanning but the resolution is often such that the data lack the geometric purity of the vector approach. Raster scanning of OS line maps is now commercially available

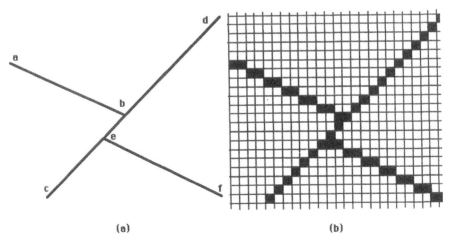

(a)   **(b)**

Fig. 9.2   (A) Vector data are held as a set of coordinates. The above figure can be stored as the six pairs of coordinates of the points – a and b, c and d, e and f; (B) the lines in (A) can be represented by squares or pixels that are given the value of 0 (blank) or 1 (filled). Thus the whole map can be stored as a set of binary numbers

Table 9.2 Comparison of raster and vector data

	Raster	Vector
Data capture	Fast	Slow
Data volumes	Large	Small
Graphic quality	Medium	Good
Geometrical accuracy	Low	High
Generalization	Simple	Complex
Point pattern analysis	Poor	Good
Network analysis	Poor	Good
Area analysis	Good	Poor
Overlay of data	Easy	Difficult

for any part of the UK (OS and Alper Systems 1989) on a 4000 by 4000 pixel basis, so that, for example, with a 1 : 2500 scale map sheet covering 1 km^2, each pixel represents 0.1 by 0.1 mm of paper or 250 by 250 mm on the ground. This should be sufficiently precise for background display on a computer screen but lacks classification into feature types. If spatial analysis is to be carried out, rather than displaying the maps as background images, then the data must be classified into relevant features. While some progress has been made in the automatic classification of such features through artificial intelligence procedures, these techniques are not yet in common usage.

When compiling both graphic and text data into an LIS, it is essential to check their integrity to ensure that no false entries have been made. Some checking may be done automatically before the data are stored within the system – the software may, for example, check for obvious mistyping at the keyboard such as the entry of alphabetic characters where a number should be recorded; or sensible limits may be set for numbers such as reasonable maximum and minimum values for the price of a property – the operator being warned of entries outside the range. In some organizations, the data are entered twice by two separate operators, the computer checking that the values agree and notifying the operators if there are any discrepancies. Such a procedure does not guarantee that the same mistake will not be repeated but it does reduce significantly the number of wrong entries. Graphic data may be checked by plotting them on a screen or by making hard copy and carrying out a visual check. After careful examination, any errors in data entry must be eliminated, hence the need for an editing facility in the system. The editing facility must also allow the data to be changed when there is a need to bring them up to date.

Data may be stored on magnetic tape, on magnetic disc or on optical disc. In GIS, magnetic disc is the preferred medium since direct access to items of data is possible by moving the reading head quickly to any point on the disc, in the manner of a gramophone record player pick-up. Magnetic tapes are often used to make back-up copies in case of disaster and for sending data to other organizations, their disadvantage being that it is necessary to wind the tape through to the appropriate point to retrieve any item of data. This sequential access takes more time than with the direct access that is possible with magnetic disc.

Relative to the past, the volumes of data that can be stored on disc are now large. This is important in land and property information systems where there is a need to store large quantities of data and to process them very rapidly. High volumes of data can also be stored on optical disc and used for graphic display but the systems only allow the user to 'write once, read many' times and cannot permit changes or corrections to the data.

Once stored, the data may be retrieved and analysed. Data are raw facts that must be processed and displayed in a form suitable for decision-making. Examples of data analysis are given in Kivell (this volume, Ch. 6) while further examples can be found in Kirkwood (1984) and Dixon (1988). The resulting information may be displayed in the form of text, tables and spreadsheets, maps or graphic images and output either to a computer screen or to a hard-copy printer. For graphics, the use of colour is advantageous not only because it can be more pleasing to the eye but also because more layers of information can be displayed – for instance differentiating more clearly between different types of land use. The resolution of colour screens is inferior to that possible for black and white although in recent years there has been a considerable improvement in quality.

An activity of growing importance is terrain visualization whereby the impact of any proposed development can be displayed. Maps portray data in planimetric form but the computer can store the three-dimensional coordinates of all points that define the terrain and display these data as they would be viewed from any perspective. Not only does this apply to constructions such as buildings but also to the surface of the earth itself so that changes resulting from excavations for new road alignments or the impact of a new reservoir on the landscape can be portrayed. Examples of such digital terrain models and the analysis of data in three dimensions are given in Raper (1989).

9.6 System selection

A number of strategies exist to guide the implementation of a property-related GIS. Many address the particular problems in local authorities (for example, Grimshaw 1988; Heywood 1989), though by implication the factors identified apply to all system selection. The LAMSAC GIS strategy, based on experience particularly with Northamptonshire County Council and Hertsmere District Council, recommended:

1. A study of current map use
2. A study of user requirements and feasibility analysis
3. Detailed analysis of needs
4. Systems design and specifications of requirements
5. Invitation to tender (ITT)

6. Bench mark/evaluation of tenders and product selection (Larner 1989: 45).

Others suggest that: 'Perhaps the greatest threat to success are unrealistic claims about the powers and benefits that a GIS can deliver . . . the best way to begin a GIS project is to raise awareness in a structured and controlled way' (Pearce 1990: 34).

At an early stage there must be a detailed study of user requirements, differentiating between what are essential components of the proposed system and those things that would be nice to include but are not of high priority. The use of spatial information within an organization is often extensive (Healey and Brett 1987) and there are many potential beneficiaries of GIS within a local authority, extending across almost all departments (LAMSAC 1989). Thus potential uses must be prioritized and systems developed capable of growth, addressing the most cost-effective applications from the start. No system is yet available that will solve all problems and hence some compromise becomes essential.

Having identified user requirements, a short list of possible systems can be prepared. Selection strategies will need to be consistent with organizational policies on IT procurement – so that, for instance, the system selected can be used with existing data, current systems operation and, possibly, maintenance contracts. Having drawn up a short list, purchasers will then issue an invitation to tender (ITT), defining precisely what is required. At this stage there is an inherent danger that there may be unseen deficiencies in the specification so that when the system is introduced it will, in effect, do what was specified but not what the purchaser had intended. There is a further risk that the system selected may satisfactorily meet the initial requirements but has neither spare capacity nor a facility for it to be expanded when required.

Having, as a result of evaluating the responses to the ITT, reduced the number of suppliers to possibly two or three, a bench-mark test should be carried out. The function of this is to confirm the claims of the vendors and that they can meet the specification (McLaren 1989). Having made the final selected may satisfactorily meet the initial requirements but has neither spare and responsibilities of all parties. Systems implementation can then begin.

9.7 Land and property information as a corporate resource

Although there are substantial costs to building a land and property information system, especially in terms of the acquisition and updating of the data, there are considerable benefits. These, however, tend to be diffuse and difficult to quantify. They can include avoided costs – the expense that would have been incurred if information were not available. The key element is, however, the use of information as a corporate resource – one that is shared within an organization and possibly with other organizations.

The sharing of data poses both technical and institutional problems. An organization's computers may be linked through a *local area network* (LAN) or to more distant locations through a *wide area network* (WAN). Linkages may also be established with other organizations through a *value added network* (VAN) – so called because value is added by combining two data sets creating information that is not available in either independently. Technical constraints may arise through difficulties with interfacing one system with another or through the use of incompatible protocols and formats (Petrie 1989). In the latter case, the National Transfer Format (NTF), administered by the Association for Geographic Information (AGI), offers a solution to the exchange of digital map data (AGI 1989). The value of land and property information depends on the extent of its use. Many factors inhibit the more general exchange of spatial information, a major factor being the lack of a suitably agreed spatial referencing system (Openshaw 1990). Whereas there are many outstanding technical problems with the hardware and software, there are many that are more dependent on human management.

The value of spatial data varies in accordance with the circumstances of their use. Factors that affect value include the context within which the data are presented, their content and degree of completeness, the characteristics, needs, purposes and opportunities of the data users, the time at which the data are available and any constraints upon their use. These constraints may be legal or social.

Legal constraints may arise where the data are owned by an individual or organization, or where matters of secrecy or security are concerned. The latter relate to much of the data held by central government, from trivial items through to data of national security as used in fighting crime or terrorism. The right of access to information may need to be restricted – for instance, although there may be every reason why individuals should know where electricity cables are buried outside their homes, it is dangerous to provide open access to information showing the whole of the National Grid. It is said that seven men with high explosives could plunge the whole country into chaos by destroying installations at seven key nodal points in the network.

The use of data may be restricted by copyright. Those who produce data have a right to reward for their labours. The Copyright, Designs and Patents Act 1988 covers digital maps as well as paper copies, and protects tables and compilations including those stored in any medium by electronic means. Protection of data relates to the particular data set. Thus one would hope that any surveyor working to a standard specification could produce a map that is apparently identical to that produced by another surveyor. Provided one had not been copied from the other, the copyright of each would be with the producer – even though each map appeared to be the same. This is unlike intellectual property where exploiting another person's idea may be in breach of the law – especially if the original idea had been patented. Patent law gives protection to inventions and is therefore not relevant to raw data. Patents must be applied for and registered whereas copyright protection is automatic.

The use of data may be constrained by social factors. In part this may be through ignorance – for instance that the data already exist or through a failure to realize the limitations of the user. Social constraints may also be deliberate, since by denying or limiting access to data one can exert power and influence. It is easy to argue that the professionalism of data users should derive from their judgement, not from their holding privileged information. In practice, those who hold information that is to their advantage tend to deny others access to it. While this may be for strictly commercial reasons, there may also be genuine concern about confidentiality or intrusion into the privacy of the individual. Currently in the UK, there is no legal right to privacy.

The development of networked land and property information systems may be inhibited through concern about the extent to which a data provider is legally liable if data are released that subsequently lead to expensive misjudgements. Professional indemnity insurance may offer some protection against the latter and is mandatory for certain professionals.

Since the function of any information system is to support decision-making, the quality of data and the way that they are processed will contribute to the quality of decision-making. The quality of information transcends matters of accuracy and precision. It is also concerned with such attributes as accessibility, appropriateness in any given circumstance, clarity, compatibility, consistency, freedom from bias, updatedness and even verifiability. Not all these characteristics are measurable quantitatively, yet the value of information in commercial and utilitarian terms must depend upon them. Since the function of land and property information systems is essentially to support the land market, much money is at risk if wrong decisions on land use and land use potential are made.

9.10 Conclusion

The techniques involved in land and property management are currently undergoing rapid change. The technology is progressing faster than most people are able to respond. New hardware and more sophisticated software are being introduced. There are short-term costs of capital outlay, training and reorganization. The cost of initial data acquisition often far outweigh the costs of hardware and software and are substantial. There are longer-term costs of systems and data maintenance, system upgrading or replacement and more far-reaching institutional change. In return, the benefits can exceed the costs by factors as high as three to four times.

The concept of spatial information as a corporate resource poses technical and institutional problems. The technical problems in general centre around standards. National agreement on such standards is beginning to emerge. The institutional issues are much more difficult to resolve. They require new thinking about the treatment of information within an organization and between organizations. Sharing of data necessitates the formation of land information policies on such matters as legal liability and pricing of products and services and the freedom of access to national data sets. The future of land and property information systems lies more in the hands of administrators and managers than technologists. What matters is good management rather than good systems.

References

AGI 1989 *The National Transfer Format* AGI, London

Andersson S 1989 Demand for access to the Swedish land data bank system – a second wave. *Mapping Awareness* 3(1): 9–12

Audit Commission 1989 *Local Authority Property – a Management Overview* HMSO, London

Bromley R D F, Coulson M G 1989 The value of corporate GIS to local authorities. *Mapping Awareness* 3(5): 32–5

Dale P F, McLaughlin J D 1988 *Land Information Management – An Introduction with Special Reference to Cadastral Problems in Third World Countries* Oxford University Press, Oxford

DoE 1975 *National Land Use Classification* HMSO, London

DoE 1987 *Handling Geographic Information. Report of the Committee of Enquiry Chaired by Lord Chorley* HMSO, London

Diebold J 1985 *Managing Information: The Challenge and the Opportunity* AMACOM

Dixon T J 1988 *Computerized Information Systems for Surveyors* The Royal Institution of Chartered Surveyors, London

Grimshaw D J 1988 Land and property information systems. *International Journal of Geographical Information Systems* 2(1): 57–65

Haines-Young R H, Doornkamp J C (eds) 1989 *Geographical Information Systems for Local Government* Seminar Lecture Notes No 4, Institute of Engineering Surveying and Space Geodesy, University of Nottingham

Healey A J, Brett R H 1987 *The Use of Spatial Information* Northamptonshire CC and OS, Southampton

Heywood I 1989 Systems selection for local authorities: criteria, guidelines and timetabling. In Haines-Young R H, Doornkamp J C (eds) *Geographical Information Systems for Local Government* - Seminar Lecture Notes No 4, Institute of Engineering Surveying and Space Geodesy, University of Nottingham pp 48–65

Kirkwood J 1984 *Information Technology and Land Administration* the Estates Gazette, London

LAMSAC 1988 *Geographic Information Systems (GIS) – Product Profiles* LAMSAC, London

LAMSAC 1989 *An Approach to Evaluating GIS for Local Authorities (Requirements Study)* LAMSAC, London

Larner A 1989 Implications for GIS in local government. In: Haines-Young R H, Doornkamp J C (eds) *Geographical Information Systems for Local Government* Seminar Lecture Notes No 4. Institute of Engineering Surveying and Space Geodesy, University of Nottingham pp 43–7

McLaren R A 1989 The art of benchmarking GIS technology. *Proceedings of National Mapping Awareness Conference Oxford*, Paper 6

Maguire D J, Goodchild M F, Rhind D W 1991 *Geographical Information Systems: principles and applications* Longman, Harlow, Essex

Mahoney R 1989 Should local authorities use a corporate or departmental GIS? *Mapping Awareness* 3(2): 57–9

NJUG 1988 *NJUG 13* NJUG publications, London

Openshaw S 1990 Spatial referencing for the user in the 1990s. *Proceedings of National Mapping Awareness Conference* Oxford, paper 25

OS and Alper Systems 1989 Note in *Mapping Awareness* 3(5): 15–16

Pearce N J 1990 Taking the risk out of system selection. *Mapping Awareness* 4(1): 33–7

Petrie G 1989 Networking for digital mapping. *Mapping Awareness* 3(3): 9–16 and 3(4): 38–42

Raper J (ed.) 1989 *Three Dimensional Applications in Geographical Information Systems* Taylor and Francis, London

Rowley J, Gilbert P 1989 The market for land information services, systems and support. Paper presented to National Mapping Awareness Conference, Oxford

Shand P J, Foster M (eds) 1990 *The Association for Geographic Information Year Book 1990* Taylor and Francis, London, and Miles Arnold

Smith P J 1989 Tomorrow's open land registry and the dawn of a national information system. Paper presented to National Mapping Awareness Conference, Oxford

US Department of the Interior 1989 *Managing our Land Information Resources* Bureau of Land Management, Washington

10

Local and regional information systems for public policy

Les Worrall

10.1 Geographical information systems: their role in decision-making

As one writer recently stated, 'unless you intend to rely on blind luck, information is clearly crucial to urban management and decision-making' (Cartwright 1989: 8). Even though everyone would accept that data and information are the raw materials of decision-making, data and information issues currently seem to be taking second place to technological issues in the current phase of the development of geographical information systems. (GIS). The factor which will determine whether or not GIS become more globally accepted will be their proven usefulness to decision-making in practical planning environments. This, in turn, will depend on how well GIS are integrated into decision-making processes and on the quality of the underlying data structures which support GIS. The issues which remain to be resolved are how can GIS most effectively contribute to the creation of more robust and more accessible processes of social decision-making, how can GIS assist decision-makers to learn how to intervene at the right time, at the right place with the right policy instruments, and what data infrastructure is needed to support GIS in a policy-analytical environment?

GIS can be defined as a configuration of hardware, software (and increasingly) communications networks and analytical procedures for the extraction of information from data to support decision-making and to achieve planning or managerial objectives. A practical, rather than esoteric, definition of a GIS raises some real concerns about the contemporary stage of development of GIS. The first concern is that the data needed to serve specific management objectives or decision-making processes are often not collected at all, or if they are collected, they are often in a form which is not readily usable. The second concern is that the type of problem dealt with by planning agencies often transcends functional and organizational boundaries and this places major demands on the ability of data from many sources to be integrated to form a coherent picture of economic and social change. The third concern is that planning processes require a continuous flow of information: this is seldom achieved. If urban planning and management are to be effective, they must be built on a firm foundation of rigorously defined data systems. This is the great challenge of the GIS movement; it should focus our attention on the underlying quality and usability of data but GIS proponents are usually too obsessed with the technology. If this necessary improvement in data quality does not take place, it will lead to a modern variant of the GIGO syndrome – garbage in gospel out.

The poor quality of nationally available data series in the UK seriously undermines the potential usefulness of GIS to public policy-making. Improving the quality of 'Database UK' is an area where practitioners, academics and GIS interest groups should collectively be allocating more of their time and energy. Because many GIS developments have been technology-driven and not planning-led, and because the pace of technological development has far exceeded the pace of development in the data infrastructures needed to inform public policy-making, GIS has become by default more an inventory-making tool than a decision-making tool. The activities of primary data collection and enhancing the integratability of existing data systems are accorded far too low a priority by both central and local government. The low status accorded to meeting legitimate data needs of public policy-makers is disconcerting given the real potential of GIS.

The purpose of this chapter is to examine some of the urban and regional information systems currently available to support public policy-making, to identify their deficiencies and to make recommendations about future developments. The stimulus to this chapter arose from earlier observations that despite a large number of putative public policy-making applications of GIS there are 'few academic texts which focus on design criteria for urban information systems to support public policy making', that there are major gaps in data availability in the UK and that there is little consensus even about the conceptual frameworks needed to structure data collection (Worrall 1991).

The importance of handling geographical information in a public policy context has long been accepted (GISP – General Information System for Planning – Department of the Environment (DoE) 1971) but has never been implemented. It seems that the Chorley Report (DoE) 1987 – the latest attempt to impose national standards for geographic referencing, classification conventions and common periodicity – is heading down the same non-implementation route as GISP. The low status accorded to developing information collection processes and meeting legitimate data needs is disconcerting because the 1980s has been marked by a massive growth in the availability, power and sophistication of computer systems. There has also been a growing realization within some, but by no means all, public authorities that strategic information systems are essential to support decision-making in an increasingly managerialist environment. This has been accompanied by a realization that many of the data sets produced primarily for administrative purposes can be used to support decision-making in the public sector if integrated within a broader framework.

While enormous claims are now being made about the potential of GIS in public policy-making, their role is essentially to assist in the efficient storage of data, to enhance their ability to be interrogated, to assist in the process of integrating disparate data sets and to provide an effective medium for presenting intelligence to decision-makers to support them in their attempt to make more informed and effective decisions. A current issue of concern is that developments in GIS are significantly in advance of the usability of the data sets available for analysis: policy analysts need to be more concerned with the quality of the data than with the way that data are packaged.

The National On-line Manpower Information System (NOMIS) demonstrates that these technical tasks can be achieved very efficiently (see Ch. 8) but the main deficiency of NOMIS is that the data infrastructure upon which it is built does not satisfy the requirements of decision-makers working at a localized strategic planning level for reasons which will be discussed later. Indeed, NOMIS was not originally designed either for this level of usage nor for use by local authorities (O'Brien 1988). NOMIS would become more generally useful if it could be interfaced with locally generated data systems (such as community charge registers). There also exist within some local authorities extensive local databases such as those developed in Telford (Worrall 1989b) and Strathclyde (Peutherer 1988): there is no technical reason why these systems could not be integrated within NOMIS to provide a prototype of the sort of system needed to support decision-making at the local level.

The development of elaborate GIS without corresponding developments in data quality also has a hidden problem attached to it: the production of highly polished output will create a spurious impression of accuracy to unreliable data. While research into methods of handling data and presenting information is essential, its value is undermined if equivalent developments do not take place to improve the quality, integratability and periodicity of the data needed to support local and regional planning and research.

The task of this chapter is not to discuss the development of GIS, but to make some recommendations primarily about the data infrastructures and data requirements which should underpin these systems. Information system development to support policy-making in the UK is currently going through a paradoxical phase: while the technology to store, manipulate and present information is improving dramatically, the quality of the nationally available data systems to support public policy-making has deteriorated. GIS will be most useful to public policy-making when the pace of development in the data infrastructure is equal to the pace of technological development. The tendency to computerize what is there needs to be replaced by a process of identifying the data sets needed to support public policy-making.

10.2 The current context in the UK

The quality of nationally available data sets in the UK is lamentable and they are generally inadequate to support public policy-making particularly at the urban or intra-urban levels (Worrall 1991). For example, the Census of Employment (CoE), which is the major national source of small area, workplace-based employment data, was originally annual but from 1987 will be biennial, is partial (it excludes the self-employed), subject to sectoral and spatial referencing errors and usually made available up to 3 years in arrears (Ch. 2). The CoE has 'long been regarded by planners and regional economists as inadequate for policy making and research' (Elton 1983: 143).

Other problems have arisen because of the politicization of statistics: the 29 (downward) revisions of the definition of unemployment have made local time-series analysis of Joint Unemployment and Vacancies Operating System (JUVOS) data yield an unreliable and misleading impression of unemployment change (Ch. 3). This is a critical deficiency given the primacy of the local unemployment rate as an indicator of local economic and social health and local labour market performance in resource allocation decisions. More specifically, data on the personal characteristics of the unemployed are not available and this limits 'the ability of policy makers to understand the nature of unemployment and respond effectively' (Elton 1983: 144).

The quality of migration data between urban areas is particularly poor and there are no data on intra-urban moves: this is particularly important given that a significant migration stream is the resultant of labour market adjustment both within and between functional urban regions. The main data source, the National Health Service Central Register (NHSCR), is an administrative system from which is produced a quarterly age/sex profile of migration between counties and metropolitan districts based on individual, but not household, migration. The system gives a partial and biased image of both the volume and profile of migration as families with young children and the elderly are more likely to reregister than young single people (Balusu 1987).

Redfern (1988) has also pointed to major problems with the availability of small area economic, social and demographic data. In particular, the decennial census is incapable of supporting the continuous modes of planning adopted in some local authorities and there is an unwillingness in central government to use existing sources of information (such as the NHSCR) and VAT registrations to develop 'Scandinavian-style' population and employer registers. To quote Redfern (1986: 422), 'the personal data held in administrative registers in the United Kingdom lack consistency, are out-of-date . . . and are unco-ordinated'. Not only are there deficiencies in the availability of aggregate data, but there are also deficiencies in, or accessibility problems to, micro-level data.

Within Scandinavia, population registers are linked to other register systems on real property and workplaces to form a comprehensive and integrated database for economic and social planning. It is possible to record-link people to the dwellings in which they live and the places in which they work for complex analysis. In addition, the property and business location registers form the basis of highly detailed, national land information systems. Integrated data systems of this type are discussed in detail at the national level in Thygesen (1984) and at the regional level in Boalt and Eriksson (1989). In Finland, there has been a move towards the concept of a register-based census which will permit both the cross-sectional and longitudinal integrated analysis of a wide range of economic and social data items (Laihonen and Myrskyla 1989). The data infrastructures used to inform public policy-making in Scandinavia are significantly in advance of those available in the UK.

Openshaw and Goddard (1987: 1425) identified several reasons why the quality of 'Database UK' is so poor. They listed the separation of computing and statistical services in central government, pressures to reduce public spending, an 'institutional departmentalized' Civil Service which 'can stifle innovation, is risk-averse, and prohibits initiative' and by an over-zealous adherence to the principle of confidentiality. The most tangible expression of central government's attitude to the UK statistical base can be seen in the comments raised after the publication of the Rayner review in 1980. The concern about nationally available statistics in the UK can be reduced to three main issues: first, a concern about the usefulness of the statistical series produced to policy analysts and social scientists after the Rayner review; second, a concern about access to statistics (this is really a concern about the cost of statistics); and third, the concern that central government is committing fraud with figures by wilfully manipulating the statistical series to disguise uncomfortable social problems (Hoinville and Smith 1982), to suppress discussion of its policies (Benjamin 1989) and to present its policies in a more favourable light.

It could be expected that the poor quality of nationally available statistical series should have prompted local government to develop its own systems to support local policy analysis and development: this has generally not been the case. Most local authorities rely solely on an increasingly out-of-date census as a source of socio-demographic information. This unwillingness to invest in the development of appropriate information systems has again been due to a cost-cutting mentality in many local authorities, the absence of people with the necessary vision and skills and the need to apply valuable resources to coping with the impact of central government imposed legislative change such as education reform and the poll tax.

There are, however, a number of exceptions to the general rule within local government: in terms of scale, Strathcylde Regional Council (Peutherer 1988; Black 1985) is the most impressive example of an authority which has taken an integrated approach to information system development. Strathcylde has focussed much of its effort on the development of primary data collection systems which have been subsequently used as an organizing framework for the integration of other, more administration-oriented, information systems such as client record systems in social services. Cheshire County Council (1983, 1985) developed a comprehensive system to monitor family stress within the county by merging administrative records from a wide range of disparate sources with the 1981 Census. An integrated information system, using Tyne and Wear as a case study, is being developed in the Regional Research Laboratory for the North (Coombes 1987). At a smaller scale, the development of a suite of integrated information systems to monitor the development of Telford New Town is described in Worrall (1986, 1988a,b, 1991). Some of these examples are discussed later in an attempt to identify current best practice.

Perhaps the most frustrating factor in all this is that central and local government are both the guardians of a growing gold-mine of data which could, if built into the appropriate frameworks for data integration (such as NOMIS), be used to improve decision-making in the public sector. Technically, there is no reason why Inland Revenue and DSS micro-data could not be used to develop small area profiles of income distribution and state dependence and why VAT registration records could not be used to develop local employment registers which could, among other things, be used as a sampling frame for local labour market surveys. These data would be used to facilitate the more efficient targeting of resources, to lay bare some of the social and economic processes which shape our urban and regional systems and to explore the causal structures of urban and regional problems. The fact that central government and most local authorities have sought to develop policies, particularly in the labour market, without an adequate empirical base raises questions about the wisdom of these policies and a more fundamental question about how the effects of these policies have been monitored and evaluated against objectives.

Concerns about data availability and quality pose questions about the potential utility of GIS to decision-makers, and it is not surprising that in two recent papers (Batty 1988; Birkin *et al.* 1987) it has been argued that GIS will suffer the same fate as the large-scale urban and regional models of the late 1960s and early 1970s because GIS proponents will be unable to demonstrate the usefulness of their systems to decision-makers. The usefulness of these systems will only be demonstrated if the systems are built on the foundation of adequately specified information systems and if the the role of GIS in the decision-making process is properly specified.

Openshaw and Goddard (1987: 1436) concluded that this will lead to an increasing emphasis on the need to build models which are 'data-rich but theory-poor'. The implications of this are that data have finally been recognized as the raw material of decision-making: this is a finding that urban researchers and model builders will ignore at their peril. The emphasis in the future will need to be directed more towards an approach to urban and regional analysis which is problem and data orientated. There will thus be a need to pay more attention to the design and development of primary data sources and to the development of techniques for estimating data which are either too costly to collect or are effectively uncollectable (Rees and Woods 1986; Rees *et al*. 1987).

10.3 Design criteria for planning orientated information systems

The greatest impediment to improving our knowledge of how urban and regional systems operate and to the development of more effective public policy is the lack of data. More important, is the lack of coherent and integrated frameworks for collecting data. Most of the economic, social and demographic information systems currently used to inform policy-making have evolved separately, have been designed for administrative and not analytical purposes, are organized on different principles and cannot be integrated. These deficiencies undermine the ability of the systems to inform policy-making and as a result much urban policy is not empirically grounded and cannot be monitored.

Our concern with urban and regional systems is an active and not a passive concern and so GIS must be designed in the context of the planning process they support. The planning mode alluded to here is a continuous process of strategic choice based upon an understanding of the causal structure of the system under control, an ability to identify nascent problems, an ability to anticipate the future, an ability to learn from intervention and an ability to adapt to the unforeseen.

Like the scientific method, a planning process can be conceived of as a procedural standard which allows planners/researchers to present their solutions/conclusions as logically consistent and valid. Such a planning process places major demands on the information systems required to support it. Consequently, GIS must meet several design criteria if they are to be a comprehensive and consistent base for urban planning and research and for the modelling and forecasting of urban change.

The information systems designed to support a continuous planning process should meet several design criteria:

1. *Periodicity*: Data collection cycles must be sensitive to the pace of urban and regional change and synchronized with the planning cycle – as local authorities produce an annual budget it is arguable that the key elements of the data system should be refreshed at least annually, either by direct observation or by data estimation procedures, to ensure, for example, that the pattern of resource allocation reflects changing needs.
2. *Timeliness*: Processing delays particularly with the Census of Population (CoP) and with the CoE mean that data from these sources often become available up to 3 years in arrears. This seriously undermines the usefulness of these data sources.
3. *Integratability*: Information synergy, and the need to relate the many dimensions of urban and regional change to each other, mean that individual information systems should be designed using a common approach

to spatial referencing and common definitions. Data should be stored at the lowest level of aggregation to facilitate later reclassification, particularly when there is a need to merge primary and secondary source material.

4. *Reliability*: While surveys like the *Labour Force Survey* (*LFS*) provide statistically reliable results at the national level, their sampling tolerances preclude their use at the subregional level. Reliable data to at least county level should be a minimum requirement.

5. *Definitional rigour*: There is an inherent risk in using administrative definitions of economic and social phenomena in an analytical context as the administrative definition may not accurately represent the phenomena which they purport to measure. For example, how should unemployment be defined? In 1971, a period when the administrative definition of unemployment was much more liberal than it is today, Worswick (1976) found that the administrative count of unemployment (792 000) was significantly (574 000 or 42%) less than the estimate derived from the 1971 Census (1 366 000). The post-1982 revisions have, therefore, translated what was then a poor measure of labour underutilization into an even poorer measure.

6. *Topicality*: The data items to be collected in an integrated data system must be the result of a conscious, usually political, decision and should not be determined by data availability.

7. *Ability to forewarn*: Much public policy has been developed as a reaction to changes which have already occurred: while it is impossible to put urban and regional systems into reverse, it is often possible to apply the brakes. The regular and systematic monitoring of urban and regional change, and the filtering of monitoring intelligence into the decision-making process, are essential for 'pro-active' policy-making.

8. *Flexibility*: The policy analyst needs to be able to switch between different (spatial, sectoral) levels of analysis, and information systems should be capable of being interrogated to expose the latent structure of problems to the requisite degree of resolution and be flexible enough to support research at a level of detail sufficient to inform policy development.

To summarize, information systems in a policy-making environment should be 'judged by [their] contribution to solving, organising, or rationalising complex choice and decision problems' (Nijkamp 1984: 4); they should 'lead to discernible action' (Steeley 1981: 8) and provide 'a comprehensive and consistent basis for socio-economic research, policy analysis, and planning' (Land and McMillen 1981: 242).

10.4 Some examples of operational systems

Strathclyde Regional Council

Strathclyde Regional Council provides services to a population of 2.3 million people. The regional council is a strategic planning authority with a wide range of functions and an interventionist approach to social and economic policy development (Strathclyde Regional Council 1980). The council needs a wide range of information about the region and its population at a variety of spatial scales to support its decision-making processes: the spatial units range from local authority districts through school catchment areas to enumeration districts (EDs) (Peutherer 1988).

A central element of the Strathclyde database is the local Voluntary Population Survey (VPS). This is described in Black (1985). The VPS is conducted in conjunction with the annual electoral registration canvass. Each dwelling is geographically referenced by unit postcode (of which there are 75 000 in the region). The availability of a range of demographic data on each of 75 000 areas clearly poses problems and so the data are aggregated to postcode sectors (410 in the region). At the postcode sector level, data from other sources can be integrated within the broader data system. These data sources include JUVOS unemployment data and vital statistics (on births by age of mother, deaths by age, sex and cause) from the Registrar-General for Scotland. In addition, departments within the regional council are in the process of geographically referencing their client-orientated record systems using unit postcodes.

The great advantage of the Strathclyde approach is that the system has been designed, albeit on a piecemeal basis, to meet the strategic planning needs of the regional council. Developments have taken place on three fronts: in developing primary collection systems, by integrating data sets from other agencies and by integrating administrative systems within a broader strategic framework. The data contained within the system are current (the population base is updated annually, hence the denominators in social indicator calculations are far more accurate than if the 1981 Census was used) and flexible, particularly in terms of geographical coverage. The systems have been used both to develop and monitor the various economic and social policies within Strathclyde Regional Council (a discussion of the Strathclyde economic policy framework is given in Young 1989).

The Telford Model

The Telford Model was designed primarily to monitor the economic, social and demographic development of Telford New Town. It has subsequently been enhanced to form the empirical basis of the Wrekin Council's annual policy planning process (Worrall 1988a). The Telford Model comprises a suite of both primary and secondary (both local and non-local) information systems (Worrall 1987b, 1989b). The primary systems are an annual population survey, an annual crime survey (until 1988), an annual employment survey, a database on the disabled (from 1989) and (until 1982) a 6-monthly unemployment survey. The secondary inputs are JUVOS ward-level unemployment data, ward-level Office of Population Censuses and Surveys (OPCS) vital statistics and data obtained from Social Services administrative systems in the county council.

The model fulfils three main roles: first, it provides the empirical basis for an annual policy planning process (Worrall 1986); second, it provides a body of information to support research and to frame additional research projects (Worrall 1987a, 1989a); and third, it supports a small area social indicator system (Worrall 1987c). One of the most tangible outputs of the Telford Model is the annual monitoring of small area social change in Telford using a wide range of social indicators. (For a full discussion of the use of territorial social indicators see Clarke and Wilson 1986.) The economic indicators produced (for 100 data zones and/or 28 wards) include measures such as employment participation rates by single year of age, algorithmically assigned economic activity rates and the distribution of employment between households, while the social measures include the percentage of children in one-parent families, the percentage of children in care or under supervision orders, household composition (by minimal household unit), housing vacancy rates, standardized mortality ratios and fertility.

The Telford Model has also permitted the local labour market to be analysed in some detail and has allowed changes on both the demand and supply sides of the local labour market to be analysed simultaneously. These include the changing sex–age profile of labour force participation, the changing household distribution of employment and the changing degree of openness in the local labour market. From the population survey, it is possible to examine labour force participation at both the individual and household levels. Trends in individual and household participation can then be examined in the context of broader unemployment and employment trends affecting the local labour force as a whole and further examined to reveal small area differences. The pattern of labour force participation at the individual level from 1980 to 1989 is shown in Figs 10.1 and 10.2, with trends in the distribution of employment among non-pensioner households being shown in Fig. 10.3.

From these figures, it can be seen that the 1980s witnessed a radical change in the profile of labour force participation by both age and sex: there was an equally radical change in the way that employment was distributed between households. At the individual level for males, the shifts in participation rates have been highest for those under 25: this reflects the well-known fact that employers often use the youth labour market as a flexible means of reacting to changes in demand. Fluctuations for prime-aged males (35–54) were less severe over the decade. Apart from the over 60s, the general trend among males has been for participation rates to return to their 1979 levels. For females the position has been radically different, with participation rates being higher in all age-groups (except the 55–59 age-group) in 1989 than in 1979. When viewed together, Figs 10.1 and 10.2 show that there were both major fluctuations in the sex–age profile of labour demand and a substantial change in the sex–age composition of the local labour supply during the 1980s.

At the household level, there were fundamental changes in the way that

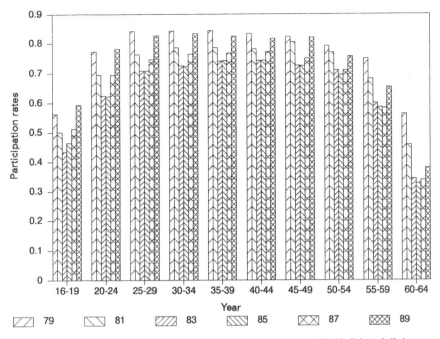

Fig. 10.1 Telford: male participation rates by age-group, 1979–89 (biennially)

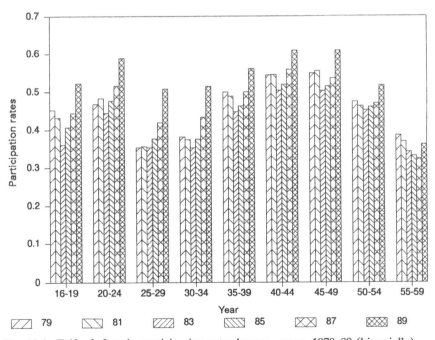

Fig. 10.2 Telford: female participation rates by age-group, 1979–89 (biennially)

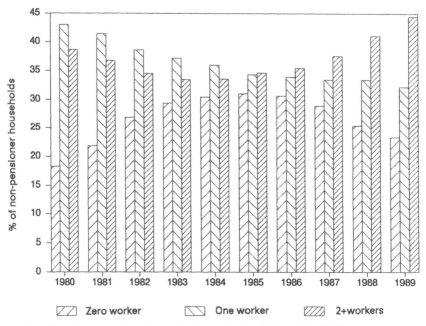

Fig. 10.3 Employment distribution by household, Telford 1980–89

employment was distributed between non-pensioner households during the 1980s. From Fig. 10.3, it can be seen that this change formed a continuous process of which the pivotal point was the year 1985 when the three household types were almost equally distributed. The percentage of households without a worker present increased from around 18 per cent in 1980 to peak at 31 per cent in 1985, and while the one-worker household was dominant in 1980, the dominant household type by 1989 was the multi-worker household. In 1989, despite the fact that almost 45 per cent of non-pensioner households contained two or more workers, around 23 per cent of households had no workers present.

Further research on this latter issue reveals that there were wide tenure differences in the household distribution of employment. For example in 1989, from a household survey, it was found that while over 61 per cent of owner occupied non-pensioner households contained two or more workers, only about 20 per cent of non-pensioner households renting from either the local authority or the New Town Development Corporation contained two or more workers. Major differences in the sex–age economic activity profile by housing tenure in Telford have also been found: this is shown in Table 10.1. From this table, it can be seen that the economic activity profile of women in owner occupation, particularly between the ages of 25 and 45, is substantially higher than for women who live in publicly rented accommodation: indeed housing tenure has been found to be a

Table 10.1 Telford: tenure differences in female economic activity rates.
(*Source:* 1989 residents' survey.)

Age	Owner occupiers	Public rented	Total
16–19	0.5000	0.5750	0.5569
20–24	0.6429	0.4848	0.5382
25–34	0.6000	0.3452	0.5339
35–44	0.6600	0.3917	0.6069
45–54	0.5882	0.5026	0.5585
55–59	0.3333	0.2671	0.3731
60–64	0.1500	0.1010	0.1210

key discriminating variable in determining labour force participation for females.

At the small area level, there are wide differences in the pattern of labour force participation at the household level and these have changed quite dramatically through the 1980s. These changes are summarized in Table 10.2. In 1980, the indicator had a range of 29.1 percentage points but between 1984 and 1986 the range increased to over 52 percentage points before declining to over 41 per cent by 1989. The table shows that the recession undergone by Telford in the early and mid 1980s did not affect all neighbourhoods within the town equally: the rising standard deviation of the indicator between 1980 and 1986 indicates that the degree of socio-economic polarization within the town increased between these dates.

The Telford labour market is an open labour market and the degree of openness has altered quite dramatically throughout the 1980s. Data on the indigenous supply of labour are derived from the annual population survey and data on the total number of jobs in Telford and in-commuting are derived from the annual employment surveys: by merging these data it is possible to identify a measure of net commuting which has been used in the past to quantify the extent to which the social gain of employment created in

Table 10.2 Telford households without economic support: small area analysis.
(*Source:* Population surveys.)

	1980	1982	1983	1984	1985	1986	1987	1988	1989
Highest	34.02	48.75	55.51	61.32	61.18	62.31	56.75	50.35	48.48
Mean	18.55	27.44	29.91	31.05	31.81	31.49	30.32	26.99	24.79
Lowest	4.92	7.33	7.66	8.16	8.48	9.87	8.31	9.54	6.96
Range	29.10	41.42	47.85	53.16	52.7	52.44	48.42	40.81	41.52
Standard deviation	6.46	10.40	11.56	12.51	11.96	12.25	11.38	9.95	9.28

Note: The data examine the distribution of non-pensioner households without workers presented across 101 standard statistical zones in Telford.

Telford has not directly benefited Telford residents. A summary of change on this net commuting inflow measure is shown in Table 10.3.

The information in Table 10.3 provides an informative insight into how a spatial labour market adjusts under differing employment creation conditions. In the period 1982–87, almost 10 500 additional jobs were created in Telford and yet the numbers of residents in employment only increased by around 5100: more simply, over this period two jobs had to be created in Telford for one Telford resident to become employed. Since 1982, the general trend has been towards increasing net in-commuting: this raises questions about the relative employability of Telford residents and indicates that commuting is probably one of the most flexible variables in the process of local labour market adjustment.

This brief and summary discussion of some of the analysis which has been conducted in Telford shows that the local labour market is highly adaptive and that significant policy-relevant change has taken place in relatively short periods of time. If reliance had been placed solely on nationally available sources these changes would not have been apparent and many important social and distributional aspects of labour market change would have escaped investigation in the more detailed survey research projects which have been nested within the Telford information systems.

The examples of information systems development and the policy-analytical uses to which these information systems have been put demonstrate the inherent weakness of the nationally available information systems when evaluated from a practical planning viewpoint. While the localized examples of Strathclyde and Telford reveal what has been achieved in terms of developing locally usable systems, they also expose the problem of the fragmentation which will necessarily occur if LAs design their own systems to fill the void of an inadequate national statistical base without a generally agreed framework for information systems design.

10.5 The potential contribution of an accounts-based approach and micro-simulation techniques

Earlier, reference was made to the fact that there is little consensus about the data collection and organizational frameworks needed to develop spatially

Table 10.3 Telford: net commuting inflows 1982–89

	1982	1983	1984	1985	1986	1987	1988	1989
Employment	38 842	39 273	40 571	43 586	45 990	49 300	52 762	57 900
Employed	35 069	34 799	35 211	36 560	37 909	40 175	44 077	47 110
Net inflow	3 773	4 474	5 360	7 026	8 081	9 125	8 685	10 790

Notes: Employment data are from the annual employment surveys. Employed refers to the number of Telford residents in employment from the annual population surveys.

orientated information systems. An accounts-based approach is a very potent device which has been underutilized in the development of local and regional information systems. An accounts-based approach imposes a rigour and a structure on information system design which is sadly lacking in many applications particularly in the UK. An accounting framework can be described as a device both for consistently organizing data and to facilitate the analysis of that data. House (1981: 441) argues that an 'integrative accounting system clearly points up the need to either generate new data bases or to make it more possible to link up existing disparate data bases'.

Because contemporary society produces vast quantities of statistical data often as a by-product of administration, the purpose of accounting systems should be to organize that information, to provide a medium for substantive analysis, to provide an integrative framework for information system development, to measure economic and social change (particularly the effect of change measured on one dimension to other aspects of change), to provide a contextual framework for economic and social indicator development and analysis, to provide a medium for testing, simulating and evaluating policies and programmes and to provide a consistent basis for forecasting. The fitting of Telford unemployment data into a time-series accounting framework for a longitudinal analysis of unemployment careers is shown in Worrall (1987a). More general applications of social accounting frameworks are provided by Casini Benvenuti and Cavalieri (1986), Madden and Batey (1983) and Ledent (1986). A discussion of the social accounting methods for forecasting occurs in Chapter 13.

RUDAP (Ruimtelijk-Demographische Aspecten van Processen), a spatial demographic information system, was designed and implemented in the Netherlands in the early 1980s (Scheurwater 1984) and is a good example of an application of an accounting framework. The system was designed to monitor the implementation of the Third Report on Physical Planning (Voogd 1982) using a process planning mode which was both 'cyclical and continuous' (Scheurwater 1984): correspondingly, the planning process had large-scale data needs. This, in turn, had major implications for the design of the information systems required to support the process of continuous decision-making. The primary data source for RUDAP is the municipal population registers which are maintained in the Netherlands (van der Erf 1984). The Dutch population registers record changes which a person undergoes during his/her lifetime including change of address. The municipal data sets are combined nationally to produce a multiregional demographic accounting framework. RUDAP illustrates two important features: first, it shows that information systems which are designed to meet the needs of a particular planning process have a higher chance of being successfully implemented; and second, it shows the advantages of having a robust national demographic accounting framework.

Integrated single- or multiregional models of the type alluded to above have major data requirements and it is in the field of data availability, reliability and integratability that most problems arise. Specifically, the main problems arise from the lack of uniformity in spatial, sectoral and temporal definition and in the relative paucity of 'flow' data. While these problems are important the RUDAP example shows that they are not insuperable. In addition, data estimation procedures can be used to simulate missing or unreliable data and to estimate flows from temporal differences in the stocks of various phenomena (Rees and Woods 1986; Rees *et al.* 1987). The constraints of data availability can be relaxed by these techniques and this will lead to the production of a disaggregate database which can be used to model and analyse more effectively the dynamics of urban and regional development.

Given that the problem of data availability is unlikely to be overcome even in the medium term, it is important to consider alternative, non-observational means of refreshing the data needed to inform public policy decisions. Techniques for overcoming the problems of data availability have been developed: one such technique based on iterative proportionate fitting (IPF) (Birkin 1987) has been used to produce synthetic micro-databases (Birkin and Clarke 1986) which have been applied to explore questions of public policy (see Clarke and Wilson 1985 on health service planning, Clarke 1986 on simulating household dynamics and Clarke *et al.* 1988 on simulating housing careers). The development of integrated macro/micro-urban and regional models is an area which could pay significant benefits when applied to practical policy analysis. The benefit of an accounts-based approach is summarized by Willekens (1984: 22) who stated that 'Good demographic accounting may . . . not only reveal data inconsistencies that may otherwise be covered up, but may also provide a proper approach for integrating the direct measurement, the indirect estimation and the analysis of demographic data.' These benefits are just as relevant to the economic domain.

10.6 Issues for the development of geographical information systems to support urban and regional planning

The purpose of this chapter has been to address four issues. The first was to identify some concerns about the current stage in the development of information systems for public policy-making: specifically, a concern that much development in GIS has been technology-driven and that the pace of technological development is far exceeding the pace of development of the data infrastructures needed to inform public policy-making. The second purpose was to describe the critical deficiencies in nationally available statistical series before identifying some of the more innovative solutions to the problem of data availability adopted in two UK practical planning environments. The third task was to identify design criteria for information system development before finally listing some of the advantages of adopting an accounts-based approach, identifying some of the methodological developments made in the field of producing synthetic data and the potential contribution of micro-simulation modelling to public policy questions.

It is important to list the practical steps needed to overcome some of the problems identified above:

1. More effort should be made to link information systems development to the needs of the planning processes used in (some) local authorities and to identify their role in practical policy-making environments: information systems cannot be developed in a vacuum and public policy must become more empirically grounded if intervention is to be effective.
2. In the arena of data collection and unlocking data from within organizations, steps need to be taken on three fronts simultaneously:
   (a) first, to apply pressure to central government to improve the quality of the nationally available statistical base, to relax its anachronistic view of confidentiality and to make more generally available data from sources like NHSCR (a population register), the Inland Revenue (ED-level income profiles), HM Land Registry and the DSS (ED-level state dependence profiles);
   (b) second, to work through organizations within local government and the public sector generally to develop a basic data model and communications standards for local authorities together with a strategy for the production of information which can be integrated with other data sources; and,
   (c) third, to identify best practice in some of the more innovative approaches to data collection and information systems design and to disseminate this more effectively.
3. There is also a need for closer working arrangements to be developed between 'academics' and 'practitioners' and also between information systems designers, planners and model builders: hopefully the Regional Research Laboratories will make a significant contribution in this area.
4. Given that a fundamental review and improvement of the UK statistical infrastructure is unlikely, more effort needs to be applied to the development of the means to produce synthetic data sets.

Finally, GIS will stand or fall by their perceived and proven usefulness in a public policy-making environment and it will only be by improving the quality of the data infrastructures on which these systems are built that the GIS bandwagon of the 1980s will be prevented from trundling down the same path to oblivion that the large-scale models of the 1960s and 1970s took.

Acknowledgements

I acknowledge the Planning Department of Shropshire County Council for their joint work on the Population Survey and Prism Research of Telford for allowing access to Employment Survey data. The views in this chapter are my own and should not be ascribed to any organization with which I am associated.

References

Balusu L 1987 Use of migration data in population statistics. Paper presented at the Institute of British Geographers Population Study Group, Oxford

Batty M 1988 Informative planning: the intelligent use of information systems in the policy making process. Paper presented at the Regional Science Association, 5th International Workshop on Regional Strategic Planning, Enschede, Netherlands

Benjamin B 1989 *Accessibility and Other Problems Relating to Statistics Used by Social Scientists* Report to the RSS Official Statistics Study Group

Birkin M, Clarke M 1986 RUIN - Really useful information for urban and regional analysis: methods and examples. Paper presented at the British Section of the Regional Science Association Annual Conference, Bristol

Birkin M 1987 *Iterative Proportionate Fitting (IPF): Theory Method and Examples* Computer Manual 26, School of Geography, University of Leeds

Birkin M, Clarke G P, Clarke M, Wilson A G 1987 *Geographical Information System and Model-based Locational Analysis: Ships in the Night or the Beginnings of a Relationship* Working Paper 498, School of Geography, University of Leeds

Black R 1985 Instead of the 1986 census: the potential contribution of enhanced electoral registers. *Journal of the Royal Statistical Society Series A* **148** 287–316

Boalt A, Eriksson M 1989 Transforming a spatial-orientated system into an easy accessible data base: some experiences from Stockholm County. Paper presented at Urban Data Management Symposium '89, Lisbon

Cartwright T J 1989 Information systems for urban management in developing countries: the concept and the reality. Paper presented at the International Workshop on Improving Urban Management Policies, Hawaii

Casini Benvenuti S, Cavalieri A 1986 Applications of a bi-regional input–output model in regional policy analysis. In Batey P W J, Madden M (eds) *Integrated Analysis of Regional Systems* London Papers in Regional Science 15, Pion, London, pp 201–16

Cheshire County Council 1983, 1985 *Areas of Family Stress in Cheshire* Cheshire County Council, Chester

Clarke G P, Wilson A G 1986 *Performance Indicators within a Model-based Approach to Urban Planning* Working Paper 446, School of Geography, University of Leeds

Clarke M 1986 Demographic processes and household dynamics: a micro-simulation approach. In Woods R, Rees P (eds) *Population Structures and Models* Allen and Unwin, London, pp 245–72

Clarke M, Wilson A G 1985 A model-based approach to planning in the National Health Service. *Environment and Planning B* **12**: 287–302

Clarke M, Longley P, Williams H 1988 Micro-analysis and simulation of housing careers: subsidy and accumulation in the UK housing market. Paper presented at the European Regional Science Association Congress, Stockholm

Coombes M G 1987 Integrated local databases: progress in the Regional Research Laboratory for the North. Paper presented at the Institute of British Geographers Population Study Group, Oxford

DoE 1971 *General Information System for Planning: Report of the Study Team* DoE, London

DoE 1987 *Handling Geographic Information: Report of the Committee of Enquiry chaired by Lord Chorley* HMSO, London

Elton C 1983 The impact of the Rayner review of unemployment and employment statistics. *Regional Studies* **17**: 143–46

Hoinville G, Smith T M F 1982 The Rayner review of Government Statistical Services. *Journal of the Royal Statistical Society Series A* **145**: 195–207

House J 1981 Social indicators, social change and social accounting: towards more integrated and dynamic models. In Juster F T, Land K C (eds) *Social Accounting Systems* Academic Press, New York, pp 422–49

Laihonen A, Myrskla P 1989 *The 1990 Population and Housing Census: General Plan* Finnish Census Bureau Report 1989 Helsinki p 17

Land K C, Juster F T 1981 Social accounting systems. In Juster F T, Land K C (eds) *Social Accounting Systems* Academic Press, New York, pp 10–25

Land K C, McMillen M M 1981 Demographic accounts and the study of social change. In Juster F T, Land K C (eds) *Social Accounting Systems* Academic Press, New York, pp 242–300

Ledent J 1986 Consistent modelling of employment, population, labour force and unemployment in the statistical analysis of regional growth. In Batey P W J, Madden M (eds) *Integrated Analysis of Regional Systems* London Papers in Regional Science 15, Pion, London, pp 25–36

Madden M, Batey P W J 1983 Linked population and economic models: some methodological issues in forecasting analysis and policy optimization. *Journal of Regional Science* **23**: 141–64

Nijkamp P 1984 Information systems: a general introduction. In Nijkamp P, Rietveld P (eds) *Information Systems for Integrated Regional Planning* North-Holland, Amsterdam, pp 3–34

O'Brien L 1988 Strategic planning using an interactive on-line GIS: the National On-line Manpower Information System. Paper presented at the Regional Science Association, 5th International Workshop on Regional Strategic Planning, Enschede, Netherlands

Openshaw S, Goddard J 1987 Some implications of the commodification of information and the emerging information economy for applied geographical analysis in the United Kingdom. *Environment and Planning A* **19**: 1423–39

Peutherer D 1988 Developing small area information systems for service and resource planning in Strathclyde Regional Council. Paper presented at the Regional Science Association, 5th International Workshop on Regional Strategic Planning, Enschede, Netherlands

Redfern P 1986 Which countries will follow the Scandinavian lead in taking a register-based Census of Population? *Journal of Official Statistics* **2**: 415–24

Redfern P 1988 *A Study of the Future of the Census of Population: Alternative Approaches* Statistical Offices of the European Community, Luxemburg

Rees P, Clarke M, Duley C 1987 *A Model for Updating Individual and Household Populations* Working Paper 486, School of Geography, University of Leeds

Rees P, Woods R 1986 Demographic estimation: problems, methods and examples. In Woods R, Rees P (eds) *Population Structures and Models* Allen and Unwin, London, pp 301–43

Scheurwater J 1984 Towards a spatial demographic information system. In ter Heide H, Willekens F J (eds) *Demographic Research and Spatial Policy* Academic Press, London, pp 69–91

Steeley G C 1981 Monitoring, information and decision making. *BURISA Newsletter* **50**: 6–8

Strathclyde Regional Council 1980 *Social Strategy for the Eighties* Strathclyde Regional Council, Glasgow

Thygesen L 1984 A national register-based statistical system and its implications for local governments. Paper presented at the 15th International Association for Regional and Urban Statistics Conference, Copenhagen

van der Erf R F 1984 Internal migration in the Netherlands: measurement and main characteristics. In ter Heide H, Willekens F J (eds) *Demographic Research and Spatial Policy* Academic Press, London, pp 47–54

Voogd H 1982 Issues and tendencies in Dutch regional planning. In Hudson

R, Lewis J R (eds) *Regional Planning in Europe* London Papers in Regional Science 11, Pion, London, pp 112–26

Willekens F J 1984 Approaches and innovations in policy-orientated migration and population distribution research. In ter Heide H, Willekens F J (eds) *Demographic Research and Spatial Policy* Academic Press, London, pp 21–35

Worrall L 1986 Information systems for policy planning in a shire district. *BURISA Newsletter* **72**: 8–10

Worrall L 1987a Information systems for urban labour market planning and analysis. In Gordon I (ed.) *Unemployment, the Regions and Labour Markets* London Papers in Regional Science 17, Pion, London, pp 139–57

Worrall L 1987b Population information systems and the analysis of urban change. *Town Planning Review* **58**: 411–25

Worrall L 1987c Urban demographic analysis: a case study of Telford. Paper presented at the Regional Science Association Population Workshop, London

Worrall L 1988a *Information System Development and the Management of Urban Change* Local Authorities Research and Intelligence Association Occasional Paper 2, London

Worrall L 1988b The measurement of urban social and demographic change. *Proceedings of the British Society for Population Studies Annual Conference* London, BSPS, pp 114–31

Worrall L 1989a The assessment of housing needs in the Wrekin District. *BURISA Newsletter* **87**: 8–11

Worrall L 1989b Urban demographic information systems. In Congdon P, Batey P W J (eds) *Advances in Regional Demography: Forecasts, Information and Models* Belhaven Press, London, pp 25–40

Worrall L 1991 Information systems for urban and regional planning in the UK: a review. *Environment and Planning B* **17**: 451–62

Worswick G D N (ed.) 1976 *The Concept and Measurement of Involuntary Unemployment* Allen and Unwin, London

Young R 1989 Lessons from Strathclyde's experience: boosting peoples' self-confidence. In Dyson K (ed.) *Combating Long-term Unemployment: Local/ec Relations* Routledge, London, pp 194–217

11

Obtaining information from businesses

Michael Healey

11.1 Business surveys and research design

Most of the chapters in this book are concerned with sources of information on economic activity and land use which are available in printed form or are accessed via on-line databases. The majority of this information comes originally from direct surveys. Clearly it is important to understand the nature of such surveys and the problems associated with them because the results of the limitations eventually find their way through to most of the published and unpublished statistics used by those interested in local and regional studies. This chapter assesses the different ways of obtaining information from businesses right across the spectrum of economic activities, from farmers and manufacturers to office managers and hoteliers. The following chapter examines land use surveys.

Surveys of business are widely used where no other sources of information exist or are accessible. They are also used to fill in gaps or update existing information. In many situations a survey of businesses is the only way in which information may be collected because the questions that planners, market researchers and academics are interested in obtaining the answers to are not asked in other surveys. Surveys can, then, be an important way of overcoming some problems of access, the length of the update cycle, and the absence of information on the topic being examined. However, there are many difficulties to be overcome before reliable data may be obtained by survey and it is not a method to adopt without considerable prior thought and preparation.

Numerous books have been written on survey research methods (e.g. Dixon and Leach 1978a, b; Gardner 1978; Hoinville *et al.* 1978; Marsh 1982; Moser and Kalton 1971), but most are concerned with social surveys rather than investigations of businesses. Although the general advice in these texts is relevant to the economic researcher, for the most part, the books lack examples of industrial surveys and discussion of the specific problems involved in such research. Indeed one of the best-known books on survey research methods, by Moser and Kalton (1971: 4), explicitly states that 'since this book is concerned with social rather than economic surveys, little will be said about surveys of shops, business firms or similar economic units'. Useful hints are also contained in the literature on élite interviewing (e.g. Dexter 1970; Moyser and Wagstaffe 1987), but most of the limited work that has been published on this topic has been written by sociologists, political scientists and anthropologists and little involves economic élites. The discussion by Kincaid and Bright (1957a, b) of interviewing the business élite is an exception. More general discussions of surveying businesses are contained in the industrial market research literature (e.g. Hague 1987a; MacFarlane-Smith 1972; Rawnsley 1978; Wilson 1968), but most of this work is, not surprisingly, concerned with industrial buying and market assessment, and not all of this experience may be directly transferable to undertaking other types of industrial survey.

With a few exceptions (Bachtler 1984; Gudgin 1976; Healey 1983a, 1986), discussions of surveys of businesses most relevant to the research worker interested in local and regional studies of economic activity are restricted to the appendices of research theses and monographs, and brief comments on survey methodology in a few textbooks (e.g. Ilbery 1985a). The lack of guidance in the literature means that most researchers have no choice in designing their surveys but to rely on a combination of common sense and previous experience. The great majority of research workers are rightly more interested in the findings of their surveys than in the survey process. However, the reliability and usefulness of the research findings and their interpretation are inextricably linked with the nature and quality of the survey methods used and the underlying research design adopted.

Sayer (1984: 221) distinguishes between extensive and intensive research designs: 'the two types of design ask different sorts of question, use different techniques and method and define their objects and boundaries differently'. **Extensive** research is primarily concerned with establishing some of the general patterns and common properties in a population (e.g. what have been the main changes in the patterns of industrial location?). It typically uses standardized questionnaire and interview surveys of representative samples or total populations, and descriptive and inferential statistics in the analysis. It also tends to concentrate on taxonomic groups such as 'urban areas' or 'new firms' (e.g. Fothergill and Gudgin 1982). Extensive research is the dominant research design used in positivist approaches in the social sciences. In contrast, in **intensive** research the main questions involve how some causal processes work out in a particular case or limited number of cases (e.g. how was industry restructured in a specific local economy during a particular period?). Intensive research is typically undertaken using less standardized and more interactive interviews and mainly qualitative forms of analysis. It is the dominant research design associated with realist approaches to social enquiry. Intensive research commonly examines groups whose members are causally related to one another, such as firms which are related 'horizontally' through competition (i.e. operate in the same market sector) or 'vertically' through linkages (e.g. Rawlinson 1990). The businesses selected may not be typical and may be selected one by one as the research proceeds and as an understanding of the membership of a causal group is being developed.

Table 11.1 Intensive and extensive research: a summary. (*Source:* Sayer and Morgan 1985: 151)

	Intensive	Extensive
Research question	How does a process work in a particular case or small number of cases? What *produces* a certain change? What did the agents actually do?	What are the regularities, common patterns, distinguishing features of a population? How widely are certain characteristics or processes distributed or represented?
Relations	Substantial relations of connection	Formal relations of similarity
Type of groups studied	Causal groups	Taxonomic groups
Type of account produced	Causal explanation of the production of certain objects or events, though not necessarily a representative one	Descriptive 'representative' generalizations, lacking in explanatory penetration
Typical methods	Study of individual agents in their causal contexts, interactive interviews, ethnography. Qualitative analysis	Large-scale survey of population or representative sample, formal questionnaires, standardized interviews. Statistical analysis
Are the results generalizable?	Actual concrete patterns and contingent relations are unlikely to be 'representative', 'average' or generalizable. *Necessary* relations discovered will exist wherever their relata are present, e.g. causal powers of objects are generalizable to other contexts as they are necessary features of these objects	Although representative of a whole population, they are unlikely to be generalizable to other populations at different times and places. Problems of ecological fallacy in making inferences about individuals
Disadvantages	Problem of representativeness	Lack of explanatory power. Ecological fallacy in making inferences about individuals

Proponents of intensive research are critical of the 'chaotic conceptions' typically found in the taxonomic collectives (e.g. degree of rurality) used in extensive research (Sayer 1984). Although intensive research is not as common as extensive research it is rapidly gaining adherents.

Both types of research design are required but they fulfil different functions, the one primarily descriptive, the other primarily explanatory (Table 11.1). 'In principle, it ought to be possible for them to be complementary' (Sayer and Morgan 1985: 150). However, the close links between theory and method (Massey and Meegan 1985) mean that the choice of survey method is not like selecting goods from a supermarket shelf, and the same method may be used in different ways in different research designs. For example, in extensive research an interview may be used to try to quantify and generalize about aspects of business behaviour, whereas in intensive research the aim of the interview may be to uncover the causal mechanisms underlying a particular process and to refine and test previously used categorizations.

The remainder of this chapter focuses on the assessment of different methods of collecting information from the owners and managers of businesses. It may be, of course, just as important to collect data from unions, workers, trade organizations and various government bodies (see e.g. Ch. 17), but the best ways of obtaining information from these groups may be somewhat different; although many of the points made here should have at least some relevance to collecting data from representatives of a wide range of organizations. The chapter begins with a discussion of the advantages and disadvantages of contacting all businesses, sampling from a population, and undertaking case studies. This is followed first by an assessment of various survey methods and the response rates associated with them; and then by an examination of the construction of questionnaire and interview schedules and the interpretation of responses. Non-standardized interviews are considered separately. The chapter concludes with a discussion of the problems created by the increased demands for business surveys. Space constraints mean that only selective coverage of these topics is possible. Examples of questionnaire and interview schedules are included in the appendices.

11.2 Populations, samples and case studies

Ideally in survey-based research all relevant businesses should be contacted, but this is rarely possible on resource grounds unless the 'population' of business is relatively small (e.g. Department of Trade and Industry (DTI) 1973; Healey *et al.* 1987; Ilbery 1985b; Massey and Meegan 1979), or it is thought necessary to try to obtain a complete coverage (e.g. Census of Population (CoP); Agricultural Census). In other cases sample surveys are usually used. Indeed, in extensive research 'samples can be more economical, less time consuming, and more accurate than large numbers such as total populations' (Gardner 1978: 79). This is because experience and sampling theory show that in selecting the size of sample the point of diminishing returns is reached fairly quickly. For some purposes, for example for in-depth investigations, it may be more appropriate to examine case studies than to analyse a sizeable group of businesses.

The starting-point for most surveys is to define the relevant **population** or **sampling frame**. This should ideally contain a complete list of the members of the population to be studied, with no duplicate entries. If a stratified sample is required the list will need to have, in addition, information for each member of the population on the variable(s) to be used in the stratification. One of the major problems in much economic research is the unsuitability of most of the sampling frames. This point is well made by Curran (1986) who, in reviewing small firm research in the UK, states:

> There are no national sampling frames for small business research in Britain in any strict sense, that is, lists of enterprises defined in some broadly acceptable way as 'small' from which statistically adequate samples might be drawn. Indeed, such lists do not exist even for particular types of small enterprise or even the small enterprise populations of specific geographical areas . . . and, in practice, researchers have continued to resort to less satisfactory alternatives.

Directory sources are one of the most common sources used for constructing sampling frames. For the purposes of the present discussion directories are taken to include not only published commercial directories, but also establishment databanks and lists compiled by local authorities and trade organizations. The problems of directory sources are well known and are touched on in several chapters of this book (see Chs 14, 16–18). They include partial coverage, bias towards larger businesses, out-of-date information and inaccurate data. Although many directories are now continuously updated, the cost of on-line searches may be prohibitive. For the unpublished lists the problems of finding out about their existence and obtaining access to them may be added. The main conclusion to come from this is that it is best to use as many different sources as possible, although their relative suitability will depend on the unit of analysis (e.g. establishment, firm or enterprise) and coverage required (percentage of units or employment) for the investigation (Aubrey *et al.* 1989; Foley 1983; Healey 1979).

The desirability of this strategy is enhanced when it is realized that the extent of overlap between different sources is sometimes small. For example, in identifying a panel of manufacturing establishments to monitor in selected London boroughs, North *et al.* (1983) found that only 5 per cent of establishments in Wandsworth, Southwark and Merton were listed in all three main data sources used – Census of Employment (CoE), Industrial Market Location Directory, and Borough Land Use Survey (Fig. 11.1). Although cross-checking between directories is essential, many inaccuracies in the information compiled may still remain, which can usually only be removed by contacting the businesses direct. This is clearly illustrated by the experience of Gould and Keeble (1984) in setting up a databank on new firms in East Anglia, between 1971 and 1981. A detailed and laborious checking procedure

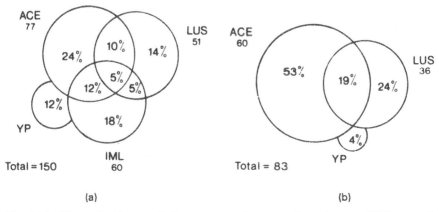

Fig. 11.1 The coincidence of data sources for manufacturing establishments in London boroughs: (a) Wandsworth, Southwark and Merton; (b) Hackney and Enfield, ACE=Annual Census of Employment; LUS=Borough Land Use Survey; IML=Industrial Market Location; YP=Yellow Pages. (*Source:* North *et al.* 1983: 122.)

yielded a list of over 2100 apparently new manufacturing firms. However, on telephoning each of these firms: 'Fully 1,400 firms, or two-thirds, were removed from the original list, because they proved not to be manufacturing firms (300 firms), not new within the time period (350 firms), not independent but subsidiaries or branches of other firms (200 firms), not indigenous but in-migrant (100 firms), or simply not traceable . . . (500 firms)' (Gould and Keeble 1984: 191).

A list of the main non-government establishment databanks held by academic institutions covering manufacturing industry in the UK is given in Appendix 16.1. Access to these is usually available for academic research, subject to the availability of resources and the safeguarding of the research interests of the compilers.

In many situations it is very difficult to construct a suitable sampling frame. For example, to study the reasons for industrial movement a list of all the firms moving into a specified area is required. Unless an establishment databank has already been developed which contains this information for the study area, recourse has usually to be made to local newspapers and industrial development and planning officers. The resulting lists are often partial and sometimes biased towards the larger firms and companies setting up on industrial estates. Unfortunately, the extent of the bias is usually unknown and cannot therefore easily be compensated for.

The issue of the minimum size of sample is not an easy one to generalize about as it depends in part on the research design and the nature of the subsequent analysis planned. Large samples are most important in extensive research where empirical generalizations are being sought about the nature of the underlying population. With small samples, of under 40 or so businesses, the standard errors may be too large to be acceptable (see e.g. Gardner 1978;

Moser and Kalton 1971). The more subcategories required for the analysis the larger the initial sample needs to be. With intensive research, detailed investigation of a small number of cases is common, although examining a larger number of businesses will give a more comprehensive coverage of causal processes in operation and the range of outcomes. Interestingly, in government-sponsored research there is pressure to reduce sample sizes. For example, in the early 1980s, as part of the 'lifting the burden on business' ethos, the DTI began to require ministerial approval if the sample size exceeded a certain level (currently 24 firms) (J. Bachtler, personal communication, 1990). It seems that at least as far as policy purposes are concerned the government is coming around to the view that smaller samples are yielding sufficiently reliable results.

If statistical inferences are to be drawn about a larger population, as is often the case in extensive research, but not intensive research, then the sample must be representative. This necessitates some form of random sampling. Many economic surveys are based on **simple random samples** from some specified population (e.g. Marshall 1983). Frequent use is also made of **stratified random sampling** (e.g. Lever 1982). The annual Census of Production is based on this method and since 1984 stratified sampling has been used in the CoE (see Chs 2 and 16). Both censuses include all the larger establishments, but only a proportion of the smaller ones. Simple random sampling is most suitable where generalizations are sought about *establishments*, while stratified sampling is better suited for making generalizations about *employment*. Similar differences arise with sampling in land use studies (Ilbery 1985a, and also Ch. 12). In farm surveys, for instance, one technique is to use random numbers to generate random grid intersections within the study area. The researcher then determines who farms the land on which the intersections fall and includes them in the sample. The probability of being included in the sample with this method increases with farm size. Clark and Gordon (1980) suggest various **cluster sampling** techniques to overcome this bias, which have the additional advantage of minimizing travel distances for the investigator. However, the techniques are not ideal and sampling errors have been estimated to be roughly twice as large as those with simple random sampling (Dixon and Leach 1978a).

Faced with the inadequacies of many sampling frames, research workers often abandon any pretence of random sampling and adopt some form of **quota sampling**, in which a deliberate attempt is made to obtain a cross-section of businesses under study. Leigh and North (1978) provide a good example in their examination of regional aspects of acquisition activity in British industry. They chose four sectors which not only had fairly high and consistent trends in acquisition activity, but also had contrasts in geographical distribution, growth rates, types of market served and structural concentration. A variant on this is to select a range of contrasting geographical areas and undertake a census of every business in the relevant categories within each area. This technique has been used in small business research, where it has been called 'saturation surveying' (Leyshon 1983; Turner 1989). Such a technique is generally only feasible where the area is small, for example, a ward or a postcode sector; for larger areas, such as conurbations, the number of relevant businesses is usually too high for a census, but this raises again the need for an adequate sampling frame. For example, in their work on new and small firms in Manchester and Liverpool, Lloyd and Dicken (1982) were fortunate to have an establishment databank which, when supplemented with other information, provided an authoritative list of manufacturing firms, from which they were able to select a quota sample stratified by location, size and industrial sector. The main problem with non-random sampling is that the findings apply to the sample and cannot be generalized to a larger population. However, the research worker often has no choice but to use this method of sampling and much can be learnt from the results as long as care is taken in interpreting the findings.

Another form of non-random sampling is to **match pairs** of establishments so as to hold some variable(s) constant. This approach has been adopted in studies of industrial linkages (e.g. Lever 1974; Peck 1985) and the performance of small firms (Hitchens and O'Farrell 1987). For instance, Peck contrasted the effect of a good linkage environment on linkage behaviour in the engineering industry, with that of a poor environment, by matching 30 pairs of plants to eliminate the influence of product type, plant size and ownership. In order successfully to implement a matched pairs design, O'Farrell and Hitchins (1988) argue that there needs to be a sufficient knowledge of the theory underlying the processes being studied to permit selection of the most important matching criteria. However, they also note that: 'A matching design is only as good as the researcher's ability to determine how to match the pairs, and this ability is frequently limited' (p. 68).

In some circumstances it is more useful to undertake **case studies** of particular businesses than to examine a group of companies. Whereas extensive surveys of populations or samples of businesses are usually undertaken to establish the range and extent of various characteristics of the businesses and the main features of their behaviour, case studies are usually used where the purpose is to examine behaviour in depth (e.g. Peck and Townsend 1984). They are particularly appropriate when research is at an exploratory stage and the lack of previous work means that the key aspects of the topic are unclear. Thus some of the early studies of the 'geography of enterprise' adopted a case-study approach (e.g. Steed 1968; Watts 1971). There is, of course, little choice but to use case studies when examining industries which are dominated by a few firms, such as the brewing industry (Watts 1980: 165–245); or where analysing local economies dominated by a few enterprises (e.g. Healey 1985). Case studies have, however, also proved helpful in other situations, for example in examining the impact of corporate reorganization on service linkages (Marshall 1989). Case studies may also be used to provide a framework for the later design of an extensive questionnaire (e.g. Hardill *et al.* 1989). Whereas in extensive research the role of case studies is usually for comparative purposes or to begin the search for generalizations, in intensive research case studies are usually used to see how general mechanisms work out in particular situations and to untangle necessary and contingent factors (e.g. Rawlinson 1990; Sayer and Morgan 1985).

Opinion on the role of case studies varies, as the following two quotes indicate:

> Case studies should normally be avoided because it is difficult to know the extent to which cases are typical . . . (Fothergill and Gudgin 1985: 104).

> The validity of the extrapolation depends not on the typicality or representativeness of the case but upon the cogency of the theoretical reasoning (Mitchell 1983: 207).

Jackson (1988) argues that case studies only need to be typical if statistical inferences rather than logical inferences are being made. This again relates back to the choice of research design.

11.3 Survey methods and response rates

There are three main types of survey method used in studies of local and regional economic activity, either on their own or in combination. They are observation, self-administered questionnaires and interviews. Response rates vary widely within the last two categories, though care is required in interpreting rates because the basis on which the figures are quoted differs. Sometimes they refer to the percentage of all businesses approached, at other times to the percentage of all relevant businesses contacted, and sometimes to the percentage which did not completely refuse, excluding those who were still thinking about it, or had not responded within the survey period. For example, in one survey the response rate ranges from 24 to 54 per cent depending on how it is calculated (Goddard 1973). The response rates cited in Table 11.2 refer, where they are known, to the percentage of all businesses approached operating in the relevant categories which provided a usable reply within the time period allocated to the survey. The extent to which the replies are usable may, of course, also vary. For example, in a survey of the tourist industry, Williams *et al.* (1989) successfully interviewed 82 per cent of the businesses approached, but for the analysis of the routes into entrepreneurship the percentage of businesses approached providing usable information fell to 63 per cent.

Observation

Although **participant observation** is frequently used in social research (e.g. Jackson 1983) its application in economic studies is far more limited. The discussions by Brannen (1987) and Winkler (1987) of the methodological

Table 11.2 Response rates to different methods used in local and regional surveys of economic activity

Author	Subject	Response rate (%)*
Postal surveys		
Aubrey *et al.*(1989)	Employment in manufacturing plants	21
Marshall (1983)	Business service firms	25
Bachtler (1981a)	Insurance companies	28
PSMRU (1986)	Ethnic minority businesses	42
Elias and Healey (1991)	Small and medium-sized employers	44
Gripaios *et al.* (1989)	High-technology firms	47
Bowler (1982)	Pick your own schemes	57
Hardill *et al.* (1989)	Adoption of computer networking	57
Keeble and Kelly (1986)	Computer electronics	57
Griffiths (1988)	Farm diversification	59
Ilbery (1985b)	Vine growers	59
Bachtler (1981b)	Bank branches	60
Pratt (1989)	Rural industrial estates	74
Fothergill *et al.* (1983)	Industrial premises	79
Contact diaries		
Goddard and Morris (1976)	Decentralized offices	25
Goddard (1973)	Office establishments	32
Telephone surveys		
Northcott and Rogers (1982)	Manufacturing firms	63
Elias and Rigg (1990)	Demand for graduates from employers	69
Watts (1982)	German multinationals	83
Marshall (1983)	Non-respondent offices to a postal survey	96
North *et al.* (1983)	Follow-up survey of manufacturing firms	97
Foley (1983)	Electrical engineering establishments	99
Face-to-face interview surveys		
Cooper (1975)	Manufacturing firms	28
Townroe (1975)	Migrant companies in London	30
Hoare (1978)	Engineering firms	35
Cooper (1989)	Employers in West Midlands	50
Diamond and Spence (1989)	Infrastructure and industrial costs	50
Leigh and North (1978)	Firms involved in acquisitions	52
Turner (1989)	Small businesses	55
Healey (1983b)	Multi-plant enterprises	62
Rhodes and Kahn (1971)	Office establishments	65
Lloyd and Dicken (1982)	New and small manufacturing firms	73
Munton (1983)	Farmers in London's green belt	74
North *et al.* (1983)	Monitoring industrial change	74
Evans (1990)	Farm-based accommodation businesses	78
Mason (1989)	New manufacturing firms post-1979	79
Marshall (1983)	Consulting engineering firms and management consultancy firms	80

Table 11.2 Cont.

Author	Subject	Response rate (%)*
Townroe (1975)	Migrant companies in the North and East Anglia	80+
Williams *et al.* (1989)	Tourist-related businesses	82
Mason (1982)	New manufacturing firms 1976–79	84
Elias and Healey (1991)	Large employers	89
Moseley (1973)	Manufacturing and service firms	89
White and Watts (1977)	Broiler industry	89
Shaw and Williams (1988)	Tourist-related businesses	90+
Ilbery (1987b)	Agricultural diversification	93
Pratt (1989)	Property professionals	95
Healey and Clark (1985)	Manufacturing establishments	95+
Healey *et al.* (1987)	Clothing firms	98
Rawlinson (1990)	Small precision engineering enterprises	98
Massey and Meegan (1979)	Electrical engineering companies	Almost 100
Watts (1991)	Multi-plant companies	100

*Response rates may be defined on a variety of bases – see text for discussion.

issues involved with 'fly on the wall' observation of directors at work are exceptions. The more common use of observation in industrial studies is as an external check to enable the exact location of the business to be plotted. **External observation** is often used in the initial stages of data collection or to update and supplement other sources. The main advantage of the technique is that it provides a quick method of identifying the businesses currently operating in a relatively small area, such as retail stores in the city centre, farms in a parish, or units on an industrial estate. Observation can also be a very useful way of constructing or establishing the accuracy of a sampling frame prior to a survey and has been used in, for example, agricultural studies (Clark and Gordon 1980), small business research (Turner 1989) and in setting up establishment databanks (Ch. 16).

External observations have also been used to note the nature and state of the buildings and local environs. For example, Fothergill *et al.* (1983) made site visits to all establishments that did not reply to their questionnaire on industrial premises and also to those factories where questionnaire replies had missing or ambiguous information. Visual inspection of the premises and site, generally from outside the perimeter, enabled them to fill in most of the information sought by their survey. Observation, however, also has its problems. Not all businesses display a name plate and where one is shown, it is sometimes out of date. It is also difficult, in many cases, to identify the business of the firm simply from observation. For example, firms described as 'electrical engineers' may be manufacturers, wholesalers, contractors or even retailers (Foley 1983). These points may usually be resolved by checking with the firm by one of the other survey methods.

Self-administered questionnaires

Most self-administered questionnaires are sent and returned by post. However, some may be left with businesses to complete and return by post and others may be collected personally. Postal surveys are an inexpensive way of administering a questionnaire. Self-administered questionnaires are most appropriate where there are only a few simple factual questions, many of which can be answered by the respondents ticking the relevant boxes. Hague (1987a) suggests that postal surveys are best suited to researching organizations where there is one key decision-maker. One of the strengths of the technique is that respondents can complete the questionnaire at their own pace, encouraging a greater chance of a more considered reply to each question. The written questionnaire also ensures consistency in the questions asked and avoids variability between interviewers. However, in the absence of a skilled interviewer there is no opportunity to probe vague answers, there is no control over the order in which the questions are completed and some questions may be missed by the respondent (de Chernatony 1988; Swain 1978).

The annual Census of Production, Agricultural Census and CoE all use postal questionnaires. The response rates to these surveys are very high and though the first two request detailed information on several subjects the respondents have a statutory obligation to complete the forms. The 1987 CoE obtained a 97 per cent response (Department of Employment (DE) 1989), while it is estimated that the Agricultural Census obtains a response in excess of 85 per cent (Ch. 7). The greater size of the government surveys, however, brings its own particular problems. Such large surveys are 'highly insensitive to many mundane sources of error – misunderstood instructions on forms, misreading of hastily written figures, misplacing a decimal point, losing one's place in copying, accidental "corruption" of data in computer files, or printing errors' (Government Statisticians' Collective 1979: 144). Such problems are not as significant in the smaller non-statutory postal surveys, but the response rates in these surveys vary widely, as can be seen in Table 11.2. The mean response rate for the postal surveys included in the table is 51 per cent and the range is from 21 to 79 per cent.

A specialized form of self-completed questionnaire is the **contact diary** which has been used in communication surveys to identify the nature of information linkages within and between business organizations. The diary aims to give a more comprehensive picture of the nature of contact patterns than can be obtained by asking managers to make estimates. Managers are asked to record in the diary details of all contacts over a given period, usually 3 days to a week (Alexander 1979; see also example in Appendix 11.1). The three studies referred to in Table 11.2, which used contact diaries, show that response rates of over 50 per cent are possible with this survey method. However, before such surveys can begin, lengthy negotiations with senior management are often necessary to obtain their agreement that managers within their organizations may be approached.

An important factor which may help to account for some of the variation in response rates obtained by self-administered questionnaires is the topic of the survey. According to a seminar of industrial market researchers lack of interest in the subject by the recipients of industrial questionnaires is one of the main reasons for non-response (Swain 1978). For example, the response rate obtained in a postal survey of skills and training (44%) was reduced by the self-employed and the owners of businesses with only one or two employees perceiving the topic of the survey to be of marginal interest and relevance to their businesses (Elias and Healey 1991). This view seems to be confirmed by the very high response (89%) obtained in a parallel interview survey of large businesses for whom the issues covered in the survey were more relevant.

A further factor may have been that the postal survey was undertaken in a city, because Gudgin (1976, 1978) reports that the response rate he obtained for a postal questionnaire was inversely related to the degree of urbanism, varying from 40 per cent in London to 85 per cent for the same short questionnaire in rural Lincolnshire. He also notes that the highest response rates are usually obtained from large establishments and the lowest from small plants. When longer questionnaires are used, Gudgin suggests that response rates may fall to a third or less. However, a recent comprehensive review of the factors affecting response rates to mailed questionnaires, used mostly in social surveys, found that the number of questions has an ambiguous impact on response rates (Harvey 1987). Indeed, it has been suggested that business respondents may equate brevity with superficiality, with detrimental results (Hoinville *et al.* 1978).

A number of other factors such as personalization, sponsorship, layout and style also seem to have an uncertain effect on response rates. Nevertheless, as Harvey (1987: 347–8) notes:

> enquiries over the last half century have suggested that there are . . . a number of fairly well established principles for researchers undertaking data collection using mail questionnaires. Researchers should enclose stamped reply envelopes and some pre-contact is likely to be advantageous. Follow up procedures will usually lead to increased response rates. . . . In general, the questionnaire should be constructed in a way that is likely to appeal to the respondents and should direct itself to arousing, rather than assuming, the interest of the respondent.

Interestingly there is evidence that stamped envelopes bring a higher rate of return than prepaid envelopes because the cash element in the form of the stamp which creates an implied obligation to reply (Wilson 1968). Sometimes a deadline for the return of the questionnaire is given, but if this is too long it may slow down or lessen the response rate as respondents may put it to one side to answer nearer the time. According to Bachtler (1984: 109) **follow-up surveys** are 'almost universally successful'. He refers to one survey in which the researcher sent six follow-up letters with a promise in the sixth that completion would spare the respondent further correspondence: result 100 per cent response! Although multiple follow-ups have been shown to increase response rates the most dramatic increase is generally noted with the first follow-up (de Chernatony 1988). For example, Fothergill *et al.* (1983) report that in a survey of industrial premises in the East Midlands the initial response rate to a two-page questionnaire was 47 per cent, and a reminder letter raised this to 70 per cent. A second reminder, worded quite differently, brought a further batch of replies, raising the response rate to 79 per cent of all establishments which were open and within the scope of the study. The best timing for a follow-up survey occurs when the replies to the first survey reach a plateau. A courteous reminder letter enclosing a second copy of the questionnaire and a stamped reply envelope is essential.

To avoid ambiguity the clearest way to express the effect that follow-up surveys have on response rates is in terms of the reduction in non-response rather than the rate of increase in response.

A 50 per cent reduction in non-response has a clear meaning regardless of response rate but what does a 20 per cent improvement in response rate mean when the initial response rate was (a) 30 per cent or (b) 80 per cent? Any effort is aimed at non-respondents and it is the effectiveness of the effort in reducing their numbers that is the main concern (Swain 1967: 15).

Interviews

An alternative way of administering a questionnaire is by interview. When there are a large number of questions, or the subject of the survey involves an investigation into the reasons for decisions or the perception of the owner or manager, an interview is usually essential. The interviewer not only has more control over who answers the questions than could be achieved with a postal survey, but he or she can also clarify any ambiguous questions, probe answers that are too brief and query discrepancies in the replies. Interviews are, of course, essential when an intensive research design is adopted, because of the need to be able to explore topics interactively.

Standardized **face to face interviews** are often used as an alternative to self-administered questionnaires sent through the post because, though more expensive in terms of time and cost, they usually achieve a higher response rate. This is partly because industrialists generally prefer being interviewed to filling in questionnaire forms (North *et al.* 1983). The mean response rate for interview surveys listed in Table 11.2 is 75 per cent, but, as with postal surveys, the range is wide, from 28 to 100 per cent. There is some indication that response rates may vary, as with the postal surveys referred to earlier, with the degree of urbanism. In one survey of migrant companies the response rate in London was 30 per cent, while in East Anglia and the North it was over 80 per cent (Townroe 1975). Other researchers have, however, managed to achieve response rates in excess of 90 per cent where the businesses were based predominantly or exclusively in cities. Such high response rates have variously been ascribed to: 'the fact that preliminary findings had been made available in advance' (Massey and Meegan 1979: 170), and to the persistence and persuasiveness of the interviewers (Healey 1984).

Telephone interviewing is growing in popularity in economic research. It is a relatively cheap method, especially if restricted to a local area, and in many cases an immediate answer to the survey questions can be obtained. Whereas a letter can be ignored or passed over a ringing telephone demands attention. Many people are encouraged to speak more openly on the telephone than face to face. However, it is also easier for the respondent to terminate the interview (Wilson 1968). Response rates also seem to be higher than for similar postal surveys (Table 11.2). For example, Foley (1983) reports that he contacted slightly over 200 establishments by telephone. The management of only two firms refused to provide answers to a few straightforward questions and in both cases they offered to respond if supplied with a postal questionnaire. Telephone interviewing has also been used to follow up non-respondents to postal surveys, to target particular groups, and to increase the overall response rate (e.g. Elias and Healey 1991; Marshall 1983). Telephone interviews would seem most suitable for obtaining answers to a small number of simple questions where it is reasonable to suppose that the person being spoken to knows the answers without having to consult any records. Hague (1987a) suggests that there is a limit of about 10 minutes for the length of time that a respondent's attention can be held on the telephone for a structured interview. However, the method has also been used successfully for longer interviews of up to about half an hour, where the managers had been visited in an earlier survey (North *et al.* 1983) or where the respondent has previously agreed to participate and had the opportunity to collect together relevant factual information prior to the interview (Elias and Rigg 1990). Several observers suggest that female telephone interviewers are more successful than males (MacFarlane-Smith 1972; Wilson 1968).

The best strategy for **obtaining an interview** seems to vary with the size of firm and the nature of the investigation. A short interview at a small firm can often be obtained by simply 'knocking on the door'. Similar tactics are also often used in studies of farmers (e.g. Ilbery 1985c). For large businesses an appointment is usually required. Identifying the best person in a large company to approach for an interview can itself be a major problem, especially in a general survey. In a specific study, for example of purchasing linkages, the section of the firm to approach can be much clearer. For many enquiries the best person to contact may be someone at the head office or divisional office rather than at a branch. A request for an interview may be made either by letter or telephone. The latter often obtains an immediate reply, but may not achieve a higher response. Indeed Forsythe (1977), who used both techniques in a study of the sources and uses of statistical information by businesses, found that firms whose chief executives were telephoned to explain the survey and request the names of users of statistics within the firms, provided names less often and gave fewer names than did those asked to provide names by letter.

Sometimes a combination of methods is needed. In one survey, of multi-plant enterprises, the research worker wrote initially to a named individual, usually the managing director, explaining the purpose of the survey and requesting an interview (Healey 1979). A stamped and addressed envelope was enclosed. Only half of the final 64 enterprises agreed within a month of sending the letter to participate in the study, but the response rate was doubled, to 62 per cent, through reminder telephone calls and letters and by querying those replies which said the company did not come within the terms of the study. In a later survey of predominantly large employers the same research worker avoided the problem of waiting for the respondent to reply by saying in the introductory letter that the interviewer would be telephoning the respondent a few days later to try to arrange a time for an interview (Appendix 11.2; Elias and Healey 1991). Although some interviews were still delayed because of pressures of work this method considerably reduced the average time between the first contact and the interview, where agreed, taking place. Indeed in some cases the respondents telephoned the interviewer on receipt of the letter to arrange a suitable time to meet. This survey achieved a response rate of 89 per cent. Using a similar approach Watts (1991) obtained a response rate of 100 per cent. He wrote to the chief executives requesting an interview with them or the person they thought most appropriate and followed this up with a telephone call a few days later. A modern variant of this for larger businesses is to send the introductory letter by fax. This gives an impression of importance and urgency. By telephoning the secretary of the person to be contacted, not only can the fax number of the business be obtained, but also the best time can be ascertained to telephone back to arrange a date for an interview.

In some cases it can be helpful to send a list of factual questions to the respondents prior to the interview to enable information to be assembled, which might not be immediately available at the interview. When this strategy was used in the Elias and Healey (1991) study the form was completed either prior to, or at, the interview in over 80 per cent of the cases. In one or two cases, however, the interview was cancelled, when the respondents received the list of questions, on the grounds that it would take them too long to assemble the information required. On the other hand, the frequent difficulties involved in persuading the respondents to return the form, when it was left for completion after the interview, provides support for sending the form out before the interview.

Polite persistence is important in obtaining an interview. A series of rejections can be dispiriting for the research worker; however, it is almost always worth querying an initial refusal to see what is the reason. For example, a company may consider itself to be too small to be of importance. Often stronger resistance comes from the secretaries of the owners and managers than from the owners and managers themselves. In one survey when telephoning a week after sending a letter to the owner requesting an interview:

the tactic used to circumvent the 'over protective' secretary was to reply when asked what the phone call concerned was to say 'It's alright, he knows what it is about'. This usually resulted in being put straight through to the owner, who even if he had not read the introductory letter would invariably agree to an interview (Rawlinson 1990: 111).

Clearly response rates vary widely not only between different survey methods, but also within. It is useful to distinguish between different types of non-response. Frequently, out-of-date and inaccurate sampling frames mean that a proportion of responses are 'not known', 'gone away' or 'not relevant'. These may be distinguished from the outright refusals and inadequate responses, the minimization of which is the aim of good survey design. Some of the firms 'no longer operating at this address' may be identified by the Post Office returning the letter to the sender.

A potentially important factor affecting the success of postal surveys and most interview surveys is the **introductory letter**. Most survey texts emphasize that the aim of this letter should be to interest the respondent in the survey (e.g. de Chernatony 1988; Swain 1978). It should give a short statement of the purpose of the study and its sponsorship, together with some indication of its value to the respondent and the use to which the findings will be put. Short, clear sentences assist in this. The overall tenor of the letter should be a courteous and tactful request for help and should specify the nature of the involvement being sought. Sometimes a separate sheet outlining the research project is appropriate. However, Dexter (1970) warns against giving too much detail about the nature of the research when seeking an interview. He suggests that élite people given a complex explanation tend to feel that the matter would probably be better handled by some specialist or that their inability to answer possibly complicated questions would be embarrassing. On the other hand, there are ethical issues involved if an attempt is made to mislead the interviewee about the nature of the research topic or the use to which the findings are to be put. Most surveys promise that the findings relating to individual businesses will be treated confidentially. It is relatively easy to hide the identity of individual businesses in a report on a large survey, but it is more difficult to achieve if a case-study approach has been used, particularly if the businesses examined are large. If interviewees known that their businesses may be identified in the report it may affect their willingness to participate and, if they do agree, it may influence the nature of the answers they give.

Given the number of requests that busy executives receive, any 'opening' in the correspondence requesting assistance that gives them an excuse for refusal may be exploited. For this reason Kincaid and Bright (1957a) advise against multiple requests in the same letter, reporting that when they asked not only for an interview but also for any literature pertinent to the study, several companies refused co-operation on the ground that since they had no pertinent literature there would be no point in conducting an interview. Although the evidence is equivocal, as few researchers are prepared to risk marring their success by deliberately sending out shoddy work, a well-designed and presented letter, typed on headed notepaper, which is personally addressed with a handwritten signature, would seem to be a sensible way of trying to persuade the owners and managers of businesses to co-operate. An example of an introductory letter is given in Appendix 11.4. Prior publicity for the survey may help to prepare the ground. For example, before sending out a postal questionnaire to establishments on rural industrial estates in Cornwall, Pratt (1989) sent material to local papers explaining the survey and persuaded local radio stations to interview him about what he was doing.

A particular problem, which tends to increase the lower the response rate, is that the participating businesses may be unrepresentative of those approached. Whether the population or a sample has been contacted the possibility of bias in the pattern of responses needs to be tested. Unfortunately much of the key information about the characteristics and behaviour of the businesses which refuse to participate is not available from other sources. The representativeness of postal surveys can be assessed by a telephone check of a sample of non-replying businesses. However, reliance often has to be put on other measures, for example, the economic sector, size, profitability and location of the non-respondents, which can be derived from other sources, such as directories and the Companies Registration Office (e.g. Healey 1983b). In some situations it is possible to weight responses to allow for variations in response rates between different groups (Elias and Healey 1991).

11.4 Standardized questionnaire and interview schedules

The nature of the responses obtained in surveys depends critically on the quality of the questionnaire and interview schedule design and the way the questions are phrased. It should be emphasized that 'the construction of a schedule, the precise framing of questions, the sequence of questions at an interview and the exclusion of bias are skilled procedures which require every attention to detail' (Jackson 1963: 83).

With standardized interviews one of the aims is to minimize the interviewer-related errors. This applies particularly where more than one interviewer is involved, though consistency of approach by a single interviewer is also important. Fowler and Mangione (1990) argue that it is essential that all interviewers:

1. Read the questions as directed;
2. Probe inadequate answers non-directively;
3. Record answers without discretion; and
4. Communicate a neutral, non-judgemental stance.

The questionnaire and interview schedules used in the business surveys referred to in Table 11.2 vary widely in **layout and question design**. At one end of the range are the standardized highly structured schedules used in self-administered questionnaires (and some interviews), in which most of the questions are concerned with factual material and there are a large number of precoded responses for the respondent (or interviewer) to tick. Two examples of postal questionnaires are included in the appendices. The first is a copy of the questionnaire which Keeble and Kelly (1986) used in their survey of high-technology industry (Appendix 11.3); the second was used by Elias and Healey (1991) in their study of skills and training among every identifiable small- and medium-sized employer in Coventry (Appendix 11.2). The amount of information requested in the two surveys was not dissimilar, but whereas the high-technology survey was originally typed on both sides of an A4 sheet of paper, the skills and training questionnaire was printed as an eight-page A4-sized booklet, all but the cover of which is reproduced in the appendix.

In many interviews the number of questions asked is often much larger than is attempted in postal surveys. For example, in the parallel survey that Elias and Healey (1991) undertook of skills and training issues among large employers the number of sheets of questions was three times that used in the postal survey. Most of the additional questions were open-ended questions with a number of probes which were designed to encourage respondents to account for various aspects of their behaviour and to discuss the strategies that they were adopting to overcome skill shortages. Space limitations permit only one example of an interview schedule to be reproduced in the appendix. The one chosen was used in a study of agricultural diversification among farmers located in the urban fringe (Ilbery 1987a). It illustrates the kind of interview schedule often used in extensive quantitative economic research, consisting of a mixture of closed and open questions which were asked, using identical wording in a fixed order (Appendix 11.5).

A **pilot survey** is essential to test the questionnaire or interview schedule. Even with a brief, simple questionnaire it may take several months to remove ambiguities from the questions and to achieve an efficient layout with concise wording (Gudgin 1978). Before the pilot survey it is useful to discuss a proposed questionnaire with a few individuals familiar with the sectors being studied, because there are important differences in the terminologies used in different industries. This pre-survey work is particularly important for self-administered questionnaires where it is not possible to clarify misunderstandings, but it is also essential preparation for interviews. Poorly worded questions may also reduce response rates for self-administered questionnaires: 'If the recipient finds it obvious that the reply will be misleading (some will answer the question as put, some will answer the question which they think the researcher meant to put) he is likely to join the ranks of non-respondents' (Swain 1967: 19). Face-to-face piloting of questions which are to be used in a self-administered survey can help refine the wording of questions (e.g. Hardill *et al.* 1989; Elias and Healey 1991).

Apparently simple questions, such as 'How many people are employed at this establishment?', are open to different interpretations. Should, for instance, part-timers, drivers and salesmen be included? People based at the establishment, but on the payroll of another, can also cause problems. As Gudgin (1976) has commented, 'the problem of gaining accurate answers to questions particularly from postal surveys is more intractable than that of getting a response at all'. This seems to be a problem especially with surveys of large firms in which the scope for error on the part of the respondent is greater than in a small firm. In one postal survey of predominantly large factories, two completed forms were returned by each of 10 establishments. The figures supplied for total employment differed by more than 10 per cent in half the cases. Of these, three differences exceeded 20 per cent and one was greater than 100 per cent (Gudgin 1976).

With less straightforward questions the problem is potentially greater. In many surveys respondents are asked to select between various categories which best describe their business or aspects of their behaviour. Where the categorization is complicated, as with, for example, the Standard Industrial Classification (SIC), it is generally better to leave it until the coding stage and encourage the respondent to describe the nature of the business or behaviour in sufficient detail to enable the classification to be applied consistently. However, where it is necessary that the respondent does the classification some guiding notes may be appropriate. For example, in collecting data on the occupational structure of businesses Elias and Healey (1991) provided examples of occupations for each of the eight occupational groups that they used (Appendix 11.4). It is usually best to leave sensitive questions, for example concerning financial matters, until near the end of a questionnaire, although Evans (1990) recommends placing them in the middle of an interview schedule in order to avoid comments from respondents such as 'I wondered when you were going to ask that'. Once these questions had been asked he found that the interviews were generally more relaxed.

Obtaining accurate answers to questions on values, motives and perceptions is even more difficult. Indeed Dean and Whyte (1958: 38) argued over 30 years ago that there is no point in asking whether the informant is telling the truth. Instead the researcher should ask: 'What do the informant's statements reveal about his feelings and perceptions and what inferences can be made from them about the actual environment or events he has experienced?' On the other hand, Marsh (1979: 304) warns that: 'We must not confuse an impossible attempt to achieve "absolute truth" through unbiased questions, with the aim of being objective in our quest for truth, through trying to be as rigorous as possible in the way we draw conclusions from observations we make about the world, what people say and how they behave.'

The answers given to questions about values, motives and perceptions are influenced by a variety of factors. For example, Gasson (1973), in a study of the goals and values of farmers, notes the danger of relying on verbal indicators of values is that the answers given are influenced by who else is present at the interview and the relationship established with the interviewer. Further at the technical level open questions tend to elicit fewer responses than closed questions (Belson and Duncan 1962). This may be illustrated by a comparison of the reasons given in two surveys for industrial movement

Table 11.3 Methods for asking the reasons for plant movement in two surveys. (*Sources:* DTI 1973; 532; Healey 1979; 163.)

	ILAG survey 'What caused you to consider opening a new plant in a new location? Was it for one of the following reasons?'	Number of times reason given as major as % of 531 moves		Textile and clothing survey 'What were the circumstances which led you to open this branch plant (transfer this plant)?'	Number of times at least 1 of group of reasons given as % of 69 moves
B1–2	To permit an expansion of output	83	1 and 5	Expand production of products presently manufactured or introduction of new products	75
B12–15	Inadequate existing premises or site	50	2, 4 and 7	Lack of space, reorganization of activity locations, plant unsuitable	48
B26–29	Unsatisfactory labour supply	40	3	Shortage of labour	25
B30–31	Inducements and facilities made available by official bodies	27	9 (part)	Assisted area grants and loans	3
B22–29	Opportunity to purchase or rent premises or site at new location	20	6	Premises available	12
B18–19	Town planning difficulties	11	8	Lease/compulsory purchase/fire	7
B16–17	Lease of former premises fell in, or good offer received	5			
B4–6	Too far from established or potential markets	19	9 (part)	Other	13

Table 11.3 Cont.

	ILAG survey 'What caused you to consider opening a new plant in a new location? Was it for one of the following reasons?'	Number of times reason given as major as % of 531 moves	Textile and clothing survey 'What were the circumstances which led you to open this branch plant (transfer this plant)?'	Number of times at least 1 of group of reasons given as % of 69 moves
B20–21	Refusal or expected refusal of IDC	12		
B33	Desire to be in more attractive surroundings	4		
B7–11	Too far from supplies, actual or prospective, of materials or services	3		
B34	More profitable to operate elsewhere, no other postulated reason being major	1		

(Table 11.3). Though the surveys differed in several respects the results are, in general terms, similar. However, the Inquiry into Locational Attitudes Group (ILAG) survey, which used a closed questionnaire, not only gave a larger number of reasons, but also obtained a higher percentage of firms citing each group of reasons. It is probable that these differences are largely due to the prompting involved in the ILAG survey, which listed over 30 possible factors encouraging industrial movement. Watts (1991) gives another illustration of the effect of prompting responses. He found that business rates were mentioned as a reason affecting the choice of plant to close only when business executives were presented with a list of possible factors.

Not using precoded responses, according to some (e.g. Hague 1987b), may be a sign of laziness on the part of the researcher; alternatively, it may reflect the desire of the interviewer not to restrict the range of responses. As Moyser and Wagstaffe (1985: 18) note, 'we have been surprised at the extent to which some élite figures will try to fit themselves and their views into "boxes" for the sake of social scientific methods'. Precoded responses are more common in self-administered questionnaires than in interviews, where respondents often find them tedious. This problem may be reduced by using a limited number of cards, with the responses printed on them, which are passed to the respondent at the relevant point during the interview. However, many élite interviewees prefer to give their own interpretations rather than be forced to choose between categories of responses which often do not seem to give an adequate summary of the situation as they perceive it.

More fundamental is the difficulty of **interpreting the answers** given. Respondents, in attempting to be helpful, may give answers which they think will please the questioner, or they may try to justify their actions (the problem of *post-facto* rationalization). Further as Dexter (1970: 144) warns, 'the interviewer cannot safely assume that the particular words he uses are in fact the stimulus to which the interviewee is responding'. For instance, Gudgin (1976) found, in a postal survey in the East Midlands, that many firms gave apparently rational reasons for choosing their locations, including labour supply, availability of premises and proximity of markets. However, in answer to subsequent questions it emerged that 90 per cent had investigated no alternative locations and 80 per cent had strong prior links with the area. A related problem in asking about reasons for locational choice is the need to distinguish the reasons for the choice of region and settlement from those influencing the choice of the site. Considerable confusion arises in many surveys where the scale element is not made clear in the phrasing of the questions and the interpretation of the results, because different factors tend to be important at different scales. For example, the availability of labour tends to distinguish one region or subregion from another, while the availability of a suitable building may make one part of a subregion more attractive than another. The ILAG survey suffered to some extent from this problem in that only a single question on locational choice was asked: 'Having decided to open a new plant in a new location which factors determined the location you chose?' (DTI 1973: 573). Further difficulties arise when questions are asked about events which occurred several years previously, or when the person answering the question had not taken part in the decision. There is also the problem of inferring the motives of a company from the responses given by a single representative of an organization.

These problems concerning questions about values, motives and perceptions can be reduced to some extent by asking only questions that it is reasonable to expect respondents to be able to answer, by careful questionnaire design, and by exploring the topic in detail. Much of value may be learnt where such surveys are carefully executed. However, the results should be interpreted cautiously and are, perhaps, best taken as simply the views of a group of managers representing their businesses. It follows that such questions should usually only be asked in interviews, where follow-up questions and various prompts can be used. An increasing number of research workers are using a non-standardized interview format for exploring business decision-making and strategies.

11.5 Non-standardized interviews

Methodologists of the standardized interview, such as Fowler and Mangione (1990), stress the desirability of socially sterile conditions in the interview situation. However, this is a normative ideal. Others, such as Brenner (1978: 138), argue that as the interview involves an interpersonal encounter, it is unrealistic to try to make 'the interviewer act *without acting*, as if the interviewer could ever *not* influence the situation of action in the interview by means of his own performance'. With non-standardized interviewing the interaction between interviewer and interviewee is emphasized rather than minimized. When interviewing the élite, such as owners and managers of businesses, Dexter (1970: 5) states that 'the investigator is willing, and often eager to let the interviewee teach him what the problem, the question, the situation, is', within the limits of the research topic. With economic research some degree of topic control is necessary, though the extent will vary with the circumstances. Non-standardized interviews are one of the main survey methods used in intensive research, but they are also used in extensive research, although their function is different (see section 11.1). Non-standardized interviews are particularly appropriate for in-depth studies, when exploring new research areas, when investigating underlying causal mechanisms and when seeking to understand and explain aspects of business behaviour. They may also be used in the early stages of designing standardized interview schedules and self-administered questionnaires.

As non-standardized interviewing is a social process, which involves both the interviewer and the interviewee, it is important to give the interviewee an *active* role and try to be as flexible as possible regarding the form and order of questions so as to accommodate for this (A. Sayer, personal communication, 1986). With this method, the research worker is able to refer to and build upon knowledge gained beforehand about the specific characteristics of the business, 'instead of having to effect ignorance . . . in order to ensure uniformity or controlled conditions" and avoid what might be taken as "observer-induced bias"' (Sayer 1984: 223). In non-standardized interviews the questions posed vary from one interview to the next, dependent on how the interview develops, the knowledge of the respondent, and the level of understanding of the interviewer at the time of interview. Consequently 'there is no way that the information can be statistically summarised to reflect the aggregate response of the group or to compare one individual's response with another's' (Gorden 1975: 61). The draft list of themes for interviewing business managers in the ESRC Changing Urban and Regional System Research Programme, included in Appendix 11.6, gives an example of an interview guide for use in non-standardized interviews (Cooke 1989). The list of themes is indicative of the topics to be covered where they are relevant to the business being examined. In this particular case the list of themes was used to suggest topics to be covered by the different research teams responsible for the various locality studies in designing their own interview schedules.

As much, if not more, skill and thought are needed in undertaking the non-standardized interviews used in intensive qualitative research as are required in constructing the standardized surveys used in extensive quantitative research. For a start, a thorough understanding of the research subject is needed to be able to make sense of the interview and to know when clarification or further probing is necessary. Moreover, it is usually essential for the interviewer to have a sound knowledge of the industrial sector of the business and background information on the business itself, particularly where it is a large concern. In a study of professional producer service industries in Britain, which used non-standardized interviews, the researchers first gathered together a large volume of published and unpublished information on the 20 largest 'target' firms in accountancy, investment banking, chartered surveying, advertising, architecture and corporate legal services (Table 11.4). The knowledge level required by the interviewer to carry out non-standardized interviews means that they must be undertaken by the principal researchers and cannot be delegated to a team of interviewers. It follows in contract research that fewer interviews can be completed for the same budget than when standardized interviews undertaken by field researchers are used. According to Bloom (1988: 55–6): 'In the right circumstances, fieldforce interviewers do an excellent job and give very good value for money. However, the tendency has been to pretend that every piece of research can be done in this way, whereas it is actually appropriate in only a minority of situations.'

There are a number of factors which contribute to a 'successful' non-standardized interview of which 'the most important is trust and rapport' (Moyser and Wagstaffe 1985: 17). Dexter (1970: 25) recommends that 'sympathetic understanding' is the attitude most likely to promote such an atmosphere and hence yield the best response. Concentration is also essential to keep the interview flowing. Kincaid and Bright (1957a: 310) note

Table 11.4 Information sources used in a survey of professional producer services. (*Source:* Thrift *et al.* 1988: 6–7.)

1.	Accountancy	Professional directories and journals (especially *The Accountant, UK Accounting Bulletin* and *International Accounting Bulletin*) Accountancy firms' reports and reviews Interviews with senior managers and partners Facilities of Institute of Chartered Accountants in England and Wales Library
2.	Investment banking	Professional directories and journals (especially *Bankers' Almanac, The Banker* and *Euromoney*) Bank reports and reviews Interviews with directors Facilities of Institute of Bankers Library
3.	Chartered surveying	Professional directories and journals (especially *RICS Yearbook, World Property*) Chartered surveyor firms' reports and reviews Interviews with directors and partners Facilities of Royal Institution of Chartered Surveyors Library
4.	Advertising	Professional directories and journals (especially *Campaign, International Journal of Marketing*) Advertising firms' reports and reviews Interviews with managers Facilities of Institute of Practitioners in Advertising Library.
5.	Architecture	Professional directories and journals (especially *Building*) Architectural practice firms' reports and reviews Interviews with managers
6.	Corporate legal services	Professional directories and journals (especially *International Financial Law Reviews*) Law firms' reports and reviews Interviews with partners
7.	General	Business magazines and newspapers (especially *The Economist, Business* and *The Financial Times*) Facilities of University of Wales Library, Cambridge University Library, City of Bristol Library Business Statistics Office, plus SDUC and University of Bristol Libraries NOMIS, University of Durham

that interviewers of executives are likely to pay for lapses of alertness as they perceived 'a decided distaste for the inefficient, uneconomical, and inconsistent'. Starting an interview on the right note is important. After explaining who you are, the sponsorship for the study, and to the necessary degree what the project is about, Dexter (1970: 55) recommends beginning 'with comments or questions where the key words are quite vague or ambiguous, so the interviewer can interpret them in his own terms and out of his own experience'. A similar method was used by Sayer and Morgan in their study of labour–management relationships in the South Wales electrical engineering industry. They generally began their interviews with plant managers with open questions so that the managers could define the problems and make their own connections, only then following up with more focused supplementary questions (A. Sayer, personal communication, 1986). Dexter advises that a question which sharply defines a particular area for discussion is far more likely than a general question to result in omission of some vital data of which the interviewer would not have thought. Non-standardized interviews need to be piloted as much as standardized ones.

The ultimate test of success, however, has to be measured in terms of the quality of the responses in helping the research worker to meet the wider goals of the study expressed in the initial research design. A further check on the value of the information provided may be made by discussing the findings with knowledgeable informants. For example, Marshall (1989) reports that the general validity of his findings about the motor vehicle aftermarket in the West Midlands was tested by discussions with industry experts and a trade association. It can be very useful to talk to such 'umbrella' organizations both before and after company interview surveys (see also Ch. 17).

At the technical level an important interviewer skill lies in **obtaining an accurate record** of the interview. There is a wide variety of ways of taking notes. However, one thing all research workers are agreed on is the importance of writing up the notes as soon as possible after the completion of the interview. Some researchers extol the virtues of interviewing in pairs as a way of eliminating interviewer bias (e.g. Kincaid and Bright 1957b). North *et al.* (1983) found this method worked particularly well in dealing with the largest firms, not only for efficient recording of the information, but also because it was common to find two or more executives attending the interview. A useful check is to send the interviewee a copy of the write-up of the interview.

An alternative to taking notes is to tape record the interviews. Gorden (1975: 275) claims that 'it is rare to find a respondent who will object to its use'. He was, however, writing in a North American context and in a recent British study the attempt to tape record interviews with the owners of small precision engineering companies had to be abandoned after the first eight owners declined (Rawlinson 1990). By contrast, in the same study managers of large manufacturing companies raised no objection to having their interviews recorded. The acceptability of taping an interview may, however, depend on the approach the interviewer takes. Gorden (1975) advises *explaining* why it is used rather than *asking* for permission. This kind of approach seemed to work for Pratt (1989) who managed to overcome any concern about taping his interviews with estate agents, property developers and planners by explaining that he just wanted to be sure that he did not misquote them. A major advantage of taping interviews is that it enables the interviewer to concentrate on the phrasing and order of questions rather than on note-taking. As well as providing an accurate record of an interview, tape recording makes it considerably easier to include extended direct quotes in the finished report. Such quotes may not only enliven an account, but may also throw some light on the interviewer–interviewee relationship. In an unstructured interview this relationship can have a critical effect on the responses given (Dexter 1970). Against these advantages have to be placed the possibility of the act of taping inhibiting the responses and the high cost involved in transcribing and analysing the interviews. If a complete transcript is needed it can take a typist from 3 to 12 hours of typing for each hour of recording (Gorden 1975).

Non-standardized interviewing is a complementary method to standardized interviewing and the industrial research worker needs to be familiar with the strengths and weaknesses of both and to know in what circumstances each is more appropriate. There are many examples where the design of standardized interview schedules is inappropriate, because insufficient thought is given beforehand to why the information is being collected and the use to which it is to be put. Consequently, although some standardized surveys are technically very efficient, the findings may be of a limited value (Bloom 1988). On the other hand, the findings from some non-standardized interviews may be vague and anecdotal because of lack of prior preparation and/or insufficient skill in obtaining information during the interview. Aspects of both standardized and non-standardized interview techniques may be used within the same interview. Thus one section of the interview may ask a common set of factual questions of all respondents, while in another section an unstructured qualitative approach may be used to explore an aspect of the behaviour of the business (e.g. Rawlinson 1990). With both standardized and non-standardized interviewing it is useful to terminate the interview by asking whether it would be all right to contact the respondent again if further queries arise.

11.6 Conclusion

Business surveys have an essential role to play in economic research and policy development. The cut-backs in official surveys have increased the need for research workers to collect their own data. However, to be done properly business surveys take considerable time and effort and should only be contemplated after all other sources of information have been exhausted. There is also a danger of over-saturation if too many surveys are attempted (see also Ch. 17). With the high concentration of employment in large organizations, these businesses, in particular, may easily be overwhelmed by requests for information. In one recent survey, in which the author was involved, the business manager told the interviewer that this was the fourth survey he had been involved with in the last 10 days and during the interview someone else rang and tried to carry out an interview then and there over the telephone. There may then be a case for discouraging schoolchildren and undergraduates from undertaking questionnaire and interview surveys of businesses, particularly large ones, although this goes against the shift in emphasis in education to more project-based teaching in schools and enterprise initiatives in higher education.

There is also a case for making the data derived from surveys more widely available. The **ESRC Data Archive** acts as a repository for surveys, though relatively few local and regional business surveys have so far been deposited there, and confidentiality constraints usually prevent names and addresses being disclosed. Collaboration in the collection, analysis and dissemination of information is another possibility, particularly where there is overlap in business surveys. This has been attempted in Sheffield where six organizations (the Employment Department Training Agency (TA), Sheffield Chamber of Commerce, Sheffield City Council, Sheffield City Polytechnic, Sheffield Development Corporation and Sheffield University) have established an independent organization, the Employment and Training Consortium (ETIC), to co-ordinate the sharing of information. A common questionnaire is used in all interviews by consortium members with Sheffield companies. Yearly visits by consortium members to companies have been reduced from approximately 7350 to 2000 (Foley 1990; see also Ch. 1).

Maintaining good relations with the business community is a long-term strategy for encouraging their continued co-operation in survey work. Sending a short thank you letter following an interview is only polite, but is often omitted. Many research workers also undertake to send interviewees a summary of the results of their projects, but these promises are not always kept. The value of resurveying the same businesses at a later date (e.g. Munton *et al.* 1988; North *et al.* 1983) suggests that keeping good relations with the businesses surveyed can be in the research worker's own interest as well as for the wider good of the research community. Professional, well-designed and organized surveys which attempt, within the limits of the research topic, to make the experience as interesting and stimulating as possible for the participants, are also likely to make business managers and owners more favourably disposed to co-operating in future. This may be easier with the increased use of qualitative non-standardized interviews which allow interviewees to talk more freely.

There is no one 'best' way of obtaining information from businesses. This is not only because of the limited amount of literature available on the topic, but also because the survey methods suitable for different situations vary, dependent on a range of factors, including the research design, the kind and amount of information required, the resources available, and the size, organizational structure, sector and location of the businesses to be approached. Often a mixture of methods may be desirable within the same research project. Much of what has been written here is pragmatic and many of the observations have not been systematically tested. There is an urgent need for a greater priority to be given to experimenting with different survey methods and reporting on specific ways which have led to increased response rates, speeded up replies, obtained clearer answers, improved the flow of interviews and so on. Until then economic researchers will have to continue to rely on 'what seems to have worked before'. But given the variability of results between studies which have used similar survey methods this is not a very satisfactory way to proceed.

Acknowledgements

The author is very grateful to David Keeble and Phil Cooke for permission to reproduce the previously unpublished interview/questionnaire schedules in Appendices 11.3 and 11.6 respectively, and to John Bachtler, Peter Elias, Nick Evans, Dave Noon, Andy Pratt and Doug Watts for their constructive comments on an earlier draft of this chapter. The usual disclaimer, of course, applies.

Appendix 11.1 A contact record sheet for face-to-face meetings. (*Source:* Goddard 1973:156.)

Meeting record

1 How long did the meeting last?

1 ☐ 2 – 10 minutes
2 ☐ 10 – 30 minutes
3 ☐ 30 – 60 minutes
4 ☐ 1 – 2 hours
5 ☐ more than 2 hours

2 Was the meeting arranged in advance?

1 ☐ Not pre-arranged at all
2 ☐ Arranged on the same day
3 ☐ Arranged the day before
4 ☐ Arranged 2 – 7 days in advance
5 ☐ Arranged more than 1 week in advance

3 Who initiated the meeting?

1 ☐ Myself/another person in my firm
2 ☐ Any person outside the firm or any other organization

4 How many people, apart from you, were at the meeting?

1 ☐ One other person
2 ☐ 2 – 4 people
3 ☐ 5 – 10 people
4 ☐ over 10 people

If there was **only one** other person at the meeting:

5 What is the work address of that person?
....
....

6 What is the nature of business of his firm?
....
....

If there was **more than one** other person at the meeting please complete the details overleaf

7 How often on average do you have a meeting with this person or particular set of people?

1 ☐ Daily
2 ☐ About once a week
3 ☐ About once a month
4 ☐ Occasionally
5 ☐ First contact

8 What was the main purpose of the meeting?

1 ☐ To give an order or instruction
2 ☐ To receive an order or instruction
3 ☐ To give advice
4 ☐ To receive advice
5 ☐ For bargaining
6 ☐ To give information
7 ☐ To receive information
8 ☐ To exchange information
9 ☐ For general discussion
10 ☐ Other (please specify).....................
....

9 What was the range of subject matter discussed?

1 ☐ One specific subject
2 ☐ Several specific subjects
3 ☐ A wide range of general subjects

10 Was the meeting concerned with the purchase or sale of goods or services?

1 ☐ Directly concerned with purchases or sales
2 ☐ Indirectly concerned with purchases or sales
3 ☐ Not at all concerned with purchases or sales

If the meeting took place outside your place of work:

11 What is the address of the meeting place?
....
....

12 What was your principal method of transport from your office or previous meeting place?

1 ☐ Walk
2 ☐ Bus
3 ☐ Private car
4 ☐ Taxi
5 ☐ Underground
6 ☐ Train
7 ☐ Plane

13 How long did this journey take?

1 ☐ Less than 10 minutes
2 ☐ 10 – 30 minutes
3 ☐ 30 – 60 minutes
4 ☐ 1 – 2 hours
5 ☐ More than 2 hours

Appendix 11.2   A postal questionnaire examining skills among small and medium-sized employers (*Source:* Elias and Healey 1991: 147–53)

EMPLOYMENT IN COVENTRY

This questionnaire is concerned with employment at your establishment and is being carried out by the Institute for Employment Research on behalf of Coventry City Council. Your response to questions will be treated confidentially. You are assured that the information on this questionnaire will not be published in a form which could identify you or your business.

For several of the questions, you are asked to classify your employees into eight groups. Typical examples of the type of jobs which fall into each of these groups are listed below for the manufacturing and service sectors. If you have no employees in a particular category, please indicate by inserting zeros in the appropriate boxes in the questionnaire.

Category	Typical Manufacturing Sector Examples	Typical Service Sector Examples
Managers and Administrators	Branch manager, general manager, managing director, production manager, stores manager.	Director, general manager, local government officer, publican, shopkeeper, transport manager.
Professional Staff	Accountant, civil/design/production engineer, R&D scientist.	Accountant, business consultant, librarian, social worker, solicitor, teacher.
Technicians and Professional Support Staff	Computer analyst/programmer, draughtsperson, laboratory technician.	Graphic designer, nurse, youth worker, photographer.
Clerical and Secretarial Staff	Clerk, computer operator, office worker, receptionist, secretary, stores clerk.	Bank clerk, computer operator, office worker, receptionist, secretary.
Craft and Skilled Service Workers	Builder, fitter, machine setter-operator, printer, sewing machinist.	Baker, bar staff, butcher, chef, electrician, hairdresser, plumber, waiter, waitress.
Sales Staff	Buyer, credit agent, sales representative.	Buyer, check-out operator, merchandiser, sales assistant.
Operatives	Assembly worker, line worker, machine operator.	Bakery operative, bus driver, HGV/taxi driver.
Unskilled Workers	Factory hand, labourer, mate.	Cleaner, domestic, helper, mail sorter, shelf filler.

Many of the questions require a tick in the appropriate box. Such boxes are marked with a small tick beside them e.g. ☐ ✔. Sometimes we have asked for numbers. Such information should be written inside the box marked 🖎 ☐. If you are unable to be exact, please give a good estimate in such cases. Some questions ask you to write information on the questionnaire in the space marked with a dotted line.........................

If you have any queries regarding the completion of this questionnaire, you can contact the IER research team at the telephone numbers shown below.

Helen Abraham   ☎ 0203 523531 Peter Elias   ☎ 0203 523286

Thank you very much for your co-operation

Institute for Employment Research
Skills in Coventry: An Enquiry of Employers

1. Name(s) of establishment(s)

....

....

2. When did your business set up in Coventry?

pre 1960 ☐ ✔   1960-69 ☐ ✔   1970-79 ☐ ✔   1980-85 ☐ ✔   post 1985 ☐ ✔

3. (a) Is this firm an independent company   ☐ ✔

part of a group   ☐ ✔

(b) If this establishment is part of a group, what is the name of the parent group?

....

Is the registered head office of this group

Coventry based?   ☐ ✔

based elsewhere in the UK   ☐ ✔

based abroad?   ☐ ✔

4. (a) What is (are) the principal product(s) and/or service(s) that your establishment provides?

....

....

....

(b) What is (are) the principal function(s) undertaken by your establishment?

Head Office Functions ☐ ✔	Manufacturing/ Production ☐ ✔	Warehouse/Distribution Repair & Maintenance ☐ ✔
Research & Development ☐ ✔	Sales, Marketing and Retailing ☐ ✔	Other Services ☐ ✔

5. (a) How many people did your establishment employ in October 1989? ✎ ☐

(b) Approximately how many people did your establishment employ in October 1988? ✎ ☐

IF YOU DO NOT EMPLOY ANYBODY BUT YOURSELF, PLEASE GO TO QUESTION 18

RECRUITMENT IN THIS ESTABLISHMENT

For each of the employment categories outlined in the table on the opposite page, please supply the following information:

6. How many people are currently employed by your establishment in these categories?

7. How many of these people are:

    a) Male full-time employees

    b) Male part-time (less than 30 hours) employees

    c) Female full-time employees

    d) Female part-time (less than 30 hours) employees

8. How many of these people are of Asian or Afro-Caribbean ethnic origin?

9. How many people were recruited during the last 12 months in each category?

10. How many of these recruits joined your workforce direct from full-time education?

11. How many vacancies do you currently have in each category?

12. Please write in the title(s) of current vacancies which are proving difficult to fill.

INSTRUCTIONS FOR COMPLETION OF THIS TABLE

Please fill in the appropriate boxes for the job categories relevant to your establishment.

Managers/ Admins	Prof. Staff	Technicians etc.	Clerical/ Secretarial	Craft/ Skilled	Sales Staff	Oper- atives	Unskilled Workers	Q
▱	▱	▱	▱	▱	▱	▱	▱	6
▱	▱	▱	▱	▱	▱	▱	▱	7a
▱	▱	▱	▱	▱	▱	▱	▱	7b
▱	▱	▱	▱	▱	▱	▱	▱	7c
▱	▱	▱	▱	▱	▱	▱	▱	7d
▱	▱	▱	▱	▱	▱	▱	▱	8
▱	▱	▱	▱	▱	▱	▱	▱	9
▱	▱	▱	▱	▱	▱	▱	▱	10
▱	▱	▱	▱	▱	▱	▱	▱	11

Managers and Administrators	....	Craft & Skilled Service Workers	....
Professional Staff	....	Sales Staff	....
Technicians etc.	....	Operatives	....
Clerical and Secretarial	....	Unskilled Workers	....

13. Thinking about the next three years, do you expect that you will be employing fewer, about the same or more employees?

Fewer ☐ ✔   About the same ☐ ✔   More ☐ ✔

(a) Jobs where you expect to employ *fewer* people. Please be specific about the titles of jobs most affected:

....
....
....

(b) Jobs where you expect to employ *more* people. Please be specific about the titles of jobs most affected:

....
....
....

14. Again, thinking about the next three years, do you think that:

recruitment will become more difficult ☐ ✔   recruitment will become less difficult ☐ ✔   recruitment difficulties will stay the same ☐ ✔   recruitment will present no difficulty ☐ ✔

If you think you will experience recruitment difficulties will you consider:

	Yes	No	
Recruiting more mature workers (over 45 years old)?	☐	☐	✔
Recruiting people with lower qualifications?	☐	☐	✔
Increasing wage levels?	☐	☐	✔
Improving working conditions?	☐	☐	✔
Increase training of existing staff?	☐	☐	✔
Increasing overtime?	☐	☐	✔
Increasing use of part-timers?	☐	☐	✔
Increasing use of temporary workers?	☐	☐	✔
Sub-contracting to outside organisations?	☐	☐	✔

Taking any other steps? (PLEASE SPECIFY)

....
....
....

TRAINING IN THIS ESTABLISHMENT

15. Which of the following types of training do employees normally receive?

Tick all that apply

Guidance from colleagues when needed ☐ ✔

Planned on-the-job training by superiors ☐ ✔

Full-time courses lasting up to 1 week ☐ ✔

Full-time courses lasting over 1 week ☐ ✔

Day release courses ☐ ✔

Other part-time courses ☐ ✔

Other (PLEASE SPECIFY) ☐ ✔

....

....

....

16. (a) In the past 12 months, have any outside organisations provided training for you?

Yes ☐ ✔   No ☐ ✔

(b) If yes, please indicate by whom this outside training was provided.

Tick all that apply

Skill Centre? ☐ ✔

Consultancy firms offering training services? ☐ ✔

Courses run at colleges of further education? ☐ ✔

Polytechnic/University short courses? ☐ ✔

Training Board? ☐ ✔

Other (PLEASE SPECIFY) ☐ ✔

....

....

....

(c) Do you have any comments on the training provided by outside training organisations?

....

....

....

....

17. Do you have any particular training requirements that are not being met?

   ....
   ....
   ....
   ....

18. Do you anticipate requiring additional land/premises during the next three years?

   Yes   ☐ ✔

   No   ☐ ✔

Finally, if you have any other comments regarding recruitment and training issues, please use the space below, or attach an additional sheet.

   ....
   ....
   ....
   ....
   ....
   ....

Thank you very much for your valuable assistance. Please return the completed questionnaire as soon as possible to the Institute for Employment Research in the FREEPOST envelope provided.

Name ....

Date ... / ... /19 ..

Position in establishment ....

Telephone Number ....

Appendix 11.3   A postal questionnaire examining high-technology industry (*Source:* Used in survey reported in Keeble and Kelly 1986.)

CAMBRIDGE UNIVERSITY SURVEY OF HIGH-TECHNOLOGY INDUSTRY

Name of firm: **Please return to:**   Mr. T. Kelly
Address: Dept. of Geography,
                                                      Downing Place,
                                                      Cambridge CB2 3EN

Name and position of respondent   ....
....

1/   Date of formation
   i/   In what year was the firm registered, or establishment opened?   ....
   ii/ In what year was the firm first involved in computer-related activities?   ....

2/   Legal status
   i/   Is this establishment: Please tick
      1/   An independent single-plant firm?
      2/   An independent headquarters of a multi-plant firm?
      3/   A branch/subsidiary set up directly by parent firm?
      4/   A branch/subsidiary acquired by parent firm through takeover or merger?
      5/   Other (please specify) ....

   ii/ If branch or subsidiary, please state location of headquarters of parent firm
       ....

   iii/ If taken over, please state year of acquisition by present owners   ....

3/   Employment
   i/   How many people do you currently employ at this site? ....
        Of which: a/ Male   ....   c/ Part-time   ....
                       b/ Female   ....   d/ Full-time   ....

   ii/ What number (or approximate percentage) of your employees at this site are:
      1/   Management, sales and clerical staff?
      2/   Research and development staff?
      3/   Skilled employees?
      4/   Semi-skilled and unskilled employees?
      5/   Other (please specify) ....

   iii/ What was the approximate employment at this site: (at mid year)
        a/ In 1975?   ....   b/ In 1980?   ....

4/   Product Range
   Roughly what percentage of your establishment's turnover can be classified   %
   to the following categories? (to the nearest 10%).
      1/   Production/assembly of microcomputers and workstations
           (less than £10,000 per unit)
      2/   Production/assembly of mini-computers (£10,000 – £100,000 per unit)
      3/   Production/assembly of 'super-mini' or mainframe computers
      4/   Production of computer hardware peripherals including terminals,
           VDUs, printers, disc drives, data communications equipment, etc.
      5/   Production of computer accessories including discs, tapes, paper, etc.
      6/   Distribution and servicing of computer products from other firms
      7/   Provision of computer software for general use
      8/   Provision of computer software for use by specific clients
      9/   Provision of computer facilities/trained personnel
      10/ Provision of educational or other computer consultancy services
      11/ Other computer-related business (please specify) ....
      ....
      12/ Non-computer-related business

Please turn over

9/ Entrepreneurship

i/ How many founders established the firm?

ii/ How old were the founders when the company was formed? (Please tick boxes)

	Founder 1	Founder 2	Founder 3
a/ Less than 25 years old			
b/ 25–34 years old			
c/ 35–45 years old			
d/ More than 45 years old			

iii/ Did the founders possess technical qualifications?

a/ Postgraduate degree			
b/ Degree in sciences/computing			
c/ Other degree			
d/ Other technical qualifications			

iv/ What was the founders' previous employment?

a/ Academic study/research			
b/ Computing firm			
c/ Other electronics firm			
d/ Other type of firm (please specify)			

v/ Where did the founders live and work prior to setting up the firm?
   (please list town or locality)

a/ Place of residence			
b/ Place of work			

10/ Investment

i/ In seeking start-up capital, what sources of funds were important?

	Major	Minor
1/ Personal savings or previous assets		
2/ Bank loan/second mortgage		
3/ Venture capital from private company		
4/ Venture capital from a public source*		
5/ Other (please specify)		

ii/ In subsequent expansion, what sources of capital have proved important?

1/ Personal savings or retained earnings		
2/ Bank loan/second mortgage		
3/ Venture capital from a private company		
4/ Venture capital from a public source*		
5/ Equity capital/share issue		
6/ Other (please specify)		

* Please state particular source and nature of public sector finance:

....
....
....

Thank you very much for your co-operation.

Appendix 11.4 An introductory letter requesting an interview. (*Source:* Used in the survey reported in Elias and Healey 1991.)

Director of Centre
Clive M Collis
BA MA

Coventry Polytechnic
Priory Street Coventry CV1 5FB
Telephone 0203 631313
Telex 9312102228 (CP G)
Fax 0203 258597

Centre for
Local Economic Development
Direct Line 0203 838463/838605

Our ref

Your ref

Date

Dear

EMPLOYMENT SKILLS AND TRAINING IN COVENTRY

The Centre for Local Economic Development at Coventry Polytechnic is conducting a survey of employers in the City on behalf of the Coventry City Council (Department of Economic Development and Planning). We are interested in collecting information and views on the employment structure of your organization, potential skill shortages and training needs. This is vital information for the local provision of education and training in the 1990s. It will also help in the planning of essential services and in the long-term development of industry and commerce in the City.

I am writing to request a short interview with you, or one of your colleagues, to discuss these matters. The interview, which should take only about 50 minutes, will be concerned with employment at all the Coventry establishments for which you/your department is mainly responsible for personnel matters. Your response to the questions will be treated in the strictest confidence. You are assured that the information you provide will not be published in a form which would identify you or your business.

I hope that you will agree to participate in this survey because it is very important that we obtain an accurate picture for all the major employers in the City. In anticipation of your co-operation I shall telephone you in the next few days to arrange a time when I can visit you.

Yours sincerely

COVENTRY
POLYTECHNIC

Appendix 11.5   An interview schedule examining farm diversification in the urban fringe. (*Source:* Ilbery 1987a.)

Agricultural Diversification in the Rural–Urban Fringe of Birmingham

1. Farm number   3. Grid reference

2. Name   4. Total holding size (acres)

   Address

Section A: Farm and farmer characteristics

1. Is your total holding in one block?   Yes/No
   If Yes, is it divided by any major roads?   Yes/No
   If No, how many blocks/separate units does your total holding comprise?
   (including farmstead) ....
   Please state location, distance from farmstead, size, ownership and
   land use of each block/separate unit:

Block location	Distance	Size	Ownership	Land use
1. Main farmstead	–			
2.				
3.				
4.				
5.				
6.				

2. Has the size of your farm tended to increase/decrease or remain unchanged over time?

   If increase or decrease, please state the following:

Location	Size	Method of transfer	Date	Present use

3. What crops/livestock are on your farm? Please state whether area/number increasing, decreasing or unchanged. Place enterprises in rank order according to profit.

Rank	Crops/livestock	Acres/numbers	Increase/decrease/unchanged
1.			
2.			
3.			
4.			
5.			
6.			
7.			
8.			

4. Please outline main changes to the farm system over the past 10/12 years and the reason(s) for these changes.

Description	Reason(s) for change(s)

5. How long have you personally been in charge of this farm?
   ....

6. How long have you lived in this area? ....

7. Was/is your father a farmer?   Yes/No
   If Yes, was it on this farm?   Yes/No

8. Age–group: <35 ☐   35–44 ☐   44–55 ☐   55–65 ☐
                 >65 ☐

9. Education: Secondary ☐   Grammar ☐   Public ☐   College ☐
                 Others ....

10. Practical training: None ☐   Father/family farm ☐

                           Farm pupil/apprentice ☐   Day–release ☐

                           College ☐   Others ....

11. Status: (i) Owner occupier ☐   (ii) Tenant ☐

           (iii) Combination of (i) and (ii) ☐

           Please state acreage owned ....

                                   rented ....

           (iv) Lease ☐   (v) Manager ☐

           (vi) Others ....

12. If land not owned please state who is owner, and length of let. (Probe – where is owner, does he own other land in area, occupation of owner.)

13. Is the farm your only source of income?   Yes/No
    If No, what is/are the other sources?
    (If manager ask about owner)

14. Labour force:

		1986	**10/12 years ago**
(i) Total labour force	F/T	......	....
	P/T	......	....
	Casual	......	....
(ii) Number of family workers	F/T	......	....
	P/T	......	....
(iii) Number of hired workers	F/T	......	....
	P/T	......	....
	Casual	......	....

15. Please can you explain the reasons for these changes over the past 10/12 years?

Section B: Diversification

1. Do you have any of the following activities on your farm?

    Pick your own ☐ (main crops?)

    Farm gate sales ☐ (main crops?)

    Farm shop ☐

    Farm catering ☐

    Bed and breakfast ☐

    Holiday accommodation ☐

    Camping/caravan sites ☐

    Horse riding/stables ☐

    Water sports (fishing, etc.) ☐

    Shooting ☐

    Informal recreation (car parks, picnic sites, etc.) ☐

    Special attractions (museums, sports, zoos, etc.) ☐

    Indoor events (dances, clubs, etc.) ☐

    One-day/occasional events ☐

    Educational visits ☐

    Others (please specify) ....

2. When did you start these activities? ....

3. Have there been any changes in your choice or scale of these activities?   Yes/No. Please specify.

4. What proportion of your total income is derived from these activities?
    (Please try to specify for each activity.)

5. Did you seek any advice before establishing these activities on your farm?   Yes/No.
    If Yes, what sources?

237

6. Please select the **three** statements, in rank order, which most closely explain your decision to participate in these activities.

Distance of the farm from Birmingham/Coventry ☐

To make more productive use of land/buildings ☐

The proximity of the farm to major roads ☐

To make more productive use of labour ☐

To increase profit margins on enterprises already established
  on the farm ☐

To generate income from new farming activities ☐

To generate income from non-traditional agricultural activities ☐

To expand your own hobbies/interests ☐

To take advantage of grants/loans from official organizations ☐

Other reasons (please state)...
....

7. Do you envisage an expansion or reduction of any of these activities?

Yes/No.

If Yes, in what way and why?

Section C: Attitudes to urban pressures

1. Can you identify any urban pressures on farming in this area?   Yes/No
  If Yes, please specify (positive and negative).

2. Has/is your farm (been) affected by any of the following, and if so how?

    Trespass ☐

    Dumping ☐

    Vandalism and theft ☐

    Threat of development ☐

    Labour problems ☐

    Prevention of farm improvements ☐

    Traffic problems ☐

    Pollution ☐

    Land problems ☐

    Security of tenure ☐

    Compulsory purchase ☐

    Others ....

3. Do you consider that urban pressures have influenced:

    (a) your farm system Yes/No

    If Yes, in what ways?

    (b) your farm-based activities Yes/No

    If Yes, in what ways?

4. What do you consider to be the relative importance of urban and non-urban forces on your farm?

5. Do you feel that present land use controls (i.e. Green Belts, etc.) are sufficient to secure the future of farming in this area? Yes/No

    If No, why not and what controls would you prefer?

Section D: Goals and values

1. How important are the following farming attributes to you?

		Very Important	Important	Not very Important	Irrelevant
(i)	Meeting a challenge				
(ii)	Being creative				
(iii)	Belonging to the farming community				
(iv)	Pride of ownership				
(v)	Doing the work you like				
(vi)	Independence				
(vii)	Making maximum income				
(viii)	Working close to family and home				
(ix)	Self–respect for doing a worthwhile job				
(x)	Being able to arrange hours of work				
(xi)	Continuing the family tradition				
(xii)	Expanding the business				
(xiii)	Leading healthy outdoor life				
(xiv)	Safeguarding income for future				
(xv)	Gaining recognition as a farmer				.
(xvi)	Earning respect of workers				
(xvii)	Making satisfactory income				
(xviii)	Exercising special abilities				
(xix)	Purposeful activity				
(xx)	Control in a variety of situations				

Out of the twenty which are the **three** most important, in rank order?

1.   2.   3.

2. Farming in close proximity to urban areas is said to be a hard, frustrating and time-consuming occupation; can you please tell me why you continue with it?

3. Are there any other comments you would like to make about farming in the rural–urban fringe of Birmingham, and/or farm-based non-agricultural activities?

Thank you for your co-operation.

Appendix 11.6   Draft themes for interviewing business managers (*Source:* Used in planning the research programme reported in Cooke 1989)

ESRC CHANGING URBAN AND REGIONAL SYSTEM RESEARCH PROGRAMME

DRAFT INTERVIEW THEMES

BUSINESS MANAGER INTERVIEWS

1   CORPORATE STRUCTURE

1.1 STATUS OF ORGANIZATION

    (a) HQ/branch/affiliate/subsidiary/independent
    (b) Functions of local office/plant(s) (processing, production, design, etc.)
    (c) Autonomy of local office/plant(s) (purchasing, product development, investment, etc.)

1.2 SPATIAL STRUCTURE

    (a) Duration at existing location (reasons)
    (b) Proportion of total corporate employment/output
    (c) Performance of office/plant(s) compared with other:
        (i) parts of corporate organization
        (ii) competitor offices/plants (UK/overseas owned)

1.3 PRODUCT SPECIALISM(S)

    (a) Nature of product/service provided in local office/plant(s)
    (b) Market for product/service (local, regional, EC, etc.)
    (c) Product/service subcontracting relations
        (i) stock-holding (high or low inventories)
        (ii) buying in (e.g. components, R&D services)
        (iii) putting out (parts of production, processing or selling activity; franchising)

1.4 RECENT CHANGES

    (a) Changes in status/function/autonomy of office/plant(s)
    (b) Substantive changes in intra-corporate spatial relations
    (c) Substantive changes in extra-corporate market relations

2 **WORKPLACE CONTROL CHANGES**

2.1 OFFICE/SHOPFLOOR CONTROL

(a) Increased use of information technology (e.g. keyboard or other central monitoring)
(b) Direct control (e.g. reducing chains of command, quality circles)
(c) Flexibility
    (i) flexible manufacturing systems
       (deskilling/reskilling, e.g. CAD/CAM)
    (ii) flexible working practices (deskilling/reskilling)

2.2 EMPLOYEE RELATIONS

(a) Trade union presence/strength/change in role
(b) Staff association(s) growth/decline
(c) Provision of fringe benefits – type – to whom (e.g. health; share schemes)
(d) Collective bargaining
    (i) national versus plant-based
    (ii) representation (e.g. multi-or single-union)
    (iii) conditions (e.g. binding arbitration; no-strike)
    (iv) payment methods (piecework/measured day, salaried, etc.)

2.3 EMPLOYMENT

(a) Workforce –Nos decreasing/increasing
    Job loss   – (i) redundancy (type)
               (ii) freezing posts
               (iii) retirement
               (iv) other
    Job gain   – see section 4 Recruitment

(b) Occupational categories – changing ratios *(in situ* and intra-corporate)
    (i) professionals, scientists, etc
    (ii) managerial and administrative
    (iii) technicians and supervisors
    (iv) craft/qualified staff
    (v) operator grades

(c) Social composition of workforce
    (i) occupation
    (ii) gender   recent changes
    (iii) age   in proportion
    (iv) ethnicity

(d) Income – changes in relationship between income bands
    (i) by occupation
    (ii) by gender
    (iii) other

(e) Full-time/part-time employment
    (i) by occupation
    (ii) by gender
    (iii) changes in shift-length
    (iv) temporary, casualized, short-term contracts

3   INVESTMENT

3.1   FORMS OF CAPITAL INVESTMENT/DISINVESTMENT

    (a) Sources of investment finance
    (b) Type of local investment/disinvestment
        (i) labour-saving technology (nature and type)
        (ii) product/process-enhancing technology (nature and type)
        (iii) workforce control technology (e.g. central monitoring)
        (iv) buildings, etc.
        (v) plant closure (complete or partial)
        (vi) scrapping machinery
        (vii) workforce redundancies
        (viii) workforce training (apprenticeships versus MSC schemes)
        (ix) marketing, distribution, design, etc.
    (c) Time-perspective of investment strategy
    (d) Importance of geographical location to present and future investment strategy

3.2   IMPACT OF FOREIGN COMPETITION ON INVESTMENT

    (a) Changes in product/service lines/functions (e.g. type, quality)
    (b) Relocation or decentralization to other countries
    (c) Establishment of overseas plant(s)/office(s) in locality
    (d) Exporting of parts of production/service functions
    (e) Importing of parts of production/service functions
    (f) Concentration of domestic production/service activity in specific locality(ies)
    (g) Increased/decreased investment in R&D, design, etc.
        (i) in-house
        (ii) bought-in
    (h) Increased/decreased investment in marketing/sales, etc.
        (i) in-house
        (ii) bought-in
    (i) Increased/decreased investment in distribution; e.g.
        (i) automated warehousing
        (ii) telecommunications

4   RECRUITMENT

4.1   RECRUITMENT PROCESSES

(a) Methods of recruitment
  (i) agencies (including jobcentres, etc.)
  (ii) marketing (adverts; 'milk-rounds', etc.)
  (iii) informal (local community network)

(b) Local sourcing
  (i) numbers (tight or slack labour market; turnover rates; local wage pressure, etc.)
  (ii) skill/qualification type (demand and supply)
  (iii) changing requirements

(c) Training
  (i) in-house (grades, time taken, etc.)
  (ii) apprenticeships
  (iii) external training (MSC; private; school; higher education)
  (iv) elsewhere (e.g. other firms with training schemes)

4.2   RECRUITMENT TARGETING

(a) Gender
  (i) occupational specificity
  (ii) income levels
  (iii) unionization
  (iv) recent changes

(b) Ethnicity
  (i) occupational specificity
  (ii) positive discrimination
  (iii) recent changes

(c) Age
  (i) positive discrimination
  (ii) income levels
  (iii) recent changes

4.3 PROMOTION

(a) Channels (career structure; internal labour market versus external)
(b) Job evaluation
(c) Unionization (seniority, membership versus non-membership)
(d) Professionalism (also staff associations, etc.)

5   GENERAL THEMES

5.1 WORK–COMMUNITY RELATIONS

   (a) Changes in links between plant/office and local community
   (b) Parts of locality differentially tied to parts of plant/office
   (c) Plant/office and community identity (e.g. 'corporate culture')
   (d) Importance of community resources/amenities (including housing) to employer
   (e) Importance of specific employer to community

5.2 LOCAL AND NATIONAL GOVERNMENT

   (a) Local planning authority (development policy; land use, etc.)
   (b) Local aid packages (grants, rent-free, training grants, etc.)
   (c) Local costs (e.g. rates)
   (d) Agency assistance (e.g. BSC (Industry) ECSC; EC)
   (e) National government contracts (e.g. Defence, Energy, Health Service)
   (f) Exchange rate
   (g) Interest rates
   (h) Government assistance (e.g. regional aid, DTI programmes – microelectronics, biotechnology, robotics)

5.3 THE FUTURE

   (a) Product/service diversification
   (b) Ownership changes
   (c) Workforce size/character/recruitment
   (d) Occupational structure
   (e) Workforce organization (unions, staff associations)
   (f) Organizational control
   (g) New technology
   (h) Competition
   (i) Investment

References

Alexander J 1979 *Office Location and Public Policy* Longman, London

Aubrey J, Clarke G, Stillwell J 1989 *The Use of Establishment Data to Examine Employment Structure: Two Studies in West Yorkshire* Working Paper 529, School of Geography, University of Leeds

Bachtler J 1981a *A Survey of the Banking, Insurance and Finance Sector: II The Insurance Sector* Department of Geography, University of Birmingham

Bachtler J 1981b *A Survey of the Banking, Insurance and Finance Sector: I The Insurance Sector* Department of Geography, University of Birmingham

Bachtler J 1984 Wherefore art thou, questionnaire? *Frontier* (The Magazine of the Department of Geography, University of Birmingham) **3**: 105–13

Belson W, Duncan J A 1962 A comparison of the check-list and the open response/questioning systems. *Applied Statistics* **11**: 120–32

Bloom N 1988 The limitations of unqualitative research. *Industrial Marketing Digest* **13**: 49–56

Bowler I R 1982 Direct marketing in agriculture: a British example. *Tijdschrift voor Economische en Sociale Geografie* **73**: 22–31

Brannen P 1987 Working on directors: some methodological issues. In Moyser G, Wagstaffe M (eds) *Research Methods for Elite Studies* Allen & Unwin, London pp 166–80

Brenner M 1978 Interviewing: the social phenomenology of a research instrument. In Brenner M, Marsh P, Brenner M (eds) *The Social Contexts of Method* Croom Helm, London, pp. 122–39

Clark G, Gordon D S 1980 Sampling for farm studies in geography. *Geography* **65**: 101–6

Cooke P (ed.) 1989 *Localities: The Changing Face of Urban Britain* Unwin Hyman, London

Cooper H 1989 *The West Midlands Labour Market: A Study of Employers and the Unemployed in the West Midlands Conurbation* Employment Department, HMSO, London

Cooper M J M 1975 *The Industrial Location Decision Making Process* Occasional Paper No 34, Centre for Urban and Regional Studies, University of Birmingham

Curran J 1986 *Bolton Fifteen Years on: A Review and Analysis of Small Business Research in Britain 1971–1986* Small Business Research Trust, London

Dean J P, Whyte W F 1958 How do you know if the informant is telling the truth? *Human Organization* **17**: 34–8

de Chernatony L 1988 *Getting the Most from Postal Research* Occasional Paper No 3, Industrial Marketing Research Association, Litchfield

DE 1989 1987 Census of Employment: results for the United Kingdom. *Employment Gazette* Oct: 540–59

Dexter L A 1970 *Elite and Specialized Interviewing* Northwestern University Press, Evanston

Diamond D, Spence N 1989 *Infrastructure and Industrial Costs in British Industry* HMSO, London

Dixon C J, Leach B 1978a *Sampling Methods for Geographical Research* Concepts and Techniques in Modern Geography 17, Geo Abstracts, Norwich

Dixon C J, Leach B 1978b *Questionnaires and Interviews in Geographical Research* Concepts and Techniques in Modern Geography 18, Geo Abstracts, Norwich

DTI 1973 Memorandum on the inquiry into location attitudes and experience. In Expenditure Committee (Trade and Industry Sub-Committee) *Regional Development Incentives, Session 1973–74, Minutes of Evidence (from July 1973), Appendices and Index* The House of Commons, HMSO, London pp. 525–668

Elias P, Healey M 1991 *People and Work in Coventry: A Survey of Employers* Coventry City Council, Coventry

Elias P, Rigg M 1990 *The Demand for Graduates* Policy Studies Institute, London

Evans N 1990 Farm-Based Accommodation and the Restructuring of Agriculture in England and Wales unpublished PhD thesis, Coventry Polytechnic

Foley P 1983 A comparison of four data sources. In Healey M J (ed.) *Urban and Regional Industrial Research: The Changing UK Data Base* Geo Books, Norwich pp. 91–104

Foley P D 1990 Growth in the collection of labour market information. *Regional Studies* **24**: 367–71

Forsythe J B 1977 Obtaining cooperation in a survey of business executives. *Journal of Marketing Research* **XIV**: 370–3

Fothergill S, Gudgin G 1982 *Unequal Growth: Urban and Regional Employment Change in the UK* Heinemann Educational, London

Fothergill S, Gudgin G 1985 Ideology and methods in industrial location research. In Massey D, Meegan R (eds) *Politics and Method: Contrasting Studies in Industrial Geography* Methuen, London pp. 27–50

Fothergill S, Kitson M, Monk S 1983 *The Industrial Building Stock and its Influence on the Location of Employment Change* Working Paper No 5, Industrial Location Research Project, Department of Land Economy, University of Cambridge

Fowler F J, Mangione T W 1990 *Standardized Survey Interviewing: Minimizing Interview-related Error* Applied Social Research Methods Series 18, Sage, London

Gardner G 1978 *Social Surveys for Social Planners* Open University, Milton Keynes

Gasson R 1973 Goals and values of farmers. *Journal of Agricultural Economics* **24**: 521–42

Goddard J 1973 Office linkages and location: a study of communication and spatial patterns in Central London. *Progress in Planning* **1**: 109–232

Goddard J B, Morris D 1976 The communications factor in office decentralization. *Progress in Planning* **6**: 1–80

Gorden R L 1975 *Interviewing: Strategy, Techniques and Tactics* The Dorsey Press, Homewood, Illinois

Gould A, Keeble D 1984 New firms and rural industrialization in East Anglia. *Regional Studies* **18**: 189–201

Government Statisticians' Collective 1979 How official statistics are produced: views from the inside. In Irvine J, Miles I, Evans J (eds) *Demystifying Social Statistics* Pluto Press, London pp 130–51

Griffiths A 1988 *Postal Survey of Farm Diversification in Devon: Summary of Findings* Department of Geography, University of Exeter

Gripaios P, Bishop P, Gripaios R, Herbert C 1989 High technology industry in a peripheral area: the case of Plymouth. *Regional Studies* **23**: 151–7

Gudgin G 1976 Establishment based data in studies of employment growth and location. In Swales J K (ed.) *Establishment Based Research: Conference Proceedings* Discussion Paper 22, Centre for Urban and Regional Research, University of Glasgow

Gudgin G 1978 *Industrial Location Processes and Regional Employment Growth* Saxon House, Farnborough

Hague P 1987a *The Industrial Market Research Handbook* Kogan Page, London

Hague P 1987b Good and bad in questionnaire design. *Industrial Marketing Digest* **12**: 161–70

Hardill I, Taylor J, Williams H, Wynarczyk P 1989 *PICT 1989 Computer Networking Survey: Methods and Questionnaire* Working Paper No 8, Programme on Information and Communication Technologies, Centre for Urban and Regional Development, University of Newcastle upon Tyne

Harvey L 1987 Factors affecting response rates to mailed questionnaires: a comprehensive literature review. *Journal of the Market Research Society* **29**: 341–53

Healey M J 1979 Changes in the location of production in multi-plant enterprises, with particular reference to the United Kingdom textile and clothing industries, 1967–72. Unpublished PhD thesis, University of Sheffield

Healey M J 1983a The changing data base: an overview. In Healey M J (ed) *Urban and Regional Industrial Research: The Changing UK Data Base* Geo Books, Norwich pp 1–29

Healey M J 1983b Components of locational change in multi-plant enterprises. *Urban Studies* **20**: 327–41

Healey M J 1984 *The Coventry Region Industrial Establishment Databank I: A Guide to Sources, Methods and Definitions used for the City of Coventry* Industrial Location Working Paper 2, Department of Geography, Coventry Polytechnic

Healey M J 1985 Industrial decline, industrial structure and large companies. *Geography* **70**: 328–38

Healey M J 1986 Collecting data from industrial firms: directories, interviews and questionnaires. Paper presented to Institute of British Geographers research workshop on industrial geography and area development: theory, method and data, London, Oct

Healey M J, Clark D 1985 Industrial decline in a local economy: the case of Coventry, 1974–1982. *Environment and Planning A* **17**: 1351–67

Healey M J, Clark D, Shrivastava V 1987 *The Clothing Industry in Coventry* Industrial Location Working Paper No 10, Department of Geography, Coventry Polytechnic

Hitchens D M W N, O'Farrell P N 1987 The comparative performance of small manufacturing firms in Northern Ireland and South East England. *Regional Studies* **21**: 543–53

Hoare A G 1978 Industrial linkages and the dual economy: the case of Northern Ireland. *Regional Studies* **12**: 167–80

Hoinville G, Jewell R and Associates 1978 *Survey Research Practice* Heinemann Educational Books, London

Ilbery B W 1985a *Agricultural Geography* Oxford University Press, Oxford

Ilbery B W 1985b Factors affecting the structure of viticulture in England and Wales. *Area* **17**: 147–54

Ilbery B W 1985c Factors affecting the structure of horticulture in the Vale of Evesham: a behavioural approach. *Journal of Rural Studies* **1**: 121–33

Ilbery B W 1987a *Farm Diversification in the West Midlands Urban Fringe: Main Report* Agricultural Change Unit Report, Department of Geography, Coventry Polytechnic

Ilbery B W 1987b The development of farm diversification in the UK: evidence from Birmingham's urban fringe. *Journal of the Royal Agricultural Society of England* **148**: 21–35

Jackson J 1963 *Surveys for Town and Country Planning* Hutchinson, London

Jackson P 1983 Principles and problems of participant observation. *Geografiska Annaler B* **65**: 39–46

Jackson P 1988 The case for case studies. Paper presented to ESRC Research Training Workshop on putting research into practice: concepts, values and empirical research, Loughborough

Keeble D, Kelly T 1986 New firms and high technology industry: the case of computer electronics. In Keeble D, Wever E (eds) *New Firms and Regional Developments in Europe* Croom Helm, Beckenham pp 75–104

Kincaid H V, Bright M 1957a Interviewing the business elite. *American Journal of Sociology* **63**: 304–11

Kincaid H V, Bright M 1957b The tandem interviews: a trial of the two-interviewer team. *Public Opinion Quarterly* **XXI**: 304–12

Leigh R, North D J 1978 Regional aspects of acquisition activity in British manufacturing industry. *Regional Studies* **12**: 227–45

Lever W F 1974 Manufacturing linkages and the search for suppliers and markets. In Hamilton F E I (ed.) *Spatial Perspectives on Industrial Organization and Decision Making* Wiley, London pp. 309–33

Lever W F 1982 Urban scale as a determinant of employment growth or decline. In Collins L (ed.) *Industrial Decline and Regeneration: Proceedings of the 1981 Anglo-Canadian Symposium* Department of Geography, Edinburgh and the Centre of Canadian Studies, University of Edinburgh pp 109–25

Leyshon A 1983 *Saturation Surveying Techniques* Working Note 2, Small Business Research Trust, London

Lloyd P E, Dicken P 1982 *Industrial Change: Local Manufacturing Firms in Manchester and Merseyside* Inner Cities Directorate, Inner Cities Research Programme 6, DoE

MacFarlane-Smith J 1972 *Interviewing in Market and Social Research* Routledge & Kegan Paul, London

Marsh C 1979 Problems with surveys: method or epistemology? *Sociology* **13**: 293–305

Marsh C 1982 *The Survey Method: The Contribution of Surveys to Sociological Explanation* Allen & Unwin, London

Marshall J N 1983 Business-service activities in British provincial conurbations. *Environment and Planning A* **15**: 1343–59

Marshall J N 1989 Corporate reorganization and the geography of services: evidence from the motor vehicle aftermarket in the West Midlands region of the UK. *Regional Studies* **23**: 139–50

Mason C M 1982 *New Manufacturing Firms in South Hampshire: Survey Results* Discussion Paper No 13, Department of Geography, University of Southampton

Mason C M 1989 Explaining recent trends in new firm formation in the UK: some evidence from South Hampshire. *Regional Studies* **23**: 331–46

Massey D B, Meegan R A 1979 The geography of industrial reorganisation: the spatial effects of the restructuring of the electrical engineering sector under the Industrial Reorganization Corporation. *Progress in Planning* **10**: 155–237

Massey D, Meegan R (eds) 1985 *Politics and Method: Contrasting Studies in Industrial Geography* Methuen, London

Mitchell J C 1983 Case and situational analysis. *Sociological Review* **31**: 187–211

Moseley M J 1973 The impact of growth centres in rural regions – II an analysis of spatial 'flows' in East Anglia. *Regional Studies* **7**: 77–94

Moser C A, Kalton G 1971 *Survey Methods in Social Investigation* Heinemann, London

Moyser G, Wagstaff M 1985 *The Methodology of Elite Interviewing* Report to the ESRC, London, Contract No H00250003

Moyser G, Wagstaffe M (eds) 1987 *Research Methods for Elite Studies* Allen & Unwin, London

Munton R 1983 *London's Green Belt: Containment in Practice* Allen & Unwin, London

Munton R, Whatmore S, Marsden T 1988 Reconsidering urban-fringe agriculture: a longitudinal analysis of capital restructuring on farms in the Metropolitan Green Belt. *Transactions of the Institute of British Geographers New Series* **13**: 324–36

North D J, Leigh R, Gough J 1983 Monitoring industrial change at the local level: some comments on methods and data sources. In Healey M J (ed.) *Urban and Regional Industrial Research: The Changing UK Data Base* Geo Books, Norwich pp 111–29

Northcott J, Rogers P 1982 *Microelectronics in Industry: What's Happening in Britain* Research Paper 603, Policy Studies Institute, London

O'Farrell P N, Hitchens D W N 1988 Inter-firm comparisons in industrial research: the utility of a matched paris design. *Tijdschrift voor Economique en Sociale Geografie* **79**: 63–9

Peck F W 1985 The use of matched pairs research design in industrial surveys. *Environment and Planning A* **17**: 981–9

Peck F, Townsend A R 1984 Contrasting experience of recession and spatial restructuring: British Shipbuilders, Plessey, and Metal Box. *Regional Studies* **18**: 319–38

Pratt A 1989 Towards an explanation of the form and location of industrial estates in Cornwall 1984: a critical realist approach. Unpublished PhD thesis, University of Exeter

PSMRU and Ethnic Business Research Unit 1986 *Ethnic Minority Businesses in Birmingham* PSMRU, Aston University, Birmingham

Rawlinson M B 1990 Subcontracting relationship between small engineering firms and large motor vehicle firms in the Coventry area. Unpublished PhD thesis, Coventry Polytechnic

Rawnsley A (ed.) 1978 *Manual of Industrial Marketing Research* John Wiley, Chichester

Rhodes J, Khan A 1971 *Office Dispersal and Regional Policy* Cambridge University Press, Cambridge

Sayer A 1984 *Method in Social Science* Hutchinson, London

Sayer A, Morgan K 1985 A modern industry in a decline region: links between method, theory and policy. In Massey D, Meegan K (eds) *Politics and Method: Contrasting Studies in Industrial Geography* Methuen, London pp. 144–68

Shaw G, Williams A 1988 Tourism and employment: reflections on a pilot study of Looe, Cornwall. *Area* **20**: 23–34

Steed G P F 1968 The changing milieu of a firm. *Annals of the Association of American Geographers* **58**: 506–25

Swain G R (ed.) 1967 *Techniques Committee One Day Seminar on Postal Questionnaires* Industrial Marketing Research Association, Lichfield

Swain W 1978 The postal questionnaire. In Rawnsley A (ed.) *Manual of Industrial Marketing Research* John Wiley, Chichester pp. 32–42

Thrift N, Daniels P, Leyshon A 1988 *Growth and Location of Professional Producer Service Firms in Britain* Final report to ESRC, London (Grant D00232194)

Townroe P M 1975 Branch plants and regional development. *Town Planning Review* **46**: 47–62

Turner W J 1989 Small business data collection by area censusing: a field test of 'saturation surveying' methodology. *Journal of the Market Research Society* **31**: 257–72

Watts H D 1971 The location of the beet sugar industry in England and Wales, 1912–36. *Transactions of the Institute of British Geographers* **53**: 95–116

Watts H D 1980 *The Large Industrial Enterprise* Croom Helm, London

Watts H D 1982 The inter-regional distribution of West German multinationals in the United Kingdom. In Taylor M J, Thrift N J (eds) *The Geography of Multinationals* Croom Helm, London pp. 61–89

Watts H D 1991 Plant closures, multi-locational firms and the urban economy: Sheffield, UK. *Environment and Planning A* **23**: 37–58

White R L, Watts H D 1977 The spatial evolution of an industry: the example of broiler production. *Transactions of the Institute of British Geographers New Series* **2**: 175–91

Williams A M, Shaw G, Greenwood J 1989 From tourist to tourism entrepreneur, from consumption to production: evidence from Cornwall, England. *Environment and Planning A* **21**: 1639–53

Wilson A 1968 *The Assessment of Industrial Markets* Hutchinson, London

Winkler J T 1987 The fly on the wall of the inner sanctum: observing company directors at work. In Moyser G, Wagstaff M (eds) *Research Methods for Elite Studies* Allen & Unwin, London pp 129–46

12

Land use surveys

Terry Coppock

12.1 Introduction

In Chapters 6 and 7 the main sources of information on urban and rural land use in Great Britain were reviewed; the emphasis in this chapter is on obtaining such information *de novo*, whether by observation on the ground or by the interpretation of imagery obtained by aerial photography and/or remote sensing from satellites. Such information will normally be sought by central or local government. Although there is a commercial interest in the data and commercial agencies may be employed by governments, the collection of information on land use does not seem to have been regarded as a commercial venture in its own right. There will inevitably be some overlap between this chapter and Chapters 6 and 7 since the evaluation of existing data necessarily involves some consideration of methods of collection. The chief difference is that the methods discussed here can be directly related to the purposes for which the data are required, whereas Chapters 6 and 7 were primarily concerned with data collected with other objectives in mind.

The method or methods chosen for land use surveys will depend primarily on those objectives and, in particular, the level of aggregation at which information is needed, the urgency with which results are required, the desired minimum levels of accuracy and the resources likely to be available. In practice, the last, which should be determined by the other requirements, will often be the deciding factor in the choice of method.

Given the sectoral structure of government at both national and local levels, it has been rare to ask who else might require such information and there has been little recognition of the importance of information on land use as a public good. More recently, there is a growing appreciation through the Tradeable Information Initiative that there may be a market in other agencies of central government, in local government and in the commercial sector for the information collected by government agencies. Indeed, one of the requirements in several officially sponsored inquiries into the collection of land use data has been a review of the suitability of existing source data. There also appears to be increasing recognition, especially among smaller agencies, that the costs of collecting information might be shared, although this often presents problems of conflicting objectives and priorities. A growing interest in geographical information systems (GIS), discussed in Chapters 8 and 9, is also focusing attention on the value of data to other agencies; the Rural Wales Terrestrial Database Study (WALTER) is an example of an initiative by a consortium of agencies to pool their resources of data (Mather and Haines-Young 1986).

It is difficult to know what weight to attach to the potential value of land use data in the future, when unforeseen needs may arise, whether for data at different levels of aggregation or to throw light on problems which may have arisen subsequently. There are, however, numerous instances when use has been made of data collected for other purposes. For example, Coleman (1985) lists 12 national agencies that have made use of data from the *Second Land Utilization Survey*.

As Chapter 7 showed, information on land use may be required for a wide variety of purposes, but the aim of most data collection will be either to establish the situation at any one time, i.e. a snapshot or as near-instantaneous a view as possible, or to measure change occurring over a period. Although interest in both central and local government appears to be primarily focused on change, the two approaches are closely related, since measuring change of use will be of limited value unless changes can be assessed against the relative and absolute importance of the different categories. Further, comparison of snapshots taken at different periods in time is one method of measuring change.

Achieving such a snapshot is very difficult in practice. Questionnaire surveys can, in theory at least, provide an instantaneous snapshot, as the Agricultural Census shows, recording the situation on one day; satellite imagery, too, has the potential to provide information for the whole country over a very short period, given cloud-free conditions. All other approaches (including in practice remote sensing) provide data which are necessarily collected over a period. As the subsequent discussion will show, the different approaches impose their own limitations. Field survey is necessarily time-consuming, since surveyors must first reach the survey area and then move about it; aerial photography is limited by the number of days on which weather suitable for photography occurs; and satellite imagery depends both on cloud-free conditions and whether sensors are switched on during the relevant transits. Remote sensing imagery in practice will obtain information on land cover rather than land use, only ground surveys may accurately identify land use.

12.2 Sampling

Timeliness and levels of accuracy are also relevant considerations in deciding whether a complete or a sample survey is required; whether sampling is appropriate will also depend on the level of generalization at which the information is required. The collection of information is costly so that no more should be collected than is required. Although sampling may also apply at the subsequent stages of intepretation and analysis, the emphasis here is on collection. Acceptable national and regional totals may be achieved by sampling and most recent official surveys have adopted such an approach. Sampling of land use has been the subject of extensive debate since it is not straightforward; many categories of use are highly intercorrelated and are non-randomly distributed, and land use parcels also vary greatly in size and shape. Random sampling, which satisfies the conditions of statistical theory for the calculation of variance and hence of the size of sample needed, is not generally appropriate.

The aim of sampling is to provide a reliable estimate of the whole population and the optimal sampling strategy is that which generates information at the required level of accuracy and at the least cost. Cost is not the only consideration since another advantage of sampling over a complete survey is that it will economize on the use of scarce resources of expertise, for example in field survey or air photo interpretation. Designing a sample survey requires a decision on four issues: what sample unit to choose; whether to collect a systematic or a random sample, whether to stratify and, if so, on what basis; and what size of sample to collect.

There are three types of sample unit – areas, lines and points. Area sampling in the field is the most time-consuming and hence the most costly; it involves measurement of all uses within the sample area, commonly a grid cell, and so minimizes the possible effects of small parcels. Line sampling involves the measurement of proportions of straight-line traverses occupied by parcels in different uses but has rarely been used for land use studies other than in forestry: following a straight line on the ground is complicated by the presence of obstacles and by difficulties of access. The principal alternative to area sampling is point sampling, strongly advocated by Dickinson and Shaw (1978) and involving the identification of use at each sample point. This approach is economical and requires no measurement of area other than that of the survey area itself since the number of points in each category provides the basis for estimating the proportion it occupies of the total area. This approach does, however, present two problems, that of identifying the sample points accurately on the ground or on aerial photographs, a difficulty minimized by proponents of the techniques but requiring a high level of skill in map reading/air photo interpretation, and whether to record the actual use at that point or the dominant use in the parcel in which it lies. Thus, a factory includes not only buildings devoted to manufacturing, but also storage space, offices and vehicle parking; a sample point within the factory area might fall on any of these activities, but the normal practice is to identify the dominant activity in the parcel, in this case, manufacturing. This is particularly a problem of ground surveys, since the level of resolution of photographic, and more especially, satellite, imagery will set lower limits on what can be interpreted.

The choice of a systematic or random survey is influenced largely by the non-random nature of land use. The advantage of a random survey is that sampling errors, and hence the required size of sample to achieve a desired minimum level of accuracy, can be precisely defined; but by definition, a random sample tends to create a tighter clustering of sample units than a systematic sample, chosen at regular points along a grid. While ideally the alignment of the grid should be chosen at random it is common practice to use intersections of the National Grid printed on Ordnance Survey (OS) maps; a disadvantage of the approach is that the size of samples can only be determined empirically. A possible compromise approach is to choose random points within each grid cell, since sampling theory can then be used to calculate expected errors.

The third issue is whether to stratify or not. The aim of stratification is to reduce within-group variance and to maximize that between groups, and it requires prior knowledge of the characteristics of the survey area. In rural areas, stratification is generally based on topography, soils or some combination of these in land classes, since rural uses are generally highly correlated with such parameters. No ready basis for stratification in urban areas exists although social area analysis, based on the Census of Population (CoP), is one possibility. At a minimum, any sample covering both urban and rural areas will benefit from a stratification into urban and rural. A further complication may be the need to provide reliable estimates not only for the survey area as a whole, but also for administrative or other policy subunits within it, and this will require that a minimum number of samples be drawn from each stratum within each subunit.

The last major consideration is that of sample size. This is determined ultimately by the required level of accuracy – the smaller the level of acceptable error, the larger the sample. It will also vary with the level of importance of the use being sampled; the smaller the proportion of the area occupied by each category, the larger the sample needed to give the same level of accuracy. The size of the overall sample will also be determined by the desired level of aggregation; acceptable national estimates can be achieved with much smaller samples than will be required to give the same level of accuracy for counties and, *a fortiori*, for districts. The size of the systematic samples commonly used for land use surveys can only be determined empirically in pilot study, which will also give some guidance on the treatment of rare categories.

Decisions on the size of sample should also take note of the distinctions between relative and absolute error. While the expected absolute error associated with the minor use, for example one occurring at 5 per cent of the points of a sample, may be smaller than the expected absolute error for one occurring at 50 per cent of the points, the relative error would be much greater. Empirical evidence by Emmott (1981) indicates that, at the 1 per cent level, the minimum sample for a use occupying 2 per cent of the survey area is more than twice as large as that required for a 5 per cent use. Decisions will thus be needed on whether those categories occupying only a small proportion of the area are sufficiently important to justify the cost of a larger sample. Because of the major differences between urban and rural land uses it may also be necessary to have different sampling fractions for the two types of area.

12.3 Questionnaire surveys

Questionnaire surveys require a knowledge of the population of users since respondents must be contacted and, if a sample is used, a sample frame devised (see Chapter 11). Postal questionnaires characteristically achieve a response rate of 25–30 per cent, even where addresses are accurately known. Administered questionnaires may secure a better rate of response and provide opportunities for clarification, but they are considerably more expensive and time-consuming and they will also face the major problem of estimating for non-response. Identifying respondents in the field is difficult and, once identified, they may prove difficult to contact or unwilling to co-operate. Given these problems, it is not surprising that such an approach has not been used either for measurement of the stock of land use at any one time or for estimates of change.

12.4 Field surveys

Chapter 7 demonstrated that the *First* and *Second Land Utilization Surveys* (Stamp 1948) were the only complete records of land use that had been compiled and that both had a rural bias. The sample surveys undertaken by the Institute of Terrestrial Ecology (ITE) in 1978 and 1984 (Bunce and Heal 1984) did provide comprehensive estimates for Great Britain as a whole and these too had a rural bias. What all four surveys had in common was a reliance on field observation, and there have been numerous other surveys at subnational level using the same technique. At the time of writing ITE is about to commence the *Countryside Survey 1990*, the third of its 6-yearly national field surveys.

Field surveys, in fact, serve two purposes: the collection of primary data and the validation of interpretations of aerial photographs and satellite imagery, so-called *ground truthing*. For the latter, the scheme of classification will already have been determined by the demands of sponsors and the limitations of what can be interpreted from remotely sensed imagery. For the former, there will be a choice, also governed by the requirements of sponsors and constrained by what can be reliably collected. The choice is also likely to be influenced by the desirability of securing comparability with other sources of land use data.

Since the time taken to collect data in the field represents a large proportion of the total time (and hence of the cost) of the whole survey, it has been argued that a complete inventory of land uses should be made (Coleman, 1985). She has also suggested that those recording data in the field will find a simple scheme boring and will produce better results if faced with the challenge of a comprehensive survey. No information is available about the relative costs of survey using simple and complex classifications, although the latter can be expected to be more expensive, requiring closer inspection and more detailed recording. It also seems likely that the larger the number of categories the greater the risk of errors of interpretation and of variations from surveyor to surveyor. In practice, once reasonably familiar with a classification, many surveyors are likely to rely on memory, consulting detailed descriptions only when they are in doubt, a procedure that may militate against consistency. These difficulties underline the importance of clear, unambiguous definitions.

Field enquiry clearly has a number of advantages. Interpretation is under-taken at the same time as observation and, as Coleman (1985: 121) has indicated, 'the ground surveyor examines land uses in perfect close-up resolution, in full colour, in three dimensions and from familiar angles'; she also points out that the ground surveyor has access to a number of sources of supplementary information not available to those interpreting remotely sensed imagery, such as printed notices. It is also claimed that local people, including land users themselves, can be consulted, a point also made by the organizers of the ITE survey; but such enquiries are likely to be very time-consuming, and Sinclair (1985), in his proposals for a national land use stock survey, noted the need for a minimum of enquiry at site level.

Land use surveys are necessarily labour intensive, particularly if a complete survey is being attempted, but there are differences of opinion about the need to employ highly skilled manpower. The *First* and *Second Land Utilization Surveys* made extensive use of school pupils and students in higher education and other surveys have made use of graduates and senior secondary-school pupils. Both Coleman and Sinclair seem confident that they could recruit sufficient numbers of suitable surveyors. Coleman (1985) estimated that a survey of England and Wales recording the 230 categories of her classification, would require a total number of between 520 and 600 surveyors and supervisors to complete the field-work in one season; Sinclair (1985), employing a simpler classification of 24 categories, envisaged a total of 10 surveyors for an enquiry extending over 3 years, although he believed that the work could be completed in one year with a more than proportionate increase in staff. Whatever the level, a preliminary period of training will be necessary to familiarize surveyors with the scheme.

Field survey, especially of a large number of categories, requires a high level of skill to interpret them accurately, and since it will be necessary in most surveys to employ a team, consistency of interpretation must be a major objective. Competence in map reading is also a necessity, sometimes taken for granted but too often difficult to achieve, especially where map features have changed since the field map was printed, for example by the removal of field boundaries. The field surveyor is also at a disadvantage in interpreting very large features and in mapping ill-defined boundaries, such as those between scrub woodland and rough grazing, tasks that are much easier with remotely sensed imagery. The problem of change in the base map can partly, if expensively, be overcome by using copies of the OS's master survey drawings under the scheme for the supply of unpublished information. These drawings record the latest mapped information and are well up to date in respect of built structures in urban and urban-fringe areas, although not in the countryside generally; their characteristics are discussed in more detail below, in the context of the project for Monitoring Land Use Change. Alternatively, as Lukehurst (1985) has recommended in respect of West Sussex, the printed maps can be updated by transferring information to them from a variety of sources, notably recent aerial photography. She is also an advocate of surveyors working in pairs, at least at the beginning of a survey, to minimize the risk of error. It would be foolish to pretend that differences of interpretation do not occur and all practicable steps must be taken to minimize the risk.

While it is true that a surveyor in the field often has supplementary information, the observer is, for the most part, interpreting use in terms of land cover. What the surveyor observes will thus often depend on the time of year, for most rural activities are remarkably seasonal in character. Both weather conditions and length of day suggest that field-work be undertaken in the summer months in rural areas, while summer evenings and weekends probably represent the most appropriate time for urban surveys. The main problem, other than securing consistent and accurate records, is that of access, whether to private land in rural areas or to the insides of buildings in urban areas. Unless surveyors have statutory powers, their movements are confined to rights of way and areas over which the public has rights of access. Although in practice landowners may be fairly tolerant of unauthorized access, this would not be acceptable in an official survey and the alternative of seeking permission is exceedingly time-consuming. In any case, there are often areas which are difficult to reach, even with permission. In urban areas, the network of roads and footpaths provides easy access to the exterior of most buildings, but there are generally only limited indications, from external appearances, of what activities are actually taking place within them. It is true that this is a problem for all kinds of surveys and the ground-based observers may have clues that are unavailable to those interpreting imagery, although the plan view available to the latter may likewise provide clues of a different kind.

It is primarily the labour-intensive character of field surveys that has led to a growing interest in sampling, although advocates of field surveys point out that a total survey may be sampled for purposes of measurement and estimates of totals yet provide a potential source of data for the study of problems that were not foreseen when the survey was planned. None the less, considerable economies are possible with sample surveys and the feasability study undertaken for the Department of the Environment (DoE) in 1985 into methods of conducting a national survey of land use, the so-called National Land Use Stock Survey (NLUSS), came out firmly against complete field survey, although the consultants recognized that field survey as such was a necessary component of any strategy (Tym and Partners, 1985).

In addition to problems of timing and seasonality (common to all other techniques for recording rural land uses), consistency of interpretation and accuracy of location, field survey also poses problems of measurement. Although this technique combines immediate interpretation with data collection and the resulting maps may be regarded as sources of information in their own right, the analysis of that information will normally require that areas under the different categories be measured. In most studies, use has been recorded as a code on each parcel on the field sheet and each parcel coloured appropriately. Planimetry of each parcel or each block of parcels with the same code is very time-consuming, as is digitizing the boundaries for input into a land use database. The alternative, strongly urged by Coleman (1985) and by Dickinson and Shaw (1978), is the use of point sampling, based on intersections on the National Grid (see above). While land use data have generally been recorded directly on field sheets, thus minimizing the risk of incorrect location, the space available on each parcel in any area of complex use is often small and the scope for adding information limited. Point sampling on the other hand, allows land uses to be recorded on a schedule, giving the grid reference of the point, and so offers scope for additional observations. Such an approach does, however, carry the risk of that location being incorrectly recorded, whether through simple error of transcription or through incorrect identification of the grid reference.

All stages in field survey require checking for consistency and accuracy. The risk of error can be minimized by a training programme, adequate briefing and a strong organization. Although it is argued that the matching of uses at sheet boundaries provides a convincing test of validity, this is true only where sheets are surveyed by different surveyors. It is also the case that field staff may have difficulty in locating themselves at the margins of the sheets where they lack information about the topography beyond the sheet boundary. Ideally, independent sample checks should be made in the field and when data are measured or transferred to printed schedules. Once data are in machine-readable form, a variety of plausibility tests can be undertaken to guard against any gross miscoding of use or location.

It is paradoxical that the main emphasis in land use surveys has been in rural areas, where patterns of use are two-dimensional and relatively simple, but where problems of access are most acute and the time required for travel most demanding, rather than in urban areas, where the travel component is smaller and the network of public ways much finer (Chapter 6). Although accurate interpretation of activities in urban areas presents immense difficulties to the observer on the ground, in general observation has a considerable comparative advantage in such areas over the interpretation of remote sensing imagery.

12.5 Aerial photography

The third approach, the interpretation of aerial photography, has been extensively employed in identifying land uses, particularly in countries lacking large-scale topographic maps, and is one that is attracting increasing attention in both central and local government (especially the latter). The availability of past air cover was discussed in Chapter 7 and the emphasis here is on commissioning aerial photography with the collection of data on land use as the main objective (see especially Kirby 1985; Coppock and Kirby 1987). Such imagery is capable of providing direct evidence only on land cover with use inferred from that cover.

Air photographs as a source of data on land use have both advantages and disadvantages. First, they obviate problems of access to land, and interpretation takes place under comfortable conditions in a laboratory. Secondly, interpretation may be facilitated by the fact that the interpreter can see each parcel in a wider context; he will also find it easier to draw boundaries between indeterminate categories, especially where topographic clues are lacking for the ground observer. Third, air photographs provide a permanent and comprehensive record, so that interpretation can easily be checked by supervisors. Such a record can also provide information on problems which were not foreseen at the time of the survey and can be resampled at higher densities to provide estimates for smaller different areas than those for which the information was required. Given suitable weather, aerial photography may also be quicker than ground survey, although a period will then be required for interpretation and this possible advantage may disappear in a national survey because of the paucity of days suitable for photographing large areas.

Photography is generally undertaken by a commercial agency specializing in aerial photography and the normal requirement is for full-frame photography with a degree of overlap between prints to permit stereoscopic viewing. The two principal considerations in commissioning such photography are the scale of the resulting photographs and the film emulsion to be used.

Although scales of 1 : 50 000 and smaller have been used, interpretation in urban areas requires a much larger scale, usually 1 : 10 000 or greater; when both urban and rural areas are being photographed, a scale of 1 : 25 000 seems an appropriate compromise. While it is true that the larger the scale, the greater the resolution, so too will be the number of photographs to be handled (a logistical problem) and the greater the cost. Nor is there any simple relationship between scale and accuracy of interpretation, for it appears that the benefits of increased scale vary considerably from topic to topic. The scale will also determine the minimum size of parcel for which the use can be reliably interpreted. The larger the scale, the lower the altitude of the aircraft and hence the greater frequency of days suitable for photography, i.e. free of cloud and haze. There appears to be a fair degree of agreement that the probability of sufficient suitable days for a national survey at 1 : 25 000 in one flying season is low, and 2–3 years would be a more realistic estimate.

Recent photography of Scotland at 1 : 25 000 was virtually completed in two flying seasons.

Most photography has been black and white, using a panchromatic film, but it is also possible to use both natural colour and false colour infra-red. Interpretation is facilitated by the use of natural colour, which represents the ideal medium, and false colour is particularly valuable in the identification of natural and semi-natural vegetation. Both are, however, more expensive than black and white photography, although the difference is much less in commissioned photography than when prints are provided from existing negatives. Since the cost is higher and the availability of suitable flying weather is lower for colour photography, black and white panchromatic film remains the most commonly used medium.

Photography is restricted not only by weather conditions, but also by time of day, owing to the effects of shadow, and by season, which greatly affects ease of interpretation, especially of rural cover. There is no ideal season for all purposes, but late summer/early autumn seems the best compromise for interpreting rural landscapes.

Interpretation of the photography is affected by the choice of scale, film and season, but depends primarily on the skill of the interpreter, with more experienced interpreters achieving a higher level of accuracy. Interpretation can be undertaken either with a mirror stereoscope, with the results plotted on a large-scale map of the same area, or with a photogrammetric plotter which can produce both a plot and a digital record of the boundaries. Mirror stereoscopes appear to be most commonly used in land use surveys since using a plotter is both slower and requires access to the appropriate equipment, although it does subsume the processes of measurement and digitization.

Interpreting air photographs requires the preparation of a key, with a detailed description of each category and fully annotated sample photographs of the categories described. Interpreters require a period of training and, preferably, experience in interpreting land use categories. It is also desirable to provide ground checks of the validity of the interpretations. The accuracy of the interpretations will vary with the topic, with urban uses providing particular difficulty. In the Developed Areas study, involving a simple sixfold classification, levels of accuracy of between 92 and 99 per cent have been claimed for the different categories (Smith 1978), and although these figures seem high, Rhind and Hudson (1980) suggest that accuracies of between 80 and 90 per cent are regularly obtained with skilled personnel. Comparable figures are rarely available for ground cover and, despite claims to the contrary, it seems unlikely that higher levels of correct identification can be achieved by such surveys, especially given the much larger number of field surveyors usually involved and their generally lower level of expertise. Whatever the problems of interpretation, air photography seems widely accepted as the preferred approach.

12.6 Satellite imagery

Whereas neither ground survey nor aerial photography can provide an instantaneous record of land use, each transit of Great Britain by Landsat satellites takes approximately 6 minutes and the whole country can be scanned in a succession of passes, assuming it is free of cloud. This is a rare occurrence and a colour composite image of the UK has been made from a total of 52 scenes spread over a much longer period (Latham 1985). Coverage by the Satellite Probatoire d'Observation de la Terre (SPOT), with its narrower sweep and hence more orbits to cover the whole country, would take longer for complete coverage, although the possibility of cloud-free coverage of any area is some eight to nine times greater with this imagery. As with aerial photography, some seasons are more suitable for the interpretation of satellite imagery of rural areas and the possibility of providing suitable imagery for any one area during the preferred season is correspondingly reduced.

As was noted in Chapter 7, the four kinds of imagery currently available, that from the multispectral scanner (MSS) and thematic mapper (TM) carried on Landsat and the multispectral and panchromatic sensors on SPOT, all have a coarse resolution, ranging from 80 m with the MSS, through 30 m with TM and 20 and 10 m respectively with the SPOT imagery. The resulting pixel size imposes limitations on what can be interpreted and such imagery is most useful for mapping land cover occurring in large blocks. The evidence from studies undertaken in the various parts of the UK all using Landsat imagery and concerned with rural topics, suggests an accuracy of 80 per cent or less, with the most promising results being obtained from TM imagery (Coppock and Kirby 1987). No studies using SPOT imagery for the UK have yet been reported, although with its greater resolution it should permit the discrimination of streets, large buildings and lines of buildings (Wooding *et al.* 1983).

In addition to the cost of acquiring such imagery, marketed commercially by the Earth Observation Satellite Company (EOSAT) for Landsat and by SPOT Image for SPOT, heavy computing costs are involved in processing and classifying the data. Furthermore, each scene must be classified separately and, although it is possible to improve the accuracy of classification by using imagery from other dates, by field observation (ground truthing) and by integrating such imagery with a digital terrain elevation model to take account of the effects of slope and aspect, these processes add to costs.

In view of the small-scale landscapes in Great Britain and the complexity of its land use, present satellite imagery has neither the resolution nor the reliability of interpretation to make it more attractive for land use studies than field mapping or the interpretation of air photographs. Hunting Technical Services Ltd (1986) have used TM imagery, recorded between April and October 1984 when conditions for its acquisition were particularly favourable. The TM data were classified using a maximum likelihood algorithm and 12 of the 47 categories were consistently identified, including one of developed land and one of derelict land and quarries.

The indications are that there will be considerable potential in satellite imagery for monitoring change in the main rural categories and in the extent of developed land, especially given the greater proportion of cloud-free imagery from SPOT and the variable geometry of its sensors, although the level of resolution is likely to remain well below that available from aerial photography and ground survey for the foreseeable future. Nevertheless, given the fact that satellites are in orbit and are a continuing source of imagery, whereas aerial and ground surveys must be specifically commissioned, there will inevitably be applications where satellite imagery can be used. Even where its resolutions and reliability are not acceptable, it can play a useful supplementary role in conjunction with other approaches. All investigations of the applications of remote sensing in the land use field have shown the importance of keeping developments in satellite imagery under review.

12.7 Hybrid systems

When accuracy, resolution, timeliness, organization and personnel with the required skills are set against the costs of securing the categories of information sought, it is unlikely that any one approach will be sufficient. In particular, whereas aerial photography cannot yield detailed information about land use in the urban environment but can provide data of the kind required for rural areas, ground survey is comparatively more efficient in urban areas, not only because of the limited powers of discrimination of remote sensing in urban areas but also because of the time taken to reach rural survey areas, the problems of access and the difficulties of ensuring consistency between surveyors. One solution for achieving the desired standards of accuracy and discrimination between uses is to employ different methods for different purposes, most commonly by combining ground survey and aerial photography. This was the solution tentatively advocated by the feasability study for the NLUSS (Tym and Partners 1985), although this judgement required confirmation from pilot studies to determine the least-cost solutions, the strata and the sample sizes.

In the Scottish study undertaken by Coppock and Kirby (1987) four different combinations of methods were evaluated along with the primary methods themselves, each having different merits and disadvantages. None of the hybrids was in fact recommended in the expectation that aerial photography could provide the data required; but they recommended that, if aerial photography was incapable of doing so, it should be supplemented by field visits. The preferred alternative, which does not permit estimates of past change to be made and would require both further research and development and a pilot study, was for a combination involving TM imagery supplemented by aerial photography.

12.8 Monitoring change

The emphasis in the preceeding discussion has been on establishing as near instantaneous a picture of land use as possible, the so-called stock survey, but the emphasis in most studies has been on change and local authorities are recorded as having a greater interest in studies of change (Champion and Markowski 1985). As noted earlier, there are two possible ways of measuring change: comparison of snapshots at different dates and monitoring change as it occurs, preferably through some administrative process. The Scottish survey was in fact intended to provide a bench-mark in the mid 1980s against which the situation in the late 1940s and 1970s could be compared. The requirements of comparability over time will affect both the methods and the classifications chosen in order to ensure comparability.

Comparisons with the past must obviously be shaped by the availability of data for past periods. In practice, this has meant reliance on past surveys and aerial photographs. There are few sources which permit comparison across the whole country (Ch. 7) and existing data are likely to cover a range of dates and leave gaps. Assumptions have also to be made about the compatibility of the classifications used and the consistency of their interpretation. Two approaches are possible in such cross-sectional studies: a parcel-by-parcel comparison and a comparison of estimates for the whole area or for subdivisions of it. A comparison of the land use survey over the Chilterns and surrounding areas, undertaken in 1951, and the maps of the *First Land Utilization Survey* highlights some of the problems of the first approach (Coppock 1954). It proved impossible to know, for all categories, how far the changes identified by superimposing one set of maps on another represented actual changes; some categories of change, for example from farmland to residential, were undoubtedly real (always assuming that the locations were correctly identified in both surveys), but it is impossible to know how far changes in rural land use were simply the product of differences in interpretation at different dates. Such a comparative approach is necessary, however, to examine features which are linear or where the pattern of distribution is important, for example reclamation and reversal along the moorland edge. Only careful evaluation of the sources and the use of supplementary information can help to minimize the difficulties.

Where estimates of change are required for the country as a whole, major subdivisions or other large areas, these have been computed by comparing samples based primarily on the interpretation of aerial photography or field surveys. Examples of the former include the Monitoring Landscape Use Change project, undertaken by Hunting Technical Services Ltd (1986), between 1984 and 1986, and the National Countryside Monitoring Scheme, being undertaken by staff of the Nature Conservancy Council (NCC) for the council (Budd 1988). The former was based on aerial photography for 1945–49, 1968–72 and 1978–82, with a random sample of sites stratified by major soil types and by counties. Photography for each period was interpreted by means of mirror stereoscopes and the boundaries drawn on overlays of the most recent photography or on maps at the 1 : 25 000 scale. The different categories of land cover were then measured to provide acceptable estimates at county level.

The National Countryside Monitoring Scheme is based on photography from the late 1940s and the early 1970s. The aim is to secure photography within 2 years of the main survey dates of 1947 and 1973, and estimates for counties are derived from stratified random samples, the basis for stratification being an unsupervised classification of satellite imagery to give three strata. A random sample of five squares was then selected within strata in each category and the available photographs interpreted either by mirror stereoscope (for lowland areas) or by photogrammetric plotter (for the uplands). The resulting boundaries are compiled on to 1 : 10 000 base maps and digitized to provide estimates at county level of changes in the areal extent of the categories monitored. Both surveys have been handicapped by the variable quality of past photography, the impossibility of establishing ground truth for earlier periods and the spread of dates.

Only one national study of change is based on ground survey undertaken with the specific aim of providing estimates of change, the Changes in the Rural Environment project of the ITE (Bunce and Heal 1984). A survey was undertaken in 1978 of a sample of 1 km squares, identified by a two-stage sampling procedure, with the central square of each 15×15 km block first being selected and a random sample of the 1 km squares from each land class being chosen to provide a total of 256 squares. These were then mapped in detail, using a complex classification with 473 categories. These same 256 squares were resurveyed in 1984. The areas occupied by each category were measured at both dates and the results applied to the known frequency percentage of land classes to provide national estimates. Although the small size of these samples means that these estimates are reliable only for Great Britain as a whole, the methodology is capable of being applied to produce satisfactory estimates for smaller areas, using larger samples. In the 1990 survey satellite imagery is to be combined with ground survey.

The alternative approach, of monitoring changes as they occur, requires a continuing mechanism. Planning departments in a number of counties have used records of development control for this purpose, usually supplemented by other sources, (Champion and Markowski 1985); for example, an analysis of land use changes in Cheshire between 1974 and 1980 was based on planning applications and local monitoring of development control. The inadequacy of data on transfers of land between uses has led the DoE to commission a project based on the process for revising the basic-scale map of the OS, the Monitoring Land Use Change project. This project is unusual in that it represents the first occasion on which one government department has commissioned another to collect land use data on its behalf. The project began in January 1985 and employs the machinery described in Chapter 7 for the continuous revision of the basic-scale maps. It was favourably reviewed in 1988 (Gould 1988) and is to be reviewed again in 1991.

It was noted in Chapter 7 that the OS divides topographic change into primary and secondary, with primary change mainly concerned with urban features and secondary with rural. Surveyors, of whom there are some 500 based in section offices throughout Great Britain, become aware of change partly through their own observation and partly through intelligence provided by others, particularly local planning authorities. When sufficient change has accumulated in an area to justify a visit, surveyors map the changes which are then recorded on the master survey drawings held in the section office; any secondary change in the same locality is likely to be mapped at the same time. Surveyors are thus accustomed to survey those land uses that are recorded on the maps and these have been extended in this project to a total of 24 categories with a strong emphasis on urban and urban-related change. Surveyors visiting an area to record primary change are asked to identify the new and former use (i.e. that already recorded on the map), to estimate the area of change to the nearest 0.1 ha on the basis of the master survey drawings and to give an estimate of the date at which the change occurred; they are also required to give a grid reference of the centroid of each parcel so identified to the nearest 10 m and the local authority area in which the parcel lies. These details are compiled in the section office on to a schedule which is then transmitted to the DoE where the returns are consolidated. These surveyors are skilled staff who have a good local knowledge of their section area, the size of which will depend on the amount of change taking place within it. As with all field surveys, the greatest difficulties occur over categories which are transitional, such as agricultural or bare urban land which awaits development; particular difficulty seems to have been experienced over rough grass and bracken, and natural and semi-natural vegetation, and over vacant, derelict and urban land (categories which are partly descriptions of land cover and partly of land use). The dates at which changes occur are likely to be known accurately only for built development, and to be progressively less accurate the further back in time the changes occur. In recognition of the problem, changes older than 5 years are now recorded in 5-year blocks. It follows that what are recorded are not changes in land use within a defined period, but the changes that have occurred since the features were last surveyed; these may range from less than 12 months in respect of residential developments to more than 20 years in areas where little built development has occurred. More than 30 000 changes have been identified within a single year, but it will be some time before sufficient knowledge has accumulated to estimate annual rates of change, particularly in relation to changes in rural areas.

This initiative is unusual among those discussed in these chapters in that it has been subject to independent evaluation (Gould 1988). A sample of some 5 per cent of changes was visited independently and the land use recorded, although almost a quarter of those locations were regarded as almost too remote or inaccessible; of the remainder, there was broad agreement in 80 per cent of the sites, although the level of agreement was very much higher in respect of the main urban uses, the prime objective of the study. The consultants concluded that, although adjustments should be made to the scheme, the data to be collected were of the highest quality, particularly those resulting from the most frequent occurring changes, and were not available elsewhere. Mislocations are likely to be rare and misclassifications are probably most common in those categories which are most difficult to identify by any methodology. Estimates of areas and identification of grid references are also likely to achieve a high level of accuracy, but start dates can only be a rough approximation. Although only data for the whole country and the regions have been published (e.g. DoE, personal communication, 1989) the potential for studying individual categories of change within any district or other defined area in the country is being investigated in a pilot computer-based project.

12.9 Conclusions

Each of the methodologies described in this chapter has limitations in respect of reliability of interpretation, precision of location, completeness and timeliness. Each will require some administrative machinery to organize the survey, including any necessary preliminary work and training, and to co-ordinate the results and to provide these in a form in which they can be analysed. In respect of regular enumerations, such as those described in Chapter 7, agencies have established a census branch or similar organization and these will also be required for the kinds of survey described here, whether by ground survey or by the interpretation of remotely sensed imagery or some combination of these, particularly if such surveys are to be repeated at intervals or are intended to provide information on a continuing basis. The limitations of a one-off survey, unless very well documented, are likely to be soon forgotten once the team which undertook the work has been disbanded. Unfortunately, all too few such surveys have been the subject of independent evaluation and documentation.

Except in relation to statutory planning, the public interest in land use is hard to define; different agencies have different interests and different priorities. Whether any general land use survey comparable to the CoP can be achieved depends on the extent to which sufficient common ground can be found between agencies without the necessity for a classification which is so complex that it is liable to misinterpretation by those collecting/interpreting the data. The concept of a periodic land use survey (by whatever means) is nevertheless appealing, not only because the collection of data is expensive but also because possible applications cannot always be foreseen. The idea of information as a corporate resource is finding increasing acceptance and is particularly attractive to smaller agencies which cannot afford the cost of collecting it. Consultants reporting on the Rural Wales Terrestrial Database project (WALTER) have identified a need for consultation between agencies before surveys are planned and undertaken (Mather and Haines-Young 1986) and those evaluating the Monitoring Land Use Change project recommended that the DoE should improve its liaison with other organizations in order to improve the quality of its rural databases (Gould 1988). To an outside observer, it seems strange that four different initiatives, with common elements, have been undertaken in the 1980s by different agencies, all with the aims of identifying and measuring changes in land use/land cover; it is true that each has a different objective, but there does appear to be greater scope for co-operation. One of the attractions of air photographs, and in the future, satellite imagery, is that they represent permanent archives which can serve a variety of purposes both in the present and in the future. Even so, it is probably true to say that, when undertaken by skilled and well-trained staff, ground survey remains the most reliable procedure for the recording of the uses to which land is put.

Note: Changes to the availability of OS master survey drawings are referred to in the Editor's preface (p ix).

Acknowledgements

This chapter draws heavily on two team projects with which the author was involved, that commissioned by the DoE from Roger Tym and Partners to evaluate the feasibility of a National Land Use Stock Survey, and a study with his colleague Dr R P Kirby, commissioned by the Scottish Development Department, of approaches and sources for monitoring change in the landscape of Scotland. In writing it, he has drawn heavily on the expertise of his fellow researchers, much of it unpublished. He is particularly grateful to Dr Kirby's knowledge of remote sensing and of statistics, although he alone is responsible for interpreting this expertise. He also draws on 40 years of experience in the conduct and evaluation of surveys and censuses of various kinds.

References

Budd J 1988 The National Countryside Monitoring Scheme. In Bunce R G H, Barr C J (eds) *Rural Information for Forward Planning* ITE, Grange-over-Sands, ITE Symposium **21**: 66–76

Bunce R G H, Heal D W (1984) Landscape evaluation and the impact of changing land use on the rural environment: the problem and an approach. In Roberts R D, Robert T M (eds) *Planning and Ecology* Chapman and Hall, London pp 164–88

Chapman A G, Markowski S 1985 Land use information held by local authorities. In Roger Tym and Partners (eds) National land use stock survey, appendices, unpublished report to the DoE, London pp 86–97

Coleman A 1985 Lessons from comprehensive ground survey. In Roger Tym and Partners (eds) National land use stock survey, appendices, unpublished report to the DoE, London pp 121–31

Coppock J T 1954 Land use changes in the Chilterns in 1931–51. *Transactions of the Institute of British Geographers* **20**: 113–40

Coppock J T, Kirby R P 1987 *Review of approaches and sources for Monitoring Change in the Landscape of Scotland* Scottish Development Department, Edinburgh

Dickinson G C, Shaw M G 1978 The collection of national land use statistics in Great Britain. *Environment and Planning A* **10**: 295–303

Emmott C 1981 *Computer Aided Interpretation of Land Use* Survey and Mapping Paper F5

Gould Laurence Consultants Ltd 1988 Land use change statistics research project. Unpublished report to the DoE, London

Hunting Technical Services Ltd 1986 Monitoring landscape change. Unpublished final report to the DoE, London

Kirby R P 1985 Remote sensing: aerial photography. In Roger Tym and Partners (eds) National land use stock survey, appendices, unpublished report to the DoE, London pp 145–67

Latham J S 1985 Remote sensing: satellite imagery. In Roger Tym and Partners (eds) National land use stock survey, appendices, unpublished report to the DoE, London pp 168–84

Lukehurst C T 1985 The Sussex land use inventory: lessons for a national land use survey. In Roger Tym and Partners (eds) National land use stock survey, appendices, unpublished report to the DoE, London pp 98–106

Mather P, Haines-Young R 1986 Rural Wales terrestrial database (WALTER). Unpublished final report to the Welsh Office, Cardiff

Rhind D, Hudson R 1980 *Land Use* Methuen, London

Sinclair G 1985 Broad cost estimates for land use stock survey. In Roger Tym and Partners (eds) National land use stock survey, appendices, unpublished report to the DoE, London pp 132–44

Smith T F 1978 Thematic cartography and remote sensing. In *Proceedings of European Seminar on Regional Planning and Remote Sensing* Council of Europe, Strasbourg pp 154–71

Stamp L D 1948 *The Land of Britain: Its Use and Misuse* Longman, London

Tym R and Partners 1985 National land use stock survey. Unpublished report to the DoE, London

Wooding N G, Tarran A E, Jones I P 1983 *Evaluation of the SPOT satellite Simulation Data for the UK Test Site in Derbyshire* Agricultural Development and Advisory Service, MAFF, London

13

Forecasting local and regional economic change

David Owen

13.1 Introduction

Forecasting of the national economy is well established in the UK, and is carried out by both public- and private-sector bodies. Given the importance of forecasting in national and corporate policy formulation, significant resources have been devoted to the development of sophisticated models of the national economy and population. However, forecasting at the subnational scale is far less advanced, despite the efforts devoted to developing models over the last 20–30 years, and the need for local forecasts for planning the spatial allocation of resources by central and local government, stimulated by the requirement for strategic planning by county councils. Operational local forecasts also treat the economy in a much less sophisticated fashion, with less disaggregation and fewer interrelationships considered, than national models. Moreover, methodological developments in the UK have tended to lag well behind those in the USA, where model-building began earlier, models have become more sophisticated, and the number of forecasting models in operation is far greater. However, differences in size between the USA and the UK are such that in some instances national forecasting in the UK is like state forecasting in the USA.

Nevertheless, a considerable amount of local forecasting is carried out in the UK, by central government agencies, local authorities and private companies. The methods adopted tend to be pragmatic rather than following current academic practice, due to the limitations of the spatially disaggregated socio-economic data sources in the UK, and constraints on computing and manpower resources.

This chapter is concerned with the most common approaches to forecasting socio-economic phenomena at the local and regional scales in the UK during the past two decades. It has as its explicit focus models concerned with forecasting the labour market, in particular employment and unemployment, and associated developments in population forecasting, since population projections form both an output from, and an input to, many forecasting models. Before considering the most common models used in local and regional socio-economic forecasting in this country, the development of modelling approaches is discussed.

13.2 Approaches to local and regional forecasting

Numerous regional economic models have been devised in the past four decades, motivated by the development of regional science and the desire to obtain an understanding of the working of regional and subregional economic systems. In addition to this academic interest, an important driving force underlying their development has been the needs of national and subnational economic planning, in order to help assess the impact of national economic trends and policies upon particular regions, and to illuminate the structure of local economies. Regional economic models have been of three basic types, becoming more sophisticated over time.

The simplest approach to regional economic forecasting is derived from the 'economic base' model, which treats certain industries (extractive and manufacturing industries and increasingly some 'producer' service industries) as the 'basic', propulsive sector of the economy, generating regional income growth through the production of exports. Growth of the basic industries increases employment, attracting migrants and generating demand for the products of service industries. These models have generally been most successful in representing the impact of primary industries in developing economies and least successful in highly developed economies in which all regions are highly integrated into the national economy and interregional trade is substantial. Nevertheless, the 'basic/non-basic' categorization has been retained as the basis of many attempts to account for regional and sub-regional employment growth (e.g. Fothergill and Gudgin 1982) in advanced economies, though practical applications have experienced severe difficulties in actually identifying the basic sector.

The most sophisticated regional models are full demo-econometric models; these aim to replicate the main features of a regional economy, and attempt to forecast change in employment and population through their interrelationships with production. Employment and income change is linked to population change through intersectoral and interregional flows of production and population, using input–output (I–O) or social accounting matrix (SAM) formulations. The main practical application of these models has been to assess the impact of economic change upon a region or subregion, such as the employment and income effects of a major plant closure upon a locality (see Lewis 1988 for a review of some British examples). While the I–O model is conceptually simple, such models are extremely data-intensive; furthermore, the input data on commodity flows must be collected either through local surveys, or estimated (which introduces considerable unknown errors into the estimation of the model). The SAM method, which aims to trace changes of state for all aspects of the economy and population, is even more data-hungry, and data requirements increase rapidly as interregional linkages are built into multiregional models; for N industries and R regions, a full specification of an interregional model requires a matrix of N^2R^2 data items (see Ch. 10 for a local application of SAM). While these approaches have been extremely influential in model-building, their application in practical forecasting has been limited because the need for generating input data on a time-series basis, and for the periodic updating of these databases, make them prohibitively costly in financial and labour terms.

A parallel theme of regional economic modelling has been the formulation of 'miniature' macroeconomic models at the regional scale, using sets of time-series simultaneous equations to represent the relationships between (a) output, wages, prices and employment, (b) the outputs of industrial sectors, and (c) employment and population. The earliest example was probably the econometric model of Philadelphia (Glickman 1977), which has proved highly influential on other North American model builders. The distinguishing feature of an econometric model is not an underlying theory, but the way that a model is specified (in accordance with an underlying theoretical framework) and the way that coefficients are estimated (Nijkamp *et al.* 1986). Recently, regional models have tended to merge elements of all three approaches (Glennon *et al.* 1986), with econometric techniques applied to elements drawn from I–O and economic base approaches.

Operational models are far less common than theoretical formulations of ideal regional forecasting models. The main constraints on their development has been the lack of spatially disaggregated I–O tables, lack of data on migration and transitions from one part of the labour force to another and the short time series for which the available data exist. Thus most operational models are greatly simplified in comparison with theoretical models (Klein and Glickman 1977), and generally focus upon a particular city or state for which the relevant data are available, or which have been obtained through a local survey.

Among the various approaches to model-building, three main types of regional forecasting strategy have been adopted:

1. *Top-down:* the levels of the relevant variables are first determined at the national scale, using a national forecasting model, from which the corresponding regional values are derived using a statistical procedure, and constrained to sum to the national total;
2. *Bottom-up:* here the regional values of the labour market variables are determined at the regional scale, and the national forecasts are simply the sum of the regional forecasts;
3. *Regional–national models:* in these models, the levels of regional and national variables are determined simultaneously.

In the UK, the development of regional economic models has been even more limited than in the USA and Europe, due to the poor quality, and limited availability, of regional and subregional economic data. It was not until the 1960s, with the revival of regional policy by central government and the emergence of interest in the possibility of regional government that the Central Statistical Office (CSO) starting publishing detailed regional and subregional statistics and the estimation of regional economic accounts, though the published reports of the Census of Production presented regional output data for earlier years (see Ch. 4). In general, the amount and quality of socio-economic data, and the regularity with which they are collected, decline as the spatial scale decreases (as illustrated by the other chapters in this book). Indeed, it is only in the last decade that the CSO has provided estimates of income below the regional level, while statistics on production are still only available at the regional scale (and since they are estimated from employment in many cases, are somewhat dubious). Furthermore, the complete lack of data on interregional trade, and the poor quality of migration data, have precluded the development of multiregional labour market models based on I–O or SAM frameworks. Finally, the lack of regionally disaggregated price data, and the unreliable nature of regional and subregional wage data, hamper the estimation of employment functions (which relate employment in an industry to its output) at the regional and subregional scales.

Subnational economic forecasting has thus by necessity tended to be fairly naive, since it is not possible to represent explicitly the interrelationships between production and employment, and between regions. Forecasting thus tends to concentrate upon employment and unemployment – the two main indicators of the state of the local economy which are readily available, modelling these quantities in a black-box fashion. The lack of comprehensive subregional national data sets means that forecasting tends to be carried out either 'bottom-up' for individual localities (generally by local authorities applying time-series forecasting models to data collected themselves) with no attempt to aggregate up to the national scale, or 'top-down' for regions (the smallest spatial unit for which much of the necessary data are available) by national forecasting bodies using various econometric or statistical methods to disaggregate forecasts made for the UK economy as a whole.

13.3 Operational methods in local and regional forecasting

Local economic forecasting in the UK has concentrated upon employment and unemployment, due to the constraints on data availability outlined above, and because much forecasting has been carried out by county councils as part of their statutory strategic planning activities, for which purpose employment and unemployment represents a direct indicator of the impact of change in the local economy upon the local population. The forecasting process has usually been driven by the assumption of a linkage between employment and population; population and employment change are treated as positively correlated, but the direction of causality between them has often not been clearly specified.

In order to estimate the implications of change in these two quantities for unemployment rates, the forecasting of local labour supply has to be undertaken. Since local authorities are concerned with physical land use planning, the implications in terms of pressure for housing development also have to be assessed, hence they have also been concerned with forecasting levels of migration and households consequent upon change in employment and population.

In the majority of cases, the forecasting methods used have been pragmatic rather than utilizing advanced models. Openshaw (1978) argued that the growth of academic interest in mathematical model-building and the switch of emphasis from empirical to theoretical modelling in the 1970s deprived local government planners of models they could use, and which could use the data available. Furthermore, Department of the Environment (DoE)/Welsh Office (1973) guidance on the use of quantitative techniques in structure planning argued that the use of sophisticated techniques should be considered in the light of the contribution they could make to decision-making, balanced against the need for economy in terms of time and money. Many local authority practitioners would question the use of many forecasting models developed in an academic context both on the grounds of lack of suitable data, lack of attention to examination of causal relationships underlying key assumptions (Sayer 1976), and the high degree of closure implied in many of the models, contrasting with the openness of most local economies. For further details of the changing nature of links between academic and practical developments at the academic–practice interface with changes in the policy-making context see Breheny (1987, 1989).

Local authorities have become increasingly involved in the production of local population estimates and forecasts, due to the reduction in the number of areas for which central government produces forecasts following local government reorganization in 1974, the need for population estimates below the district scale, and the cancellation of two mid-term censuses (in 1976 and 1986). Population estimates are generally produced using an apportionment method, using a proxy for population change, with the results constrained to sum to the Office of Population Censuses and Surveys (OPCS) mid-year estimates for the district. Roberts and Fairman (1984) describe such an application for Gloucestershire, where small area population estimates are derived from the electoral register, using a multiplier for groups of parishes representing the ratio of the electorally eligible population to the population in private households, calculated from Census of Population (CoP) data. The parish-level multipliers are updated annually, by comparison of the population estimated from the electoral roll with the OPCS mid-year estimate, multipliers for each parish being adjusted pro rata until the two totals agree.

The most common approach to population forecasting is the single-region 'cohort survival' or 'incomplete accounting' model; Woodhead (1985) noted that in 1979, 80 per cent of major local authority models were of this type. This method demands population data disaggregated by age and gender, and information on birth and death rates. Population change between two dates (commonly calculated over 5 years as the population data are disaggregated in the census into 5-year age bands as used in the census) is estimated by first calculating births by applying birth rates to the female population of child-bearing age, while the population in each of the older age-groups is the population in the previous age-group multiplied by the probability of persons of that age surviving over the forecast period, plus net in-migration over the forecast period for the age-group. For examples of this method see McLoughlin (1969) and Field and MacGregor (1987). Most of the key assumptions used in these projection models are based on nationally projected birth and death rates and projected local migration probabilities estimated from CoP data.

Openshaw (1978) notes that the main problem with this method is the uncertainty of spatially disaggregated migration data. Extending this basic model to the multiregional situation demands the use of an interregional migration matrix, greatly increasing the complexity of the model and the potential errors in calculation, but offering the possibility of generating forecasts of in-migration to all parts of the area for which the forecast is made. The introduction of more sophisticated demographic accounting models which attempt to replicate all the possible changes of state of a population in a multiregional system has been limited by inertia, lack of computer capacity, lack of data and crucially the huge effort involved in assembling the databases required by these models.

Population forecasts are then used as the input for forecasts of the number of households and the number of economically active persons in an area. In forecasting the number of households, the simplest technique is to multiply projected population by projected mean household size. The most common approach is the calculation and projection of household headship rates (the proportion of the population who are household heads) for particular marital status, age and gender categories, which are then multiplied by the projected population in each category, to produce an estimate of the number of households.

Planning advice from central government to local authorities has recommended forecasting labour supply and demand separately, and then integrating the results to identify the implications of employment and population change upon unemployment. The established procedure for labour supply projections was outlined in a planning advice note by the Scottish Development Department (SDD 1975). The method first calculated male and female economic activity rates in the forecast area and for Great Britain from past CoPs, and the ratio of the local to the national rates was then calculated. These ratios would then be subjectively modified to reflect local development over the forecast period, and applied to the Department of Employment's (DE's) projected activity rates for Great Britain to produce projections of local activity rates. These would then be applied to the projected population in the economically active age-groups to produce an estimate of labour supply. This approach has been extended by increasing the disaggregation of the forecasts by age-group and gender. The method is simple and straightforward, but depends on the validity of a large number of assumptions for its projections to be accurate and 'it is difficult to build into this forecast in any systematic way, exogenous changes and their effects on both migration patterns and activity rates' (Bracken and Hume 1981: 14).

The methods used to forecast labour demand are more diverse. Harding (1985) identified four common approaches to local employment forecasting. The first is to apply the rates of employment change from a national employment forecast to local employment by industry data. An example of this type of forecasting for the West Midlands is described by Marshall (1986); the national employment forecasts of the Cambridge Econometrics (CE) model were first disaggregated to the regional scale using the CE 'regional converter' (based on a shift-share analysis), and these regional rates of employment change applied to county-level employment data for 1981. The advantage of this type of approach is that national or regional forecasts are perceived as 'reliable, reputable and available' (Harding 1985: 179), the main disadvantage being that they offer no explanation of why this pattern of employment change should occur locally.

The second type of approach is the trend extrapolation of local employment data. This is the most commonly used approach, generally using time-series regression models; Box–Jenkins models yield more accurate results but the lack of long time series of local employment data limits their implementation in practice.

The third type of approach has been the isolated implementation of I–O, Lowry, or shift-share forecasting models. The heavy data requirements of these theoretical models have precluded their routine adoption following early applications in the late 1960s and early 1970s. A recent application of the I–O model at the local scale is the KRIOS (Keele Regional Input–Output System) model (Proops *et al.* 1983). This used survey data for north Staffordshire, weighted by Census of Employment (CoE) data, to create a set of I–O accounts. These I–O accounts were then used to calculate regional imports and exports, and estimate labour multipliers (i.e. measures of the total change in regional employment when an individual industry sector increases its output to final demand by an amount sufficient to require the employment of an extra worker). Such models have generally been used to estimate the impact of industry closures or openings, and the KRIOS model discovered that a reduction in the workforce of the north Staffordshire tableware industry by one-tenth led to an indirect loss of a further 426 jobs in the locality. The general application of I–O techniques to subregional employment forecasting is hampered by the high degree of dependence of local economies upon external trade, and the lack of up-to-date regional trade data. An unsuccessful attempt by Warwick University to build a multiregional econometric model using an interregional trade matrix estimated by a gravity model is described by Round (1982).

The fourth approach is the modification of employment forecasts made using systematic methods by local information. This approach was advocated by the SDD in their Planning Advice Note 4 (SDD 1975). They argued for a 'broad-brush' method of projection, which was amenable to regular updating, adjusted to take into account local factors, with the adjustments and their rationale carefully recorded. Hunt *et al*. (1980) discuss the extension of simple 'relative change forecasting' (RCF) to incorporate 'soft' information about the prospects for a local economy. The RCF method applied national forecast rates of growth to employment by industry, and added an estimated 'local differential' for Coventry. This approach was fairly satisfactory during the rapid growth of the 1960s, but failed to predict the employment losses of the 1970s. The 'local prospects' method operated by collecting background information about the main industries in Coventry, and then combining this information with a set of judgements about the economic context and the likelihood of industries maintaining employment in Coventry to produce a quantitative forecast of local employment. While the method was criticized for its subjectivity, Hunt *et al*. argued that the results of RCF forecasting vary wildly according to the precise form of the model chosen, and that therefore 'judgement cannot be avoided as a key input into employment forecasting' (Hunt *et al*. 1980: 9).

In some instances individual local authorities have produced multiregional forecasting systems, in which projections for the subdivisions of a county have been interactive, rather than simply duplicating the same model in each area. Examples of such approaches are generally to be found in population forecasting, with the interaction between areas represented by the process of migration, which then feeds into the local population forecasting models. The population forecasting models devised for the Greater Manchester Metropolitan County were of this type (Dewhurst 1984). One calculates the migration to each zone of the county on the basis of a forecast of the housing stock in each area, while a second models migration into the county as a response to job opportunities in the county relative to those in the remaining regions of the UK, estimated from the Institute for Employment Research (IER) regional forecasts.

In most applications, the integration of demographic and labour market forecasts has been achieved in a linear deductive fashion, with the output of one set of forecasts used as the inputs for another, with no explicit feedback loops or constraints. The integrated forecasting system of Gloucestershire County Council (Breheny and Roberts 1978) was an early attempt at integration. This starts from an initial employment demand forecast, which is used to produce population and housing demand forecasts. The housing demand forecast is then compared with an independent housing supply forecast; if the former is greater, population consistent with supply is used to produce new age/sex projections, which in turn produce a new estimate of labour supply which is compared with the labour demand estimate. The model iterates until consistent estimates of supply and demand are achieved.

A different approach to integration has been followed by Grampian Regional Council (Cockhead and Masters 1984). In this instance, independent forecasts for population, the labour market and housing were made, and reconciled in a final stage using an 'accounting' approach. A series of key parameters for linkage of the forecasts were identified, including activity rates, unemployment rates, the net journey-to-work balance, household headship rates, vacancy rates and migration levels. In the first stage, values for these parameters derived from the forecasting models are compared with those generated by trend-based forecasts. Differences are reconciled by adjustment of these linkage parameters and other values in the forecasts, and the comparison carried out again. This iterative process continues until the differences between the estimates become negligible (nine iterations were needed for the 1983 forecasts).

Few explicitly econometric models of regional economies have been established for Great Britain – probably a reflection of the absence of a regional layer of government in England, but as a result of the more comprehensive economic and financial data available in Northern Ireland, the Northern Ireland Economic Research Centre (NIERC) has been able to develop a relatively sophisticated econometric model of the Province's economy (Tavakoli *et al.* 1988). This model produces forecasts for 25 industries grouped into 3 sectors; manufacturing, private services and public services. The model assumes that manufacturing output is determined by national demand together with local final and intermediate demands; the wages paid by this sector contribute to personal incomes and hence consumer spending, which in turn provides the demand for private services. This again links to employment and hence consumer expenditure, while government expenditure influences demand for local services both directly and via wages through consumer spending. The NIERC model uses a variant of the Cambridge Growth Project (CGP) macroeconomic model to generate UK national output forecasts as a measure of UK domestic demand, with demand in the Province represented by real consumer expenditure and exogenous government expenditure. Employment by industry is then forecast using an employment function in which the ratio of labour cost to output price, the ratio of labour costs to the prices of other inputs, capital stock and industry output are the independent variables. Real wages by industry are then estimated as a function of the ratio of capital to employment, the UK real wage rate, the ratio of sector wages to average wages, the unemployment rate and a measure of union power. Wages and other sources of income then feed into the estimation of consumption. A further block of the model is concerned with the estimation of population, using a cohort survival approach, migration (as a function of wage and unemployment differences) and unemployment.

13.4 National forecasts used as input to local economic forecasts

Local economic forecasting makes use of two types of national forecast: centrally produced projections of population and households for regions and subregions; and macroeconomic models of the entire UK economy, which produce employment and unemployment forecasts. These exogenous inputs to the local forecasting process are usually either disaggregated to the local scale using a statistical procedure of varying degrees of sophistication, or used as control totals for variables which local forecasting models aim to match. Their importance in local forecasting makes it worth while to consider the various types of national forecast available, and the manner in which they are produced.

National population estimates and projections

Annual population estimates disaggregated by age and gender relating to June of each year are produced for local authority districts and health districts by the census offices in the countries of the UK (Cockhead 1985). These estimates take as their starting-point the most recent census population (converted to a 'home population' basis), which is carried forward year by year, based on an estimate of marginal changes to the population, using the 'demographic accounting' method. This method first estimates the total population of an area and then breaks it down into age and gender groups by ageing forward the corresponding population group from the census using estimates of natural change, migration and other changes, and then constraining the results to sum to the estimated total population for the district, and the national age and gender totals. The crucial element is the estimation of net civilian migration, using census migration propensities, National Health Service Register (NHSCR) migration data and Department of Education and Science data. A comparison of the mid-year estimates and CoP data for 1981 in England and Wales found that 96 per cent of these estimates fell within 5 per cent of the corresponding census figure.

UK population projections are made by the OPCS, with subnational projections for Wales and Scotland generated by the Welsh and Scottish Offices, constrained to sum to the national totals from the UK forecast. The projections use the 'component method', which estimates the number of persons in each year of age and gender as the corresponding number one year younger one year before, less deaths in that age-group, plus or minus migrants in the age-group during the year. The results of the method thus depend entirely upon the assumptions relating to fertility, mortality and migration, and these assumptions are modified by comparison of previous projections with estimated populations, and in the light of changing trends in these three components. New projections are made every 2 years, using the mid-year estimates as the base for the projection, and projections are published for 20–30 years ahead.

Household projections

Projections of households are produced for England and Wales every 2 years by the DoE, covering a period of 15–20 years (Corner 1989). The projection method used is the well-established headship rate method. The basic method projects the household headship rates forward, and multiplies these by the OPCS population projections to produce the forecast number of households in each of 7 headship types in 3 marital-status categories for 15 age-groups disaggregated by gender. Headship rates are calculated from the 1961, 1966, 1971 and 1981 CoPs, supplemented by data from the *Labour Force Survey* (*LFS*), and extrapolated using time-series regression, with rates constrained to lie between zero and unity. Projections are made for both countries, regions, counties, metropolitan districts and London boroughs, and with such a large number of projections to be made, it is important that the model is 'robust, mechanical and computationally efficient' (Corner 1989: 97). Projections are made independently for each household category and area, and hence it is necessary to modify the projections for each subarea to ensure that they sum to the total for the area as a whole. This procedure is carried out in a top-down manner down the 'tree' of projection areas, from countries to counties and districts.

The Chelmer model (King 1987) has been developed to offer local authorities, housing associations and other actors in the housing arena access to an interactive model based upon the DoE projections for households and population, but offering the ability to use it in an interactive 'what-if' fashion, to explore the implications of differing assumptions relating to migration, employment and household formation upon household change in a given area. The model is hierarchical, operating at the region, county and district scales. The user can choose between sets of DoE household projections and alternative modelling strategies, the most significant of which relates to the choice between using constant migration age profiles (age breakdown of migrants) or constant migration propensities by age in estimating migration.

CACI forecasts of population, labour force and households

The reduction in quality, frequency of collection and incomplete topic coverage of official statistics, and the increasing interest of retailers, property developers and others in the private sector, have drawn market analysis companies into the generation of local-scale socio-economic data. One example of this is the production of annual estimates of population, labour force and households for wards and enumeration districts (EDs) in the whole of Great Britain by CACI.

The CACI model produces estimates of population by 5-year age-group and gender for June of each year, consistent with OPCS and GRO Scotland mid-year estimates for the district each ward or ED falls within. The basis of the estimates is a cohort survival model which ages the 1981 ward population from the CoP, to which is added and subtracted births and deaths estimated from ward-level vital statistics data, and migration estimated from the electoral register by applying age-specific migration propensities from the 1981 Census. The unconstrained ward estimates are controlled to the district-level mid-year estimates and converted from a 'usually resident' to a 'home population' basis. A further model estimates ED population in a similar fashion using ward-level birth and death rates, but without taking migration into account. Estimated ED populations are constrained to sum to the ward estimates. Household numbers are estimated by applying national headship rates to these local population estimates, with local corrections applied based on a comparison of actual with estimated household numbers for 1981. The labour force is estimated by applying census-based economic activity rates to the population estimates for age and gender groups, with changes in activity rates incorporated by constraining the sum of the small area labour force estimates to equal the regional totals from the *LFS*.

Macroeconomic models

A large number of macroeconomic forecasting models now exist, the bulk of which are concerned with financial indicators (such as the overall national rate of growth, balance of payments, and inflation), and only generate crude industrial disaggregations of employment and output. The PA Cambridge Economic Consultants (PACEC)/Cambridge University Department of Land Economy model is of this type (having been derived from the Cambridge Economic Policy Group's [CEPG's] model), but is used for subnational forecasting by means of a separate model which applies national employment change to individual industries on the basis of their employment trends relative to total employment.

The necessity for this extra estimation stage is obviated if the macro-economic model itself produces disaggregated employment forecasts; such models are rare. The macroeconomic model which has been most heavily used in regional and subregional economic forecasting is probably the CGP multisectoral dynamic model (Barker 1985). This is a large model, consisting of a database of time-series economic data, I–O tables, a series of 'classification converters' necessary in order to generate data on a consistent basis over time, and a set of regression relationships, all bound up in a software package which enables model parameters to be changed and alternative expectations of exogenous variables to be incorporated. The model is highly disaggregated, covering over 40 industrial sectors, and hence contains about 8000 variables and 16 000 equations, most estimated using ordinary least squares or the Cochrane–Orcutt iterative technique. It has a dynamic formulation, producing year-on-year forecasts over the medium term rather than projections for a single target year. A modified version of the model has been used by the IER as the basis of their own forecasts of industrial and occupational employment change for the UK as a whole.

The CGP model is the most disaggregated macroeconomic model operating at the UK scale, but the Fraser of Allander Institute of Strathclyde University has for some years been developing a multisectoral medium-term model of the Scottish economy. It has similarities with the CGP model, producing forecasts at a similar level of industrial disaggregation, but the database on which it is run is specific to Scotland. This is the only example of an operational macroeconomic model operating at the subnational scale in the UK.

13.5 Recent and current multiregional forecasting models

A number of organizations have published forecasts at the regional or subregional scale using integrated econometric models in the last decade or so. This section reviews some of the better-known models.

Institute for Employment Research (IER)

In the early 1980s, the IER at Warwick University published labour market forecasts for the regions of the UK, as part of the institute's medium-term forecasting of the economy (Keogh and Elias 1979). The main emphasis was placed upon the medium-term forecasting of employment by industry. The regional employment model used regression equations for the historical relationship of regional to national employment in an industry to disaggregate national forecasts generated by the IER's version of the CGP multisectoral dynamic model to the regional level, the results afterwards being scaled to sum to the national forecast total. Employees in employment in industry i in region j at time t, E_{ijt}, were estimated as a function of national employment in the corresponding industry, E_{iUKt} and time, using a mixture of arithmetic, logarithmic and logit specifications for the regression model for each region and industry. This relatively simple formulation was necessitated by the short time series of data available for estimation of the model (1965–76).

The IER model also attempted to estimate self-employment and employment in domestic service and the Armed Forces by industry. This was achieved by calculating ratios of employees to the self-employed at the regional scale, and applying these to the projected number of employees by region. The other categories of employment were estimated by assuming their regional distribution remained the same as in the 1971 CoP over the forecast period, and applying the regional shares of total employment in these activities to projected employment. The regional labour force was projected forward by combining the OPCS and GRO Scotland projections of adult population with estimates of regional activity rates. These were derived from the DE's projections of activity rates by applying regional activity rate relativities from the 1971 Census to the national projection of activity rates. (For details of more recent DE activity rate and labour supply projections see DE 1990 and Spence 1990). Unemployment forecasts were obtained as the difference between the projected labour force and projected employment, which was further adjusted to correct for 'double-jobbing'.

The IER's regional model was abandoned in the mid 1980s, because the recession of 1980–82 was perceived to have caused a structural shift in the relationship between regional and national employment, and hence the model could no longer produce plausible regional employment forecasts. However, production of forecasts at the national scale has continued (IER 1989). The IER has recently revived its regional forecasting activities in association with CE and NIERC, and plans to produce regional forecasts of employment by occupation, linked to the CE/NIERC regional model.

Cambridge Economic Policy Group (CEPG)

An alternative set of regional forecasts were produced in the early 1980s by the CEPG (Gudgin 1982). Again, the methodology was top-down, regional forecasts being driven by national forecasts of employment by industry (for eight sectors). The CEPG model was much smaller than the CGP model, containing only about 60 equations, and its national employment forecasts were less sophisticated than those from the CGP model, being predicted from forecast output or investment, or, for the smaller industries, as a time trend, with self-employment assumed to remain constant during the 1980s (Gudgin *et al.* 1982).

Regional employment in agriculture, mining and construction was treated as a function of the UK employment share of the industry and a time trend. For the service sector, public utilities and transport, regional employment was predicted from the ratio of the region's share of employment in the industry to its share of total population. A shift-share analysis of manufacturing employment was carried out, and the three components projected forward. Regional employment in the South East was estimated as the difference between the sum of forecast employment in the other 10 regions, and the national forecast.

On the demographic side, there were three behavioural equations and two identities for each region. The labour force was projected as a trend in the share of the national labour force, adjusted for pressure of demand, and the 'job shortfall' (the difference between the '*ex ante* labour force', defined as the previous year's labour force plus natural increase, minus the change in employment) in the region relative to the UK. Gudgin argues that the entire burden of labour market adjustment falls upon the labour force, and thus 'the equations in the model . . . ignore the neo-classical tradition of wage and price adjustment . . . the level of migration responds directly to the size of the job shortfall, there is no intervening movement of wages and prices' (Gudgin 1982: 13). Net migration of working-age persons was thus treated as a function of the relative regional job shortfall, together with an autonomous component representing urban–rural flows independent of local employment shortages. The regional population of working age was estimated by calculating natural increase using a cohort survival model and adding net migration, and total population was then estimated as a trend relative to the population of working age. Unemployment was obtained as the difference between the labour force and total employment.

Cambridge Econometrics (CE)

CE is a commercial firm marketing the results of the multisectoral dynamic model developed by the CGP. The regional extension of the model was developed in association with tne NIERC, and it shows strong similarities with the CEPG regional model, due to the close association of Graham Gudgin with the development of both models (CE/NIERC 1987).

The regional model takes the national employment forecasts from the CGP model and estimates separate structural employment equations for 15 industry sectors and 11 regions. For the production industries, the share of regional employment in the industry is treated as a function of the national employment share of the industry, the one-year lag for this variable, the one-year lag of the corresponding regional share, and a time trend, with a regional policy indicator added for manufacturing, a dummy for the miners' strike added for mining and quarrying and proxy for the 'troubles' included in the equations for Northern Ireland. Employment in the service-sector industries was treated as a function of the regional share of UK population, the share of the industry in UK employment and the one-year lag of the share of the industry in regional employment. The employment forecasts are then scaled to sum to the national employment forecasts from the CGP model. Regional employment forecasts, together with national output forecasts, are input to the GDP block of the model where regional productivity is estimated for nine industry sectors, as a function of UK productivity in the corresponding industry and a constant.

The labour market block is the major simultaneously estimated block of the model, containing three structural equations for each region estimating working-age migration, labour force and population. The simultaneity revolves around the concept of the 'relative employment shortfall' (RES). The employment shortfall is the difference between the *ex ante* labour force in a region and the level of employment; the RES is the difference between the regional shortfall and the national shortfall distributed between regions on a pro rata basis according to their labour forces. Two equations are used to estimate working-age migration; one treats it as a function of regional and national unemployment rates, relative regional house prices and a time trend; the other treats it as a function of the RES, relative regional house prices and a time trend. A dummy variable for the 1981 depression is included, with a proxy for the 'troubles' in Northern Ireland. Adding natural increase in the working-age population and net migration of the working-age population yields an estimate of the working-age population. The regional labour force is then estimated as a function of the working-age population, the UK participation rate, the change in the RES and a time trend. Finally, total population is treated as a function of the working-age population, the ratio of the UK population to the UK working population and a time trend. The regional population estimates are scaled to the UK totals.

PA Cambridge Economic Consultants (PACEC)

This model is the direct descendant of the CEPG regional model. It is driven by the PACEC/Cambridge University Department of Land Economy national macroeconomic forecasting model, again derived from the model developed by the CEPG. For each region, there are 20 equations, 15 of which are estimated, the rest being identities. Employment is 'modelled outside the main model and fed in using a form of shift-share analysis' (Tyler and Rhodes 1989: 124) for 33 industry sectors. The national and structural components for each region are calculated by simply applying the projected national rates of change to regional employment data. The forecast of employment change is made by adding a forecast of the differential component to the sum of the other two components. This component is projected by examining the historical trend of the differential component, and interpreting the trend in the light of shifts in urban and regional policies, the urban–rural shift of employment and infrastructure developments.

The labour supply variables are modelled in a similar fashion to the CEPG and CE/NIERC models. The central concept in labour market adjustment is again the job shortfall, the difference between the *ex ante* labour force and employment. Working-age net migration is a function of the regional and national job shortfalls; total migration is calculated as working-age migration multiplied by the ratio of working-age to total population; total population is a function of regional natural increase, net migration and lagged population; natural increase is a function of the UK rate of natural increase and regional and national population; the working-age population is a function of that in the UK and the regional and national populations; and the labour force is a function of the UK labour force and the populations of working age in the region and the UK. Unemployment is again the difference between the workforce and total employment.

The distinguishing feature of the PACEC approach is that the model has been applied at the subregional scale to forecast labour market change for the counties of the South East. The first stage was to apply the same shift-share technique to forecast employment for Greater London, and the remainder of the forecast employment change in the region was allocated to counties on the basis of their employment change between 1971 and 1981. These forecasts were then adjusted to take into account an expected improvement in the rates of growth of eastern counties (due to infrastructure developments such as the M11, Stansted Airport and the Channel Tunnel), the ending of the New Towns programme, development pressures in the west, the recovery of employment in London, and the faster growth of private services in previously slower-growing counties. County population change for 1985–95 was then estimated by applying the parameters from a regression of population change on employment change for 1971–81 at the district scale to forecast employment change. Population change was then translated into household change by the application of DoE forecast headship rates, and using DoE estimates of the number of dwellings in 1981, estimated household change was translated into estimated dwelling change.

13.5 Conclusion

This chapter has provided an overview of the current state of regional and subregional economic modelling in the UK by public and private organizations, and the evolution of such models in recent years. This review has been far from exhaustive, but the bulk of operational forecasting applications use similar methodologies to those discussed here.

The prospects for methodological improvements in modelling and the extension of local forecasting models to cover other economic phenomena are poor. The quality of local socio-economic data has declined in recent years, both topic coverage and frequency of collection being reduced in response to financial and ideological pressures (Elton 1983, and also Chs 1 and 10). Such deficiencies in data quality obviously have implications for the robustness of the output from forecasting models. However, computing power has become much cheaper, and advanced statistical applications are now becoming more common on microcomputers, thus permitting greater disaggregation and more robust estimation of forecasting models within the continuing constraints on staff time within public- and private-sector bodies making use of such forecasts.

It is also clear that with the increasing 'commodification' of information (Openshaw and Goddard 1987), commercial organizations are being drawn in to fill the gaps left in government data collection, and to provide the forecasts which other organizations need but cannot produce for themselves. Moreover, the private sector is becoming more heavily involved in spatial forecasting, often combining data from official sources with their own databases. Another trend is the adaptation of quantitative forecasts made at one spatial scale to smaller scales, often with the addition of subjective judgements about the likely trend in employment. Again, this is an area in which commercial consultancies are prominent, providing small area labour market forecasts for organizations who have to plan for economic expansion without access to objective information, one example being the need for British Rail to forecast traffic growth into London in order to plan capital expenditure. Indeed the recent rapid growth of the South East has been a major factor in the recent reawakening of interest in strategic planning, which if economic growth is maintained, may lead to a revival in regional and subregional economic modelling when the results of the 1991 Census become available.

References

Barker T S 1985 *The Cambridge Multisectoral Dynamic Model Version 6: MDM6 User's Manual* Department of Applied Economics, University of Cambridge

Bracken I, Hume D 1981 *Key Activity Forecasting in Structure Plans* Papers in planning research 26, Department of Town Planning, University of Wales Institute of Science and Technology

Bracewell R 1987 CACI's small area population, labour force and household estimates. Paper presented to the Regional Science Association workshop on Regional Demography, London School of Economics, October 1987

Breheny M 1987 The context for methods: the constraints of the policy process on the use of quantitative methods. *Environment and Planning A* **19**: 1449–62

Breheny M 1989 Chalkface to coalface: a review of the academic–practice interface. *Environment and Planning B* **16**: 451–68

Breheny M, Roberts A J 1978 An integrated forecasting system for structure planning. *Town Planning Review* **49**: 306–18

CE/NIERC 1987 *Regional Economic Prospects* CE, Cambridge

Cockhead P 1985 Population estimates. In England J R, Hudson K I, Masters R J, Powell K S, Shortridge J (eds) *Information Systems for Policy Planning in Local Government* Longman, London pp 56–75

Cockhead P, Masters R 1984 Forecasting in Grampian: three dimensions of integration. *Town Planning Review* **55**: 473–88

Corner I E 1989 Developing centralised household projections for national and sub-national areas. In Congdon P, Batey P (eds) *Advances in Regional Demography: Information, Forecasts, Models* Belhaven, London pp 91–106.

DE 1990 Regional labour force outlook to the year 2000. *Employment Gazette* **98**: 9–19

DoE/Welsh Office 1973 *Using Predictive Models for Structure Plans* HMSO, London

Dewhurst R 1984 Forecasting in Greater Manchester: a multiregional approach. *Town Planning Review* **55**: 453–72

Elton C J 1983 The impact of the Rayner review on unemployment and employment statistics. *Regional Studies* **17**: 143–6

Field B G, MacGregor B D 1987 *Forecasting Techniques for Urban and Regional Planning* Hutchinson, London

Fothergill S, Gudgin G 1982 *Unequal Growth* Heinemann, London

Glennon D, Lane J, Johnson S, Robb E 1986 Incorporating labour market structure in regional econometric models. *Applied Economics* **18**: 545–55

Glickman N 1977 *Econometric Analysis of Regional Systems* Academic Press, New York

Gudgin G 1982 The CEPG regional model of employment and unemployment. Paper presented to British Society for Population Studies conference on population change and regional labour markets, Durham University, Sept

Gudgin G, Moore B, Rhodes J 1982 Employment problems in the cities and regions of the UK: prospects for the 1980s. *Cambridge Economic Policy Review* **8** (2): 1–81

Harding C 1985 Population estimates. In England J R, Hudson K I, Masters R J, Powell K S, Shortridge J D (eds) *Information Systems for Policy Planning in Local Government* Longman, London pp 178–89

Hunt H J, Stanton P J, Hodges D 1980 Getting an accurate forecast of local employment prospects. *The Planner* **65** (Jan): 8–9

IER 1989 *Review of the Economy and Employment 1988/89* IER, University of Warwick

Keogh G T, Elias D P B 1979 *A Model for Projecting Regional Employment in the UK* Discussion Paper 1, Manpower Research Group, University of Warwick

King D 1987 The Chelmer population and housing model. Paper presented to the Regional Science Association workshop on regional demography, London School of Economics, Oct 1987

Klein L, Glickman N 1977 Econometric model-building at regional level. *Regional Science and Urban Economics* **7**: 3–23

Lewis J 1988 Economic impact analysis: a UK literature survey and bibliography. *Progess in Planning* **30**: 157–209

McLoughlin J B 1969 *Urban and Regional Planning* Faber and Faber, London

Marshall M 1986 National economic forecasting at local level: applications of the Cambridge Model in the West Midlands. *Local Economy* **3**: 49–60

Nijkamp P, Rietveld P, Snickars F 1986 Regional and multiregional economic models: a survey. In Nijkamp P (ed.) *Handbook of Regional and urban Economics* vol 1 Regional Economics North-Holland, Amsterdam pp 257–94

Openshaw S 1978 *Using Models in Planning: A Practical Guide* RPA Books, Corbridge

Openshaw S, Goddard J 1987 Some implications of the commodification of information and the emerging information economy. *Environment and Planning A* **19**: 1423–39

Proops J L R, Lee W D, Pullen M J 1983 *KRIOS: The Keele Regional Input–Output System* Discussion Paper 41, Department of Economics, University of Keele

Roberts A J, Fairman S J 1984 A practical method for small area population estimation. *Planning Outlook* **27**: 39–40

Round J 1982 *Methods of Projecting the Regional Demand for Labour using the Cambridge Growth Project Model* Final report, SSRC Grant HR5099, University of Warwick

Sayer A 1976 A critique of urban modelling. *Progress in Planning* 6: 187–254

SDD 1975 *Forecasting Employment for Regional Reports and Structure Plans* Planning Advice Note 4, Edinburgh

Spence A 1990 Labour force outlook 2001. *Employment Gazette* **98**: 186–98

Tavakoli M, Roper S, Schofield A 1988 Some issues in the modelling of small regions: the case of Northern Ireland. In Harrigan F, McGregor P (eds) *Recent Advances in Regional Economic Modelling* London Papers in Regional Science 19, Pion, London pp 83–99

Tyler P, Rhodes J 1989 A model with which to forecast employment and population change at the regional and subregional level. In Congdon P, Batey P (eds) *Advances in Regional Demography: Information, Forecasts, Models* Belhaven, London pp 124–49

Woodhead K 1985 Population projections. In England J R, Hudson K I, Masters R J, Powell K S, Shortridge J D (eds) *Information Systems for Policy Planning in Local Government* Longman in association with Burrisa, London, pp 76–95

Part Three

Economic Sectors

14

Agriculture, forestry and fishing

Brian Ilbery, Alexander Mather and James Coull

Sources of information on primary activities in the UK are diverse and widely scattered. While there may appear to be many data sources, they tend to relate to either agriculture, forestry or fishing and there is no single comprehensive published source on primary land uses and activities. Still worse, existing data sources are often deficient in some way and tend to exclude information on important items, especially concerning recent changes in primary activities and the functional elements (i.e. processes at work) of land use. Very few sources permit a detailed analysis of spatial patterns and trends as they contain information at a broad national, or at best regional, scale. The following discussion thus examines the type of data sources available for each of the three major primary activities separately. It complements the material covered in Chapter 7 on rural land use.

14.1 Agriculture

Agricultural data sources in the UK are dominated by the Agricultural Census and official publications. A wide range of other sources exist, but these are fragmentary and often deal only with particular aspects of what is a very complex industry. As with forestry and fishing, there is a shortage of information published at anything below the national scale. In the space available, it is impossible to examine all available data sources; instead discussion concentrates on what are considered to be the major ones. General reviews of agricultural data sources can be found in Burrell *et al.* (1984), Ilbery (1985) and Mort (1990).

Although beset with numerous problems, the annual **Agricultural Census** provides a major source of information which does not have an equivalent in the forestry and fishing industries. The census makes available information from national and regional scales down to county and parish (but not farm) levels. Indeed, such a scale approach – from national down to farm levels – seems an appropriate way to organize the subsequent discussion on agricultural data sources. However, such categorizations are not mutually exclusive as some data sources, particularly the Agricultural Census, transcend more than one scale.

National data sources

Three main types of data are available at the national scale.

1 Land and land use surveys

The classification of land and land use in the UK dates back to the pioneering work of Sir Dudley Stamp, who produced both a threefold typology of land quality and the first *Land Utilization Survey of Great Britain* in the 1930s (Stamp 1940, 1948); similar work was undertaken for Northern Ireland between 1938 and 1946. A *Second Land Utilization Survey* was undertaken in the 1960s by Alice Coleman (Coleman and Maggs 1962; Rhind and Hudson 1980). Such surveys produce a wealth of information for a particular period in time, but they become quickly dated and are thus of limited value for studies of agricultural change (see Chs 6, 7 and 12 for further exemplification of land use surveys).

Various other bodies have attempted to classify agricultural land on the basis of physical characteristics. The Ministry of Agriculture, Fisheries and Food (MAFF) produced its own Agricultural Land Classification in 1968, with five grades of land quality based on soil, relief and climatic parameters (the system was revised in 1988). A similar Land Use Capability Classification was produced by the Soil Survey, and the Institute of Terrestrial Ecology (ITE) has produced its own classification of agricultural land based on 26 physical attributes (Bunce *et al.* 1981). In each case, a series of maps of agricultural land quality has been published. However, no attempt was made to incorporate economic parameters into the classifications and there is not necessarily a direct relationship between land use and land quality; as Tarrant (1974: 119) remarked, 'good and bad quality farming is not restricted to good and bad quality land respectively'.

With the more widespread use of remote sensing and geographical information systems (GIS) (see Chs 9 and 12), there is a greater opportunity to monitor changes in agricultural land use. One such project has been the monitoring of landscape change by Hunting Surveys and Consultants (1986). This used aerial photography and satellite remote sensing to examine landscape change in England and Wales between 1951 and 1981; a sevenfold classification of land cover was produced (see Ch. 12 and Hooper 1988). No attempt was made to collect information on the social and economic aspects of agriculture or to explain landscape change and assess the impact of change on the quality of the landscape.

2. General agricultural trends

Each agricultural ministry in the UK (e.g. MAFF in England, the Department of Agriculture and Fisheries for Scotland (DAFS), the Welsh Office for Agricultural Development (WOAD) and the Department of Agriculture in Northern Ireland (DANI) produces annual statistical publications on the general structure of the agricultural industry; information is provided on such items as farm sizes, types and incomes, as well as on employment structure, inputs, outputs and levels of self-sufficiency. Executive summaries are provided in *Agricultural Statistics UK*, the *Economic Report on Scottish Agriculture*, *Welsh Agricultural Statistics*, and the *Statistical Review of Northern Ireland Agriculture* (Appendix 14.1).

Two major annual publications are worthy of mention. The first is the *Annual Review of Agriculture*, which from 1989 became known as *Agriculture in the UK*. This publication, which has been produced for over 50 years, provides insights into the economic condition of the agricultural industry in the UK. It includes aggregate statistics on the value of production, levels of self-sufficiency in different products, public expenditure on agriculture, agricultural land prices, input costs and outputs of different crop and livestock commodities. Information on incomes, assets and liabilities, obtained from farm management surveys (see below), can also be found in the report. Agricultural economists and geographers often make use of the annual review to identify trends in UK agriculture (Hill and Ingersent 1982; Fallows and Wheelock 1982; Grigg 1989).

Farm Incomes in the UK has been published since 1948/49. It is a government summary of various farm management or business surveys undertaken by certain universities and agricultural colleges on contract to MAFF. Each year, the trading accounts and balance sheets of approximately 2200 farm businesses in England and Wales are examined. These are used to provide detailed information, at a national level, on farming incomes, inputs and outputs, and assets and liabilities. Occasionally, some of the information is published by standard regions, as in the 1988 report which, for example, provided regional data on sales and prices of agricultural land over a 6-year period. Many of the colleges involved use these surveys as the basis for their own publications, the most famous of which is Nix's (1984) *Farm Management Pocketbook*. To come into line with MAFF, DAFS and WOAD, DANI produced its first *Statistical Review of Northern Ireland Agriculture* in 1988. The review is very comprehensive and contains information on farm incomes, inputs and outputs, gross margins, subsidies and grants, hours of work and earnings of farm labourers, farm structures and employment, and crops and livestock. Some of the material from *Farm Incomes in the UK* and *Agriculture in the UK* is reproduced in the *Annual Abstract of Statistics* of the Central Statistical Office (CSO).

In addition to these major reports, data on average earnings and hours worked by hired agricultural workers are gathered together by MAFF and DAFS and published annually as a *Wages and Employment Enquiry* and part of the *Economic Report on Scottish Agriculture* respectively. The information in the *Wages and Employment Enquiry*, for example, is based on a sample of nearly 1900 holdings visited by agricultural wage inspectors. Some regional breakdown of the data is given (Department of Employment (DE) 1987), but information is not provided at either county or smaller scales.

Other national, often unofficial, publications on agriculture are listed by Mort (1990). These include two annual publications by the National Farmers' Union (NFU): *Agriculture: Some Basic Statistics* and *Agriculture's Place in the National Economy*. The latter is based on MAFF statistics and the former produces the same kind of information on crop and meat production, size of holdings, employment, incomes and self-sufficiency that is published by the government. Rather different are Savill's *Agricultural Land Market Report* (quarterly) and *Agricultural Performance Analysis* (biannually), which provide important information on land acquired by financial institutions since 1970 and thus allow the growth of such ownership to be examined (Munton 1977).

3. Specific crops and livestock

A range of specialist bodies collects and publishes information on different crops and livestock. Foremost among these are the various marketing boards. For example, the Federation of UK Milk Marketing Boards produces a series of reports and accounts, the most notable being *UK Dairy Facts and Figures*. Produced annually since 1962, this contains information on different aspects of the dairy industry, including the number of milk producers, income from milk production, milk supplies, total sales of milk off-farm and marketing (Williams 1986). The Milk Marketing Boards divide the country into a number of regions and provide data at this scale, thus permitting some spatial analysis (Ilbery 1985; Grigg 1989). The Pig Marketing Board produces similar information, on the number of pigs, bacon prices, market supplies and feed costs, in its *Annual Report and Accounts*.

With regards to crops, both the Potatoes and Hops Marketing Boards have been producing annual reports for over 40 years. The former publishes two main documents: first, the *Potato Statistics Bulletin* which contains data on plantings (by different variety and registered growers) and monthly consumption of potatoes; and second, *Potato Processing in Great Britain* where exports and imports of processed potatoes are recorded, together with the number of raw potatoes and varieties used for processing in Great Britain. Annual information on the area under hops and total production is provided in the Hops Marketing Board's *Provisional Estimates of the Total Production of English Hops* and their *Statement of the Total Area under Hops* (Ilbery 1984). An annual *Hop Report* has also been produced by Wigan and Richardson International Limited since 1952; this outlines production totals, the main growing areas, market demand, and trade in hops and hop products.

General information on cereal crops is available from the Home Grown Cereals Authority, which collates and publishes data on cereal production, prices, supplies and trade in its *Cereal Statistics*. Indeed, an unbroken run of information on grain prices from 1771 is provided by the *Corn Returns*. If one takes more specialist crops, the Mushroom Growers' Association, for example, has provided production and manpower figures for its members since 1979 in an annual *Industry Survey*. Similarly, the Plunket Foundation publishes information on more than 600 crop and livestock co-operatives in the UK. Its annual *Directory of Agricultural Cooperatives in the UK* contains data on sales, profits, share interests, bonus payments and membership. This source has been used to map the distribution of co-operatives in the UK (Bowler 1972, 1982a).

Many of these sources of information are useful as background material and for identifying general trends in the agricultural industry. However, they are very fragmentary and often inadequate for the monitoring of structural and land use changes in agriculture. Many are infrequently used by research workers, especially when compared to the Agricultural Census which totally dominates agricultural studies at the subnational level.

Regional, county and parish data sources

While some of the data sources already mentioned permit a regional breakdown of agricultural activities, one major source has attracted the attention of social scientists for subnational studies – the annual **Agricultural Census**. This is one of the longest running series of statistics collected by central government and the only real source of information on agriculture at county and parish levels. The Agricultural Census is undoubtedly the prime source for any examination of agricultural change; it provides 'a large body of statistical information which grows in complexity and sophistication through time' (Robinson 1988: 284). Among its major attractions are its continuity (from 1866) and national coverage, although the number of items on the census schedule has increased considerably over the years, from 50 in 1919 to 249 in 1979 (Robinson 1988).

The various agricultural ministries keep a farm register, which is continuously updated; complete enumeration is difficult because there are no cadastral records or a proper system of land registration in the UK. As a consequence, up to 500 000 ha may be omitted from the census (Tym and Partners 1984). Nevertheless, a census form is addressed to all known occupiers of agricultural land in the UK, who have a statutory obligation to complete it on 4 June each year (1 June in Northern Ireland) see also Ch. 7). Information is gathered for individual *holdings*, so that a farmer with three separate holdings which are worked as one *farm* could complete three postal schedules. Over time, however, farmers have been increasingly encouraged to return just one census form where different holdings are farmed as one farm unit. Consequently, the distinction between holding and farm has become blurred, adding to the problem of inconsistency with Agricultural Census data (see below).

Farmers are asked a wide range of questions relating to areas of crops grown, numbers of livestock, farm size, the sex and type of farm workers, land tenure, farm woodland and rough grazing (excluding common land). From this information, the different agricultural ministries provide a number of tabulations and analyses, including a type-of-farming classification (based on standard man-days (SMDs), where 1 SMD is equal to an 8-hour working day), where farms have been classified each year since 1964 into 14 types according to the labour input values of their various enterprises (Church *et al.* 1968; Ilbery 1985).

Information from the Agricultural Census is not made available at the individual farm holding level. Instead, it is aggregated to parish, county, regional and national scales. The data are published in the form of different annual and longer-term reports (see MAFF 1968), except for the so-called annual parish summaries which are housed and available for examination at the Public Record Office in Kew Gardens, London and the Scottish Record Office. Summaries are not available for parishes which contain less than three farms (for reasons of confidentiality); such parishes were traditionally combined with an adjacent parish. Now, the data for such parishes are combined into a composite 'others' category which has no specific location in a particular county.

By providing data at parish and county levels, the Agricultural Census can be used to 'demonstrate the spatial variation in farming over wide areas and to study the evolution of these changes' (Clark *et al.* 1983: 155). The same authors identified four main uses of such data:

1. The simple mapping of agricultural data, as in Coppock's (1976a, b) agricultural atlases of Scotland and England and Wales.
2. The derivation of agricultural regions, as demonstrated by Anderson (1975) for England and Wales, and Ilbery (1981) for the county of Dorset.
3. The study of change over time, as in Harvey's (1963) classic study of the expanding Kentish hop industry, Bowler's (1982a) examination of regional change in UK agriculture since the 1950s, and Robinson's (1988) analysis of agricultural change from the mid nineteenth century to the present.
4. The diffusion of new agricultural developments over space and time, including, for example, the spread of maize in the UK (Tarrant 1975) and the oil seed rape revolution (Wrathall 1978).

   The Agricultural Census has also been used as a basis for the ITE's National Land Classification Database (see Ch. 7) and the economic modelling exercise on the possible consequences for British agriculture of changes in the Common Agricultural Policy (Harvey *et al.* 1986).

However, it should be emphasized that the Agricultural Census is intended mainly as a guide to agricultural production and not as a record of land use. It is, therefore, often being used for purposes other than for which it was designed. Indeed, there are two groups of problems which confront researchers when using the Agricultural Census for studies of agricultural change and land use (Clark 1982; Ilbery 1985): the first concerns the use of the parish as a statistical and geographical unit (Coppock 1960, 1965); and the second relates to the reliability and comparability of data.

Various issues are relevant with regards to the parish. First, the boundaries of civil and 'agricultural' parishes do not coincide. This is because land is often recorded as being in a parish to which it does not belong. Before 1949, a farmer's place of residence determined the parish to which farmland was returned (even if only a small percentage of the farmland was in that parish). Since then, farmland has been allocated to the parish in which the bulk of the land lies. While this helped to improve accuracy, the trend to larger, more fragmented farms, together with farmers being encouraged to complete one census return per farm rather than per holding, have exacerbated the parish boundary problem. Second, parishes vary enormously in size, shape and physical characteristics. Consequently, the level of aggregation and thus generalization will differ: the larger the parish, the greater the level of generalization, such that important enterprises may be 'hidden'. This problem is, of course, exaggerated as the data are also aggregated up to county and regional levels. Third, the number of parishes has fallen since the late nineteenth century; Clark (1982) has suggested that there has been a 26 per cent decline since 1870. This, together with the local government boundary changes in the mid 1970s, have helped to break the continuity of the data. Finally, parishes (and counties and regions for that matter) are administrative units and not fundamental units of agricultural activity.

Such problems as these are enhanced when the reliability and comparability of the data are considered. The census form is completed on one particular day of the year (4 June) and it is possible that many spring crops (already harvested) and autumn crops (not yet planted) go unrecorded; this could be a special problem when double-cropping and interplanting take place. Not only is it uncertain as to what the farmer is actually recording, but there is evidence of ignorance and a lack of co-operation by farmers (Clark 1982; Ilbery 1985; Robinson 1988). Many farmers appear to 'round up' to a convenient figure the size of farm and the number of livestock and/or area of crops. In upland areas, where use is often made of common land (which is not recorded in the census), the size of farm is not always known; similarly, over- or under-recording of numbers of hectares may occur, depending on whether a farmer is seeking a grant or trying to avoid taxation! Historically, some regions, and the south-east of England in particular, had a rather high refusal rate, and today areas of small farms and complex tenure systems (as for example in many urban-fringe areas) are unlikely to be adequately represented by the Agricultural Census. Estimates are made for 'missing returns', based either upon previous returns or national trends. In any one year, up to 15 per cent of farms/holdings do not return their census forms (Clark 1982).

The comparability of the census over space and time also varies. Much depends, for example, on farmers' perceptions of what constitutes a particular land use. This has been a consistent problem with regards to rough grazing, which means different things to farmers in different areas. One difficulty relates to the existence of rough grazing over which common rights extend rather than sole grazing rights (Robinson 1988); this has been especially problematical in Scotland with its high proportion of upland. Various amendments have been made to the definition of rough grazing, but these have only added to the confusion and reduced the ability to compare figures between different years. Comparison over time has also been made difficult by the changing definition of a farm. A major change occurred in 1968 when holdings were excluded if they had less than 4 ha of crops and grass, no full-time workers or were smaller than 26 SMDs (40 SMDs after 1973). This led to the removal of 47 000 holdings in England and Wales, although 2000 statistically significant holdings (mainly intensive livestock units) of less than 4 ha were included in the census after 1973 because they exceeded the 40 SMD threshold (Clark 1982).

During the 1980s, attempts to harmonize the UK Agricultural Census with other members of the European Community (EC) have led to the progressive replacement of the SMD by standard gross margins, another series of weightings applied to each enterprise. The total gross margin is converted into a new size classification of farms based on the European size unit (ESU) (1 ESU is equal to approximately £500 of standard gross margin). As a consequence, holdings of less than 1 ESU, under 6 ha and employing no full-time labour are 'statistically insignificant' and thus excluded from the census (Ilbery 1985). In Northern Ireland, for example, this has led to a 21 per cent decline in the number of enumerated holdings (Clark 1982).

The agricultural industry in the UK is presently undergoing considerable change and this places the authorities responsible for the census in a difficult position. It has long been known that the census does not reveal all that there is to be known about farming, especially when it comes to agricultural marketing, farm incomes and the decision-making behaviour of farmers. The real dilemma revolves around the decision of whether to change the Agricultural Census to accommodate recent developments in farm diversification and off-farm sources of income for example, or to maintain it as it is for purposes of consistency and comparability; experience suggests that the census will continue to be modified, as it has in the past. Whatever happens, and despite the problems listed here, the Agricultural Census remains the only major source of information for agricultural studies at regional and lower scales of analysis.

Farm-based data sources

Although farm surveys have been undertaken by government and other bodies, the resultant information is normally aggregated and published as regional and national summaries. However, with the inadequacies of such 'secondary' sources as the Agricultural Census, there has been a growing trend among researchers to conduct personal farm surveys and generate farm-level databases. This raises a series of issues concerning different types of questionnaire survey (e.g. postal and personal), questionnaire design and sampling, which are discussed in Ch. 11. For example, a major problem is establishing a sampling frame on which to base a questionnaire survey (Clark and Gordon 1980). Lists of farms are usually not available from the agricultural ministries. Consequently, research workers rely on such general directory sources as NFU lists or *Yellow Pages*, or on specific lists provided by specialist organizations like the various marketing boards, *Soil Association* (for organic farmers), and *Home and Freezer Digest* (as used by Bowler 1982b in his national survey of pick-your-own (PYO) farmers).

Information gathered at the individual farm scale allows an examination and greater understanding of farm systems, farm families, decision-making behaviour and farmers' attitudes and motivations. Unfortunately, surveys of farms and farmers are often undertaken on an *ad hoc* basis for specific agricultural research purposes; there is no co-ordination of surveys and very little co-operation between different research groups. Examples of such farm surveys would include a postal questionnaire of farms in England and Wales with a PYO scheme (Bowler 1982b), farm diversification on the urban fringe of Birmingham (Ilbery 1987), occupancy change and the farmed landscape (Marsden *et al.* 1986), and farm-based accommodation in England and Wales (Evans and Ilbery 1989).

One attempt to produce a national database of individual farms is the Farm Database compiled by Spa Database Marketing. The database was initially established in 1975 with information collected from a wide range of sources (including many listed in this review) and from representatives of different companies. It has been updated and improved by field and desk research each year; monitoring of change over time is thus not possible. By 1989, when the database was first marketed, information on 178 000 farms in Great Britain (over a 90% sample) was available in the form of farm type, farm size, farm location, and specific types of crops and livestock. The database, therefore, offers a potentially exciting new source of information for farm-based research. However, as with all marketing companies, a charge is made for the data, the reliability of which is still to be assessed.

A rather different approach to the gathering of agricultural information war adopted by the Rural Areas Database (RAD). Established at the ESRC Data Archive in Essex University in 1986, RAD aimed to provide a central, comprehensive information resource for rural research, planning and policy-making in the UK. This involved collecting, storing and disseminating agricultural, environmental and socio-economic data relevant to rural areas; in this sense RAD played a co-ordinating role that previously did not exist. RAD was also actively involved in the development of GIS software, in an attempt to integrate different data sets on a consistent spatial basis (Walford *et al.* 1989). The main depositors were DAFS, WOAD and MAFF, but the results of certain individual farm surveys (e.g. Bowler 1982b; Ilbery 1987) were also deposited with RAD. Unfortunately difficulties in co-ordinating the wide disparity of sources led to further development of the database ceasing in 1989.

Agriculture is a dynamic activity and changes in response to a wide range of factors. There is a need to monitor agricultural change and seek the views of the people directly involved, i.e. the farmers and supply and processing companies; both are required for effective policy decisions. However, the necessary information is either not available or in a form that is unsuitable for research purposes. This is especially the case with the increased integration of agriculture into the wider food chain and the development of agribusinesses. The Agricultural Census remains the most important source of agricultural information, but this suffers from a range of different problems and does not contain data on agricultural marketing and agribusiness activities.

14.2 Forestry

Sources of information on forestry in the UK are sparse, scattered and generally deficient. The forestry industry is complex in structure and organization, and there is no forestry equivalent of the annual Agricultural Census. Data on forest areas and their trends are relatively comprehensive by comparison with those on topics such as forest use, ownership and employment, but even these data are far from adequate. In particular, little useful information is published at scales below those of the constituent countries of the UK. One basic problem is that the overall area of forest and woodland is divided between the state and private sectors, between broad-leaved and coniferous species, and between 'high' forest and other woodland. It is not surprising that inadequacies exist when information can be broken down in these various ways, as well as by country.

General sources: structure, areas and trends

An introductory outline of the structure and organization of British forestry, including some basic statistics, is contained in the glossy publication *British Forestry,* published by the Forestry Commission (FC) (1986). Rather more detail, together with future projections, is contained in *Beyond 2000: The Forestry Industry of Great Britain* (Forestry Industry Committee of Great Britain (FICGB) 1987).

A useful basic source of data, at the reconnaissance level, is the *Annual Abstract of Statistics.* This source has in recent years contained 10-year figures on forest area in the UK, Great Britain and Northern Ireland. In addition, data for Great Britain are provided for both the FC and private sectors. For the former, total estate, plantable land acquired, total area planted, area lost by fire and volume of timber removed are covered. For the private sector, data are given for the area of productive woodland, total area planted and volume of timber removed. Separate data, similar to those quoted for the FC sector, are given for state forestry in Northern Ireland.

This source is widely available in many public libraries, and is valuable because of its accessibility. It does, however, suffer from a number of disadvantages. For example, new planting (afforestation) and restocking are not distinguished, nor is there disaggregation by type of woodland or by country (other than Great Britain and Northern Ireland). Other statistical abstracts are more helpful in this respect, although they may be less widely available. For example, the *Digest of Welsh Statistics* quotes the FC and private-sector woodland area and planting rate, as well as area of fire damage and volume of timber felled, all for 1950, 1960, 1970, 1980 and then annually from 1985. Forestry data in the *Scottish Abstract of Statistics* are more comprehensive. In recent years statistics have been listed on forest and woodland area by principal species, type and ownership (i.e. state or private sector), as well as on new planting and restocking rates. In addition, data are given on number of FC industrial staff, harvesting and consumption of Scottish timber, and FC finances in Scotland. Included in the latter data are statistics on cash receipts from sales of produce, recreational activities and so on, as well as on total expenditure. Since the forest area is relatively larger and planting rates are higher in Scotland than in other parts of the UK, the *Scottish Abstract of Statistics* is a useful and valuable basic source of information on commercial forests.

Most of the data published in the standard national abstracts are derived from the FC annual reports and accounts (Appendix 14.1). These are therefore a fundamental source, even if they are sometimes less widely available than the *Abstracts*. (Basic facts and figures from the reviews are presented in summary form on small leaflets available from the FC.) The annual reports normally contain 10-year runs of figures on private-sector and FC planting, broken down by country and by new planting/restocking. An example of the kind of trends that can readily be established from this source is illustrated in Fig. 14.1. In addition, data on FC land acquired and sold

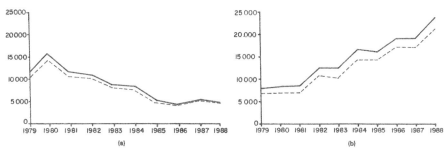

Fig. 14.1 New plantings of trees in Great Britain and Scotland, 1979–88: (a) FC; (b) grant-aided private sector. (—)GB; (---) Scotland. (*Source:* Based on data in FC *Annual Reports and Accounts.*)

are listed, together with information on topics such as timber sales and FC recreational facilities. Statistics on land use on FC land, planting and timber production are provided for each FC forest district, but this breakdown is singularly unhelpful as forest districts rarely coincide perfectly with local authority areas or with any other standard statistical unit. Furthermore, they have been radically altered through time.

Other sources of areal information include the agricultural censuses, maps, and air photographs and remotely sensed images. Care needs to be exercised in the use of Agricultural Census information: in addition to the other problems of that source previously outlined in this chapter is the fact that statistics on 'woodlands' in practice relate only to farm woodlands. Only forest and woodland forming part of agricultural holdings are covered. The use of this source is reviewed by Essex (1984). Standard map series can be useful, but as a source of information on forest area and trends they suffer from two main limitations. First, conventions for the depiction of woodland type have varied through time, so that care has to be taken to distinguish apparent trends, i.e. no more than a function of changing cartographic conventions, from real trends. Harley (1975) is helpful in this respect. Second, a time-lag exists between the establishment (or removal) of a forest and its depiction on a map. This time-lag is not necessarily constant. Major changes are usually recorded (and depicted) much more rapidly than minor ones. Satellite imagery might be expected to be a useful and up-to-date source, but in practice newly established forests may not be discernible for a number of years, until the forest canopy has developed.

A useful review of sources for assessing woodland change is presented by Watkins (1985). The same author discusses the use of FC census material in relation to woodland change (Watkins 1984a): this is a useful introduction to FC census reports and unpublished returns.

Some local planning authorities refer to forestry issues in documents relating to Structure Plans and local plans. Since the prominence of forestry issues is highly variable across the country, it is not surprising that coverage is patchy. There has been an increasing tendency, however, for strategic planning authorities in Scotland (i.e. the regional councils) to take an interest in forestry matters in general and in afforestation in particular in recent years.

Information on forest ownership is very scarce. Aggregate breakdowns into the private and state sectors are contained in standard sources such as FC annual reports and abstracts of statistics, but otherwise little information is publicly available. Research on landownership (including forest ownership) is easier in Scotland than in other parts of the UK, because of the existence of the *Register of Sasines*, housed in Edinburgh (in effect this is a public register of title to land in Scotland). This source is difficult to use, but it does offer the possibility of sketching a rough outline of the structure of ownership and how it varies across the land (see Mather 1987). Both the FC and the private forest-management companies regard such information as they have on forest ownership as highly confidential.

Economics

Forestry Commission annual reports and accounts are a basic source. Recent reports have contained useful summaries ('salient facts') of the finances of the FC (in both its roles of forestry authority and forestry enterprise) for each of the previous 5 years. More detailed financial statements, and statements of accounting policies, are also listed. This information is illuminated by the reports of the National Audit Office (1986) and of the House of Commons Public Accounts Committee (1987).

Information on the economics of private forestry is less readily available. Sample-based data on income and expenditure of private forest enterprises are available in the form of economic surveys of private forestry produced by university forestry departments (for example Todd *et al.* 1988). Timber market reports and reports of timber auctions are contained in the quarterly magazine *Timber Grower*, while reviews of forest property markets and tables illustrating timber-sale price tends are contained in the monthly journal *Forestry and British Timber*. Private forest-management companies and land agents also produce financial information in their promotional literature and house journals: one example is the *Forestry Investment Review* (e.g. Economic Forestry Group 1989, spring issue). These companies also circulate sales lists of forest properties and bare land properties for afforestation. Little comprehensive information is available on the market for forests and for land for planting, but aggregated data for sales of agricultural land for afforestation in Scotland (where most UK afforestation takes place) are available in the annual *Economic Report on Scottish Agriculture* (DAFS).

Employment

Data on forestry-related employment are especially inadequate. Some information is available from the Census of Population (CoP), but it is based on 10 per cent samples. With the small numbers employed in forestry, the sampling errors are large. Class 02 of the 1980 Standard Industrial Classification (SIC) is defined as forestry. The *Employment Gazette* lists figures from each Census of Employment (CoE), broken down by the standard regions (Ch. 2). For example, figures for 1987 are contained in the issue of October 1989 (pp. 540–8). Employment in the FC sector is covered in the annual reports and accounts, but there is no corresponding source for the private sector. Occasional estimates and informal censuses are undertaken: one is reported in the FC *Annual Report and Accounts for 1986–87*. This is a useful source for the mid 1980s, but adequate time series are lacking. Mechanization has meant that employment per unit area has decreased rapidly over the last few decades, and intensity of employment fluctuates greatly according to the stage of the forest rotation. Extrapolation and the use of average figures are therefore fraught with danger and difficulty, and the results of academic research projects, for example, rapidly become dated.

Forest recreation

Published data on forest use relate almost exclusively to timber production, but some information on forest recreation is also available, at least for the FC sector. FC annual reports and accounts contain financial data on commercial recreation (such as chalet letting), while information on facilities for non-commercial information is also contained in the same source. A report of a recent attempt to quantify the economic value of recreation in FC forests is contained in Willis and Benson (1989).

Policy, planning and grants

FC annual reports and accounts are convenient sources of policy statements and statements on current issues, which are often published as appendices. In addition, the FC publishes pamphlets outlining its policies on various issues. It also produces management guidelines, such as the recently issued *Forests and Water Guidelines*. Details of grant schemes such as the Woodland Grant Scheme are also available (usually free of charge) from the FC. The use of grant aid to encourage planting is reviewed by Watkins (1984b).

Details of consultation procedures relating to the approving of planting-grant applications and to the issuing of felling licences are also available in FC pamphlets. A useful review of public control of woodland management is contained in Watkins (1983). Until recently, the (town and country) planning system has not impinged on forestry matters in a major way. In the mid 1980s, however, the question of the impact and control of afforestation became a controversial issue, and important policy papers were produced by the Countryside Commission for Scotland (CCS) (1986), the Nature Conservancy Council (NCC) (1986) and the Convention of Scottish Local Authorities (COSLA) (1987). Since then, regional planning authorities in Scotland have begun to draw up indicative forestry strategies which outline preferred areas for afforestation. These strategies will ultimately be incorporated in structure plans.

Land capability for forestry

A physical classification of land capability for forestry in Scotland has recently been produced in the form of maps at a scale of 1 : 250 000 with accompanying explanatory booklets (Macaulay Land Use Research Institute (MLURI) 1988). No comparable source exists for the other parts of the UK, but some reference to forestry is contained in the regional bulletins produced by the Soil Survey of England and Wales (SSEW 1984: e.g. *Soils and their Use in Northern England*).

The assessments contained in these sources are purely physical in nature: other facts such as conservation, planning policies and ownership are not included. Some of them may, however, be incorporated in the Scottish regional indicative forestry strategies.

Unpublished information

In addition to the published information outlined above, a large amount of unpublished data is held by the FC and other bodies. These data are usually confidential, especially when they relate to private woodlands. Problems may persist even where issues of confidentiality do not exist or can be overcome: retrieval is often very difficult, but it may become easier as the computerization of FC records proceeds.

14.3 Fishing

While, in the fisheries of the UK, statistics are collected and published at the national level, in most cases a breakdown is available between the three parts of England and Wales, Scotland and Northern Ireland. There is also considerable regional and local detail available, especially in Scotland. The main database relates to commercial fisheries, and the amount of published information available for both recreational fishing and fish farming is still limited. There are substantial elements of consistency in the information published over the last 20 years, but there have been inevitable changes as the industry itself has changed.

The highest level of responsibility for administration and data collection rests with MAFF. However, with the great importance the industry has in Scotland, Scottish sources have always been more extensive and detailed, and this has in recent decades been continued under DAFS. MAFF publishes annually the *Sea Fisheries Statistical Tables* (*SFST*) for the UK, but the annual *Scottish Sea Fisheries Statistical Tables* (*SSFST*) are longer and more detailed. The best local detail is found in the *SSFST*, in which the creek returns give details for all landing points of significance.

An additional source are the annual reports of the Sea Fish Industry Authority (SFIA), which in the 1980s has effectively replaced the former government agencies of the White Fish Authority and the Herring Industry Board. The statistical material in the SFIA reports partly overlaps both the *SFST* and the *SSFST*, but they give some additional information on Northern Ireland and on fish consumption.

In recent decades the importance of commercial fisheries in the UK has declined in both relative and absolute terms, and the industry has experienced considerable structural changes. The volume of published material has also been much reduced, with the discontinuation of the former annual fisheries reports for both England and Wales (after 1945) and Scotland (after 1979), while statistical tables have been both reduced in number and condensed. Both the *SFST* and *SSFST* made the change from imperial to metric weights in 1975; apart from this the *SFST* has continued since 1970 with only minor modifications, but there have been several important changes in the *SSFST*.

The most useful single compilation of a list of sources of fisheries statistics for the UK is that by Scott (1981) for the SFIA, but it is now starting to be dated (see Appendix 14.2).

Landings data

Both *SFST* and *SSFST* give prominence to national totals in their data. In both publications fish landings are the main component, and are regularly given by both weight and value. In the *SFST*, the areal breakdown of most of the data is at the level of the three component parts of the UK. Landings in Scotland are now about three-quarters by weight and about two-thirds by value of the overall total, while the contribution of Northern Ireland is only about 3 per cent. For selected years from 1938 data are given for landings of the three main groups (i.e. demersal, pelagic and shellfish), and also for the 14 most important species (cod, herring and so on) individually. A separate tabulation of unit values shows that the average values have increased about 50 times since 1938 and about 9 times since 1970. For the current year there is a detailed tabulation of the landings of over 50 species.

There is considerable detail for the landings at individual ports. In the *SFST*, landings of 50 species at 13 main ports are specified (Fig. 14.2), although the grouping of the remaining landing places under 'other ports' leaves the location of over 40 per cent of the landings unclear; however, the landings at them are recorded in the unpublished annual 'Small ports exercise' (see Appendix 14.2) undertaken by MAFF. Landings for the 5 leading Scottish fishery districts are also given in the *SFST*, but data for all 20 Scottish districts are given in the *SSFST*, together with a map to show the extent of each district (Fig. 14.2). Until 1981, the creek returns in the *SSFST* gave details at 200 places or stretches of coast, but this has now been reduced to 138. Monthly returns of landings are also available in some detail for all parts of the UK, and weekly returns for England and Wales and for Scotland.

Information is given on the sea areas in which catches are made, and both *SFST* and *SSFST* include maps to show the extent of 14 recognized statistical areas around the British Isles from which the great part of catches now come (Fig. 14.3); and minor statistical coverage is given on catches from grounds as far away as the Grand Banks in the west, the Barents Sea in the north, and the Bay of Biscay in the south. Some groupings of sea areas are used, especially where the areas or catches are small. In the *SFST*, data on catches of 12 main species from the different sea areas are given, but in the *SSFST* the detail extends to over 50 species. These show that the North Sea is easily the most important single area for the national fish catch, contributing about one-half, while almost one-third comes from waters to the west of Scotland.

Fig. 14.2 Main ports of landing in Great Britain and Scottish fishery districts. (*Source:* Based on data in *SSFT* and *SSFST.*)

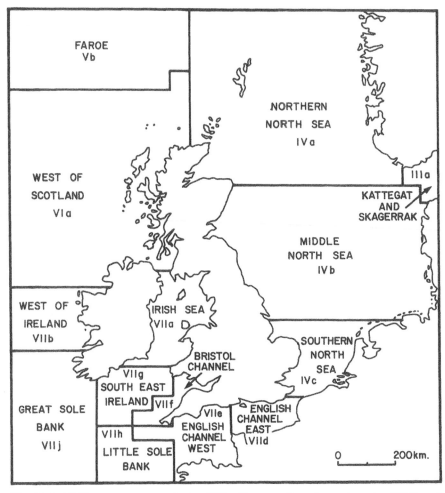

Fig. 14.3 Fishing areas around the UK. (*Source:* Based on data in *SSFT* and *SSFST*.)

In the *SFST* total catches made by different gears (e.g. trawls, seines) from different sea areas are given, but are not differentiated by species; fishing effort is also given, measured by numbers of vessel arrivals and days' absence. The *SSFST* here gives fuller data, in that 50 species are tabulated by fishing gear and sea area, while in fishing effort a measure by hours' fishing is added.

While a variety of other catch data are regularly published, there is very limited areal breakdown in them. The *SFST* for the three component parts of the UK gives total landings for four length classes of vessels, from under 24 m to over 43 m; and the *SSFST* gives total landings for five length classes of vessels, from under 12 m to over 34 m, and also distinguishes between vessels from Scotland, England and Wales, and Northern Ireland. There is also a separate tabulation in the *SSFST* of landings by 18 different fishing methods for the 5 size classes of vessel; and there is a tabulation of monthly landings for the 6 most important species. Despite the much reduced size of the distant-water fleet since the general extension of national fisheries limits to 320 km in the 1970s, distant-water landings are still given in the *SFST* for the three main ports of Hull, Grimsby and Fleetwood; and there are separate aggregate total landings for frozen fish which substantially comes from distant-water fishing.

Of considerable geographical moment is the mobility of modern fleets, whereby landings are often made at other ports than the vessels' home bases. This is often poorly reflected in published data on fishing, but there is a good overview of the situation in Scotland where since 1979 there has been a tabulation of base districts of vessels by landing districts. This is a cross-tabulation in which all districts are entered both in vertical columns and horizontal rows. The aggregate of the entries in the vertical columns gives the total of the landings in a district made by both the home fleet and visiting vessels; while the total of the entries in horizontal rows gives the total landings in Scotland of the fleet of a particular district.

Fleets and fishermen

A complete list of all Scottish boats over 10 m is now published annually by DAFS, which gives essential particulars, including ports of registry, and a similar list is now produced by MAFF for England and Wales. A digest of these data is also given annually in the *SFST*: it gives for the three parts of the UK numbers of vessels and aggregate tonnage by five length classes. Also given are numbers of vessels by five size classes for six main ports in England and Wales, and five main ports in Scotland, and the fleet is also subdivided according to the main fishing method employed. In the *SSFST* the data are fuller with numbers in 7 size classes being given for all 20 fishery districts. The SFIA reports also give summaries of grant and loan aid for vessel construction and improvement for the three parts of the UK.

The *SFST* gives numbers of fishermen, divided into regularly and partially employed, for the three parts of the UK, and employment has been fairly stable since 1970 with *c.* 25 000 fishermen after earlier reductions. These data are in effect considerably more accurate than those given in the DE's CoE, as many fishermen are self-employed; there are also significant numbers who are part-time, and tend not to feature in other enumerations. The SFIA reports give numbers of fishermen in the nine major UK ports, while the *SSFST* gives numbers in all Scottish fishery districts. The latter publication formerly published considerable detail of shore employment, but this was discontinued in the early 1970s.

The foreign contribution

The contribution made by foreign nations in the UK fish supply are shown in the *SFST*, with both data on landings by foreign vessels in the UK and imports. Details are also given for the landings of seven main species at the ports of Hull, Grimsby and Fleetwood; and landings are also subdivided into those of 10 foreign nations, including 6 of our EC partners and the Scandinavian countries other than Denmark. In the *SSFST* landings from the same 10 countries are given, but also with detail for 39 species, 10 ports of landing, the sea areas fished and the fishing methods employed.

The *SFST* data on foreign trade show that, since the enforced contraction of distant-water fishing, the contribution of imports to the UK fish supply has risen to over one-third in tonnage and almost one-half in value. Data are published on trade with main trade partners. Despite modern trade orientations, imports from EFTA still vie with those from EC countries; both are comfortably exceeded by other imports, mainly due to canned salmon and tuna from the USA and Canada. With modern increased trade flows, imports and exports for the UK are now approximately in balance in value terms, but a considerable excess tonnage is exported, largely because of the dispatch of low-value klondyked pelagic fish to the USSR and other countries. However, the EC countries take about two-thirds of exports by value.

Fish consumption

The *SFST* shows trends in per capita fish consumption over the last 50 years; there has been little aggregate change since 1970, with consumption varying mainly between 7 and 8 kg per capita annually, although the proportion of shellfish has risen and that of herring has fallen. The SFIA reports show the areal breakdown of household consumption between seven TV regions, and between fresh and frozen fish. In all regions now except the North (i.e. Scotland and Border; and Yorkshire and Tyne-Tees) the consumption of frozen now exceeds that of fresh fish. There are limited areal variations in fish consumption, but the leading areas are now Southern, and Scotland and Border – the two regions with over 10 kg per household yearly.

Freshwater fisheries

Data on salmon and trout fisheries are published annually by MAFF for England and Wales and DAFS for Scotland. These give numbers and weight of fish caught for 9 regions in England and Wales, and 11 in Scotland; and in Scotland there is a breakdown into 62 fishery districts. There are in both publications separate data for commercial and angling catches, and MAFF also give numbers of angling licences by region. DAFS data have been published since 1952 and MAFF since 1983; MAFF also have in hand the publication of retrospective data from 1951.

Fish farming

An annual release on mimeographed sheets produced by DAFS since 1979 gives data on salmon and rainbow trout production in Scotland. As well as tonnage produced this also gives information on numbers of farms and employment; and also in recent years gives some regional breakdown between Western Isles, Northern Isles and the rest of Scotland. Some additional financial data are available in the annual reports of the Highlands and Islands Development Board. There are no published data on fish farming in England and Wales.

Final remarks

The best data available for fishing are those relating to ports, especially the local level for both landings and boats. Data relating to place of catching are generally not available at local level; and data relating to distribution to the consumer are available only at the level of major regions. Data relating to fish farming and recreational fishing are still rare at the local level.

A considerable amount of data are regularly collected by the various organizations mentioned beyond what is published. It may be possible to obtain non-confidential parts of this information on request. Some of this is regularly mimeographed, and in Appendix 14.2 the main regular compilations are listed.

14.4 Conclusions

Research workers interested in the primary economic sector have access to a large number of potential data sources. However, information on agriculture, forestry and fishing in the UK is very fragmentary and suffers from such problems as a lack of accuracy, comparability and availability at geographical scales lower than national and regional. This greatly hinders studies of change over space and time, especially as data on certain important aspects and recent developments in primary activities often do not exist. In turn and as a consequence, it becomes very difficult to make effective policy decisions. It can only be hoped that better co-ordination and the development of more sophisticated retrieval and geographical information systems will improve accessibility and usefulness of the existing data sources, but many gaps will still remain in the information based on the nature and distribution of primary activities.

Note: changes to the availability of the agricultural parish summaries are referred to in the Editor's preface (p ix).

Appendix 14.1   Statistical series: agriculture and forestry

CSO annual *Annual Abstract of Statistics*
  HMSO, London
DAFS annual *Economic Report on Scottish Agriculture*
  DAFS, Edinburgh
DANI annual *Statistical Review of Northern Ireland Agriculture*
  DANI, Belfast
FC annual *Annual Report and Accounts*
  FC, Edinburgh
MMB annual *UK Dairy Facts and Figures*
  MMB, Thames Ditton
MAFF annual *Agricultural Statistics UK*
  MAFF, London
MAFF annual *Agriculture in the UK*
  MAFF, London
MAFF annual *Farm Incomes in the UK*
  MAFF, London
Scottish Office annual *Scottish Abstract of Statistics*
  Scottish Office, Edinburgh
Welsh Office annual *Digest of Welsh Statistics*
  Welsh Office, Cardiff
Welsh Office annual *Welsh Agricultural Statistics*
  Welsh Office for Agricultural Development, Cardiff

Appendix 14.2 List of sources on fishing

DAFS *Fisheries of Scotland* annual reports, 1959–79. (From 1948 to 1958
the same report was issued by the Scottish Home Department, and
from 1982 to 1987 by the Fishery Board for Scotland.)
Fresh Water Fisheries Statistical Bulletin annual 1982–
Landings weekly
Return of Sea Fisheries monthly
Salmon and Trout Farming in Scotland annual reports 1979–
The Scottish Fishing Fleet annual list 1982–
Scottish Sea Fisheries Statistical Tables annual 1887–
DANI *Return of Sea Fisheries* monthly
HIB Annual reports, 1935–1980
HIDB Annual reports, 1966–
MAF *Report on Sea Fisheries* annual 1912–38
MAFF *British Landings of Main Species at Ports in England and Wales
Other than Main Ports* (small ports exercise) annual
The English and Wales Fishing Vessel List annual
Landings of Fresh Fish weekly
Return of Sea Fisheries monthly
*Salmon and Freshwater Fisheries Statistics for England and
Wales* annual 1983–
SFIA *Annual Report and Accounts 1981–82*
*Sources of Statistics Pertaining to the British Fishing
Industry* (compiled by I. Scott) 1981
Trade Bulletin monthly. Previous to 1981 compiled by the WFA
WFA *Annual Reports and Accounts 1951–52 to 1980–81*

References

Anderson K 1975 An agricultural classification of England and Wales. *Tijdschrift voor Economische en Sociale Geografie* **66**: 148–58

Bowler I R 1972 Cooperation: a note on government promotion of change in agriculture. *Area* **4**: 169–73

Bowler I R 1982a The agricultural pattern. In Johnston R J, Doornkamp J C (eds) *The Changing Geography of the United Kingdom* Methuen, London pp 75–104

Bowler I R 1982b Direct marketing in agriculture: a British example. *Tijdschrift voor Economische en Sociale Geografie* **73**: 22–31

Bunce R, Barr C, Whittaker H 1981 *An Integrated System of Land Classification* ITE, Grange over Sands

Burrell A, Hill B, Medland J 1984 *Statistical Handbook of UK Agriculture* Wye College, Ashford

CCS 1986 *Forestry in Scotland: A Policy Paper* CCS, Perth

Church B M, Boyd D A, Evans J A, Sadler J I 1968 A type-of-farming map based on agriculture census data. *Outlook on Agriculture* **5**: 191–6

Clark G 1982 *The Agricultural Census: United Kingdom and United States* Geo Books, Norwich

Clark G, Gordon D 1980 Sampling for farm studies in geography. *Geography* **65**: 101–6

Clark G, Knowles D J, Phillips H 1983 The accuracy of the agricultural census. *Geography* **68**: 115–20

Coleman A, Maggs K 1962 *Land Use Survey Handbook* Second Land Use Survey, Isle of Thanet Geographical Association, Ramsgate

Coppock J T 1960 The parish as a geographical – statistical unit. *Tijdschrift voor Economische en Sociale Geografie* **5**: 22–5

Coppock J T 1965 The cartographic representation of British agricultural statistics. *Geography* **50**: 101–14

Coppock J T 1976a *An Agricultural Atlas of Scotland* John Donald, Edinburgh

Coppock J T 1976b *Agricultural Atlas of England and Wales* Faber and Faber, London

COSLA 1987 *Forestry in Scotland: Planning the Way Ahead* COSLA, Edinburgh

DoE 1987 Earnings and hours of agricultural workers in 1986. *Employment Gazette* July: 347–53

Economic Forestry Group 1989 *Forestry Investment Review* Economic Forestry Group, London

Essex S J 1984 The use of the annual agricultural returns to measure change in the area of woodland. *East Midland Geographer* **8**: 159–62

Evans N, Ilbery B W 1989 A conceptual framework for examining the role of farm-based accommodation in the restructuring of agriculture in England and Wales. *Journal of Rural Studies* **5**: 257–66

Fallows S J, Wheelock J 1982 Self-sufficiency and United Kingdom food policy. *Agricultural Administration* **11**: 107–25

FC 1986 *British Forestry* FC, Edinburgh

FICGB 1987 *Beyond 2000: The Forestry Industry of Great Britain* FICGB, London

Grigg D 1989 *English Agriculture: An Historical Perspective* Blackwell, Oxford

Harley J B 1975 *Ordnance Survey Maps: A Descriptive Manual* OS, Southampton

Harvey D *et al.* 1986 *Countryside Implications for England and Wales of Possible Changes in the Common Agricultural Policy* Centre for Agricultural Strategy, University of Reading

Harvey D W 1963 Locational change in the Kentish hop industry and the analysis of land use patterns. *Transactions of the Institute of British Geographers* **33**: 123–44

Hill B E, Ingersent K A 1982 *Economic Analysis of Agriculture* Heinemann, London

Hooper A J 1988 Monitoring of landscape and wildlife habitats. In Park J R (ed.) *Environmental Management in Agriculture: European Perspectives* Belhaven, London pp 21–34

House of Commons Public Accounts Committee 1987 *Forestry Commission: Review of Objectives and Achievements* House of Commons Paper 185

Hunting Surveys and Consultants Ltd 1986 *Monitoring Landscape Change* vol 1, DoE, London and the Countryside Commission, Cheltenham

Ilbery B W 1981 Dorset agriculture: a classification of regional types. *Transactions of the Institute of British Geographers* **6**: 214–27

Ilbery B W 1984 Britain's uncertain future in the international hop market. *Outlook on Agriculture* **12**: 119–24

Ilbery B W 1985 *Agricultural Geography: A Social and Economic Analysis. Oxford University Press, Oxford*

Ilbery B W 1987 The development of farm diversification in the UK: evidence from Birmingham's urban fringe. *Journal of the Royal Agricultural Society* **148**: 21–35

MAFF 1968 *A Century of Agricultural Statistics Great Britain, 1866–1966* HMSO, London

Marsden T, Whatmore S, Munton R, Little J 1986 The restructuring process and economic centrality in capitalist agriculture. *Journal of Rural Studies* **2**: 271–80

Mather A S 1987 The structure of forest ownership in Scotland: a first approximation. *Journal of Rural Studies* **3**: 175–82

MLURI 1988 *Land Capability for Forestry* MLURI, Aberdeen

Mort D 1990 *Sources of Unofficial UK Statistics* The University of Warwick Business Information Service, Gower, Aldershot

Munton R J 1977 Financial institutions: their ownership of agricultural land in Great Britain. *Area* **9**: 29–37

National Audit Office 1986 *Report by the Comptroller and Auditor General: Review of Forestry Commission's Objectives and Achievements* House of Commons Paper 75

NCC 1986 *Nature Conservation and Afforestation in Britain* NCC, Peterborough

Nix J S 1984 *Farm Management Pocketbook* Wye College, Ashford

Rhind D, Hudson R 1980 *Land Use* Methuen, London

Robinson G 1988 *Agricultural Change: Geographical Studies of British Agriculture* North British Publishing, Edinburgh

Scott I (ed.) 1981 Sources of statistic pertaining to the British Fishing industry. Sea Fish Industry Authority: Occasional Paper Series no. 3

SSEW 1984 *Bulletins 10–15* SSEW, Harpenden

Stamp L D 1940 Fertility, productivity and classification of land in Britain. *Geographical Journal* **96**: 389–412

Stamp L D 1948 *The Land of Britain: Its Use and Misuse* Longman, London

Tarrant J R 1974 *Agricultural Geography* David & Charles, Newton Abbot

Tarrant J R 1975 Maize: a new United Kingdom agricultural crop. *Area* **7**: 175–9

Todd J D, Kupiec J A, Baptie M A 1988 *Economic Surveys of Private Forestry: Income and Expenditure Scotland 1986* Department of Forestry, University of Aberdeen, Aberdeen

Tym R and Partners 1984 Monitoring land use changes. Unpublished report to the DoE

Walford N, Lane M, Shearman J 1989 The Rural Areas Database: a geographical information and mapping system on rural Britain. *Transactions of the Institute of British Geographers* **14**: 221–30

Watkins C 1983 The public control of woodland management. *Town Planning Review* **54**: 437–59

Watkins C 1984a The use of Forestry Commission censuses for the study of woodland change. *Journal of Historical Geography* **10**: 396–406

Watkins C 1984b The use of grant aid to encourage woodland planting in Great Britain. *Quarterly Journal of Forestry* **77**: 213–24

Watkins C 1985 Sources for the assessment of British woodland change in the twentieth century. *Applied Geography* **5**: 151–66

Williams R E 1986 Perspectives on milk marketing. *Journal of Agricultural Economics* **37**: 295–310

Willis K G, Benson J F 1989 Recreational value of forests. *Forestry* **62**: 93–110

Wrathall J E 1978 The oil-seed rape revolution in England and Wales. *Geography* **63**: 42–5

15

Mining, utility and construction industries

Peter Roberts and Derek Senior

15.1 Introduction

This chapter describes and assesses the major sources of information which are available on the activities of the mining, utility and construction industries. At the outset it is important to stress the very diverse nature of these industries and, as a consequence, to note that considerable variations exist in terms of the availability, quality and form of the sources of information which are discussed in this chapter.

Most of the activities which are discussed in this chapter are included within three divisions of the Standard Industrial Classification (SIC). Division 1 includes the energy and water supply industries, Division 2 the extraction of minerals and ores other than fuels, and Division 5 comprises the construction industries. Table 15.1 provides further details on the structure of the industries together with national employment totals. Other divisions of the SIC supply significant quantities of goods and services to these industries and, as is the case for a wide range of other industries, it is apparent that in some regions and localities this inter-industry relationship is of considerable significance. A good example of the strength of this inter-industry relationship is provided by the case of coal mining and the related mining machinery activities. It is also important to note the close relationships that often exist between the various mining, utility and construction industries. The best example of such a relationship is that which exists between coal mining and the generation of electricity. In certain regions, such as Yorkshire and Humberside, the strength of the coal energy cycle (Hills 1984) is an important determining factor in the continuing livelihood of the region.

Although there are a number of important conventional sources of information related to these industries, one characteristic which emerges is the variation in the spatial units which are used for the collection and analysis of information. Many of the activities which are considered in this chapter

Table 15.1 Employment in the UK mining, utility and construction industries, September 1987. (*Source: Employment Gazette* October 1989.)

		Employment (thousands)		
	Division or class	Male	Female	Total
1	Energy and water supply industries which comprise:	428.7	79.1	507.8
11	Coal extraction and manufacture of solid fuels	135.6	7.3	142.9
12	Coke ovens	2.4	0.1	2.5
13	Extraction of mineral oil and natural gas	29.4	5.3	34.7
14	Mineral oil processing	17.9	3.8	21.7
15	Nuclear fuel production	13.4	2.5	15.9
16	Production and distribution of gas and electricity	181.9	51.1	233.0
17	Water supply industry	48.1	9.1	57.1
21	Extraction and preparation of metalliferous ores	2.0	0.1	2.1
23	Extraction of minerals nes	27.0	13.9	30.9
5	Construction	904.9	122.2	1027.0

are organized in unique spatial units; the best example is that of the water supply industry which is organized in regional areas which are related to an aggregation of catchment areas. A second feature of much of the information which is available is that it is initially collected at the level of the individual production unit and it is then often presented at a relatively high level of aggregation. Thus, for example, much of the information on the coal mining industry, which is initially collected on a pit-by-pit basis, is presented for public consumption at a regional or national level of analysis. This implies that there are many gaps in the spatial coverage of information due to the spatially specific nature of many of the industries, and that much of the information which is available is only consistent with information regarding other economic activities at a relatively broad level of analysis. A third feature and difficulty arises because some activities normally present information on production at a national level, while distribution and sales information is available at a regional or local level. A final feature is that many of the industries have experienced changes in the organization of their activities through time and, as such, it is difficult to trace the evolution of production, supply and consumption over the longer term. An example is provided by the water industry where, prior to 1974, many small water undertakings existed. This industry was reorganized in 1974 and new regional water authorities were created. At the time of writing, this industry is in the throes of a further reorganization with the transfer of the ownership of the industry from the public to the private sector. This reorganization could result in further changes in the units of organization and administration.

With these general features clearly in mind, this chapter attempts to present a comprehensive review of the various generally available and other sources of information on:

(a) coal mining;
(b) other mineral and minerals industries;
(c) construction activities;
(d) utilities, both public and private.

The coverage of these four topics varies; this reflects both the absence of a single consistent source of information, and the variety of areas, regions and other spatial units which are used by the industries for the collection and presentation of information.

A useful and often neglected source of local information on many of these industries is the local planning authority. Many county authorities have, for example, produced a minerals subject plan or report of survey. Such a document frequently contains information on the distribution of mineral resources, on the organization of the industries locally and on the output of a range of minerals operations. Two problems exist in using this source of information; first, not all authorities have produced a mineral subject plan which encompasses full details of the industries using such minerals; second, the date of production of such a plan varies between authorities.

15.2 Coal mining

One of the distinguishing features of the coal industry was noted some years ago by Sir Hubert Houldsworth, then Chairman of the National Coal Board, who said, in 1953, that: 'All generalisations about the coal industry or the miner are dangerous; most of them are simply untrue . . . in a sense there is no industry: there are statistically only 900 pits' (Harris 1980). Although there are fewer pits today than there were in 1953, the industry, despite a greater degree of centralization and national control than in the past, is still very much orientated towards the individual colliery. The pit is the primary production unit and, especially under a regime that regards each production unit as a profit centre in its own right, it is the level at which much basic information is collected. Colliery-by-colliery operating results are collected by British Coal on a monthly basis. These records include information on coal production, manpower, output per man-shift and the operating costs of the colliery in terms of the production costs per tonne of coal produced (and, more recently, the cost per gigajoule). Colliery operating results provide basic information for use both by colliery management and by senior management at area and national levels who are responsible for the overall planning of the coal industry. Given that the individual pit is both the basic unit of production and the lowest level for which information can be obtained, then it is apparent that any detailed study of the coal industry should ideally commence with a consideration of the local level of employment, production and profitability.

The above source of information, together with a quarterly summary of the monthly figures, can only be obtained directly from the British Coal Corporation, but unfortunately they do not normally make the information available to individual research workers. Other important sources of information that can be obtained from the British Coal Corporation include their weekly statistical summary and their pamphlets which summarize the major operating statistics for collieries by area. British Coal's nine areas and groups differ significantly from the standard regions, in that they reflect the distribution of coal reserves and of current coal production. These regions (or areas) have been reorganized on a number of occasions since 1947, the date when the National Coal Board was formed. *Regional Trends* provides a useful summary of the major operating statistics for British Coal collieries; this information is presented by standard regions and contains details of the output of saleable coal, the average number of wage-earners on colliery books and output per man-shift.

British Coal also presents an annual summary of the major statistics on coal mining in its *Annual Report and Accounts*. This report is the most generally accessible and widely used source of information on the coal industry. It provides a number of summary tables that trace the evolution of the industry since 1947 and it also includes details of the corporation's current and future investment programme. In addition, the *Annual Report* provides information on technical developments in the industry, new prospects and exploration, the changing composition and patterns of demand in the markets for coal, coal industry personnel (including the activities of British Coal Enterprise Limited), and the organization of the industry.

As well as providing information relating to the deep mining of coal, the *Annual Report* also provides a considerable amount of information on opencast coal mining. This information is provided for the six opencast regions of the UK where opencast mining is currently occurring and, as is the case in the statistical series which are related to deep mining, details are provided of saleable output, employment and the cost of production. Additional information on opencast coal mining can be obtained from the Opencast Executive, who have made available a general summary of the major operating statistics related to this section of the industry in their publication, *Opencast Coal Mining in Great Britain* (British Coal Opencast Executive 1988). Opencast coal is mined both by British Coal and by private contractors who operate under planning and licensing procedures. A number of private contractors, such as Hallamshire Holdings, provide details of their opencasting operations in their annual company reports. Other information on opencast coal operations can be obtained from the County Planning Officers Society who publish *Opencast Coalmining Statistics* on a county-by-county basis. This digest provides details of output from both British Coal and licensed sites; and it also presents a summary of planning approvals and reserves.

There are a number of other important sources of information on the coal industry. The Colliery Guardian publishes an annual *Guide to the Coalfields*.

This digest of statistics provides valuable information on a wide range of topics, and includes a list of collieries and opencast sites, details of coal production, information on mine safety, details of suppliers to the coal industry, and information on the producers of coal. The *Colliery Guardian*, the *Mining Engineer* and a number of other trade and industry journals provide useful additional information on coal mining. Further information is available from the National Union of Mineworkers and the Union of Democratic Mineworkers who both publish fact sheets and a variety of research reports, and from the Coalfield Communities Campaign who have recently commissioned a series of research investigations. A number of local authorities, many of whom are members of the Coalfield Communities Campaign, have also published reports on coal mining in their areas. Information on coal production from small mines, employing less than 30 workers, can be obtained from the Federation of Small Mines of Great Britain. The federation's *Annual Report* provides information on coal production and other matters.

In addition to these coal industry-specific sources of information there are a number of national and international, official, semi-official and other sources of information. Many of these publications provide information on UK coal production at a national level and, while they are of great value in tracing the overall evolution of the industry and in preparing international comparisons, they are of limited value in regional or local studies. Publications such as the *Digest of UK Energy Statistics, Energy Trends*, the *Annual Abstract of Statistics* and *UK Mineral Statistics*, all provide information on national coal production and consumption. Data for two regions can be obtained from the *Scottish Abstract of Statistics* and the *Digest of Welsh Statistics*. Other sources of information on national coal production and consumption include the various Eurostat statistical series, a number of statistical series published by the International Energy Agency (IEA), the specialist reports published, on a regular basis by, for example, the Financial Times (*International Coal Report*) and by stockbrokers James Capel and textbooks such as that by Manners (1981).

Finally, a number of other sources of information exist which are of value. The Association of British Mining Equipment Companies publishes much helpful information. The House of Commons Select Committee on Energy and the Monopolies and Mergers Commission have both published a number of reports with detailed statistical appendices on coal mining. Many local authorities in their minerals subject plans and other documents publish information on coal and the problems facing mining areas. Census of Production information, although generally published at a national scale, can also provide useful insights.

15.3 Minerals

A wide range of minerals occur and are currently produced in the UK. Year-after-year totals for the major minerals produced, and totals for building materials, are produced in the official *Annual Abstract of Statistics*. These are global figures with little applicable use.

A vital and valuable source is *UK Mineral Statistics*, an annual publication by the Natural Environment Research Council (NERC) and the British Geological Survey. This brings together in one volume a wide range of mineral data, from a variety of original sources. It includes data on production and consumption, and also on imports and exports.

The information which is provided covers almost all of the wide range of minerals, including common sand and gravel (for the construction industry) and the more specialist sands (for the foundry, moulding and glass-making). Other common minerals are also included, such as common clays and shales, the specialist clays, limestone and sandstone. A comprehensive guide to the mineral resources of Britain was published in Blunden (1975).

The information is presented for a variety of geographical areas. There are national figures, but an important set of data is that presented for individual counties (and regions in Scotland). These figures give a clear indication of the importance of each county area for individual minerals. Information is also shown on the end use of minerals produced which further indicates the role of each area of the country in the construction industry.

The Crown Estate Commissioners produce figures on marine sand and gravel extraction. They control the licensing of such extraction below high-tide level.

Some of the data in *UK Mineral Statistics* are, however, aggregated into larger totals or larger areas in order to maintain confidentiality because a major producer would otherwise be identifiable, or because a small number of returns have been made. The key source for all such information is an annual census of mineral extraction, currently collected by the Central Statistics Office (CSO) – previously the Business Statistics Office (BSO). Such an annual return from minerals operators has been required since 1895. The commercial and competitive nature of the minerals industry means that such detailed information is highly sensitive.

The *Business Monitor* (PA 1007) published annually by the CSO contains statistics on minerals extracted both by county and by region, and on employment and plant and machinery used.

More detailed studies of individual minerals can be found in the series of *Mineral Dossiers* produced by the Mineral Resources Consultative Committee. There are currently 26 of these, ranging from fluorspar to gold, and from salt to china clay.

The uses made of minerals is fully documented in the above publications, and more up-to-date quarterly figures are included in *Housing and Construction Statistics*. This includes information on building materials and component production, deliveries and stock. The data are wide-ranging and include bricks, tiles, ready-mixed concrete, building blocks and sand and gravel. These data on materials are often used indirectly as a measure of activity in the construction industry, and indeed of the economy as a whole.

Much detailed work on mineral production and sales in the context of the need for future mineral extraction has been done in the last decade. This has been carried out and published by regional aggregates working parties. The 10 bodies were set up by government following the Report of the Verney Committee in 1976. The working parties include representatives from mineral operators, central government and mineral planning authorities. They have produced a series of surveys and updates, both of production and usage, but also of areas of reserve with planning permission. The 1985 survey and forecasts have now been officially incorporated by the Secretary of State for the Environment in his *Minerals Planning Guidance Note 6* which was published in 1989 (Department of the Environment (DoE) 1989a).

The individual working party reports contain details on assessed demand, likely supply from permitted reserves, and other potential and possible reserves. The culmination is the presentation and the assessment of the choices that are available in order to maintain the balance between supply and demand. Strategic planning for the minerals industry has been discussed by Roberts and Shaw (1982).

Further detail about individual mineral workings can be gleaned from the growing number of mineral subject plans, prepared by the mineral planning authorities. Some 15 counties are currently covered by such plans. The White Paper, *The Future of Development Plans* (DoE 1989b), suggests that in future all areas will be expected to produce such subject plans and these will be an increasingly useful source of mineral data. The existing plans have included maps of mineral resources, locations of individual workings, and a series of constraints regarding future development. In the main the plans have also indicated areas of search where it is most likely that future mineral workings will be found. In addition, there are consultation areas where mineral deposits need to be protected from the pressures of other development which could result in the sterilization of the resource.

There are, in addition to the official statistics and plans of central and local government, a number of valuable documents published by the minerals industry itself. Many of these documents are available to researchers on request from the various organizations and the individual member firms. There are a number of associations for the industry; BACMI (British Aggregate Construction Materials Industries) is one of the largest and produces regular statistics from its members on crushed rock, ready-mixed concrete, coated roadstone, and sand and gravel. The association also publishes briefing notes on numerous mineral issues where its members' interests are involved. This includes separate demand estimates which challenge the official forecasts of DoE and the regional aggregates working parties.

Other data are published by SAGA (Sand and Gravel Association) and BQSF (British Quarrying and Slag Federation), mainly reflecting production rates of member companies. In addition SAGA have been pioneers in the making of awards to companies which show good practice in the skill of reclaiming and restoration of former workings. Such awards are indicative of some of the best and most innovative levels of current practice.

323

15.4 Construction

There is a plethora of statistics produced about the construction industry. The industry itself both by its structure and activities is very varied, and a single source of statistics for the industry is virtually impossible to identify. Similarly, since the industry is so central to the national economy as a whole, there are many different and varied interests involved in the collection and publication of data.

The data collected refer to financial activities, to costs and expenditure, or they may concentrate upon employee-related activities such as manpower, earnings, training or accidents. Data also refer to particular sectors of the industry, ranging from the provision of construction services to commerce and business to residential development, leisure provision and transport facilities, and all of these activities are underpinned by the involvement of the construction sector in infrastructure development.

In the case of housing there are five main areas of statistics–physical data, data on value, characteristics of new housing, forward indicators and analyses at the level of the individual local authority of tender stage data. The chief source of official statistics is the DoE's *Housing and Construction Statistics* which is published both in the quarterly and annual volumes. The base for these statistics are the monthly returns made by local authorities. This statistical series concentrates primarily on giving a picture of new housing starts, housing under construction and housing completions on a regional level. More detailed geographical data are provided by the DoE in *Local* Housing Statistics, a quarterly return which presents similar figures for the latest quarter and the latest year by region, county, local authority or new town in England and Wales. Monthly figures are published in the *Monthly Digest of Statistics* which provides data for starts, under construction and completions on seasonally unadjusted bases for the last 6 months, and for recent quarters and years.

These data are widely used in housing and planning forecasts, but are dependent upon local authority returns. There are omissions in some authority returns, and occasionally the published figures are revised because of underestimates on completion.

One of the main sources of unofficial data is that which is produced by the National House Building Council (NHBC), especially in their *Quarterly Private House-building Statistics*. This includes records of intention to build, recorded some 21 days before start date. This information contains some overestimates, as some registrations are cancelled, but it is widely accepted as an early warning of market trends. Other useful information which is produced by the NHBC includes regional data on new house prices, house types, and the construction of traditional and timber-framed housing. The council also publishes general statistics concerning the structure of the industry, including data on the size of companies and on output by the various size categories of building firms. A number of building societies and the Building Society Association also publish useful material on housing.

The construction industry includes a wide variety of activities but, unlike many manufacturing industries, firms which carry out the work are not always easily identifiable. A variety of bodies are involved in many construction projects, such as contractors, public agencies and builders working privately on their own account. The range of work which is undertaken by the construction industry provides an indication of the difficulty of collecting data. Construction activity, as defined by the SIC, is an indicative definition and includes general construction and demolition work, the construction and repair of buildings, civil engineering, the installation of fixtures and fittings, and building completion work.

Official statistics are produced by the DoE and the CSO. The DoE produces quarterly and annual returns which indicate output from contractors and direct labour organizations; these returns indicate values (at current and constant prices) for all work and an indication of new work (at present prices) for contractors. Since 1979 data have been collected for individual projects rather than general returns, and this has improved the reliability of data. The use of valuation certificates within the industry has assisted this method. However, repair and maintenance work is still collected on a firm-by-firm basis, which leads to a lack of any comprehensive definition of the information which is included in these returns. The DoE publishes a widely used index of output at constant prices.

Since 1907 a Census of Production has been published for all production industries including construction. This information covers a wider range of activities than the DoE data but the two are considered to be complementary. The annual results are published in the *Business Monitor* (PA 500). Data prior to 1980 are more detailed; a slimline census has subsequently been introduced in order to reduce the burden on industry. This has been achieved by reducing the number of questions asked and by increasing the use of sampling. The figures produced on gross output are considered to be better than the DoE's firm-based method because they include work done through subcontractors.

A number of unofficial statistics are available related to the many and diverse activities included within the construction industry. These range from the demand forecasts prepared by material producers, such as the Brick Development Association, to the activity studies undertaken by the Construction Plant Hire Association. Such information ranges from data on the costs of production of individual components, to details of associated employment. In addition to these sources of information, many other reports, fact sheets and information packs can be obtained from individual construction firms. In recent years some of these firms have collaborated, mainly in order to enhance their marketing and development activities, through the establishment of organizations such as British Urban Development.

15.5 Utility industries

A wide range of utility industries currently operate in the UK and many of these industries, both directly and indirectly, are of special significance to the economic health of regions and localities. This section examines, in particular, three of these industries (gas, electricity and water) and, in addition, it provides an overview of the major sources of information that relate to the consumption of energy.

The gas industry in the UK developed during the nineteenth century and was initially organized on a local basis. The production of town gas (from coal) was undertaken both by private companies and by municipal enterprises and gas was distributed through a local grid system. In 1949 the industry was nationalized; however, it was not until after the discovery and subsequent exploitation of offshore reserves of natural gas that a truly national industry emerged. British Gas, as the major operator of the gas production and distribution system, is organized at two levels: a national production activity and, through 12 regional boards, a regional supply activity (Roberts and Shaw 1984). As is also the case for other energy production and supply activities, the regions which are used by British Gas do not fully reflect the boundaries of standard regions or of counties. British Gas was privatized in 1986.

Information on the production and transmission of gas is available from a number of sources. British Gas PLC publishes an *Annual Report* which contains a wide range of information on the quantities of gas which are produced and purchased. This information is also provided, in greater detail, by the Department of Energy in the annual report, *Development of the Oil and Gas Resources of the United Kingdom* – the Brown Book (Department of Energy 1989). The Brown Book contains details of both onshore and offshore production on a field-by-field basis, and, in addition, it provides a valuable commentary on gas reserves, exploration, transport systems, certain environmental aspects of gas production, the economic impact of gas production, and the relationship between the gas industry and the development of the UK economy. Other sources of information on gas production include the *Digest of UK Energy Statistics, Energy Trends*, Census of Production reports and the annual reports of a wide range of offshore operators including Shell, BP and other major oil and gas companies. A summary of information on gas is published annually by British Gas in their *Facts and Figures* report.

Many of the above-noted sources also provide information on the supply and consumption of gas. In addition to the information which is provided at national level, the regional gas boards publish material on their operations; this information is summarized in *Regional Trends* (CSO 1989). The Gas Consumers' Council and the Office for Gas Supply produce annual reports which include information on gas consumption. The *Digest of Welsh Statistics*, the *Scottish Abstract of Statistics* and the *Northern Ireland Annual Abstract of Statistics* are other important sources of region-specific information.

Electricity production and supply statistics generally mirror the activities of the gas industry in that there is a broad division of responsibility between national and regional levels. At national level the Central Electricity Generating Board's (CEGB) *Statistical Yearbook* contains a variety of information on the fuelling and operation of power stations and on the transmission of electricity. This information is mainly provided at a national level; however, some local and regional information is contained within the yearbook. The CEGB is responsible for the production of electricity in England and Wales; details of production in Scotland can be obtained from two sources; first, the report of the South of Scotland Electricity Board and second, the North of Scotland Hydro-Electric Board *Report and Accounts*. The Northern Ireland Electricity Service *Annual Report and Accounts* provides details of production and consumption in Northern Ireland. Other information on electricity production and consumption can be obtained from the standard national sources as listed above for the gas industry.

Other sources of information on the operation of the electricity industry include the *Annual Report* of the UK Atomic Energy Authority, reports and statements published by the Electricity Council and the Electricity Consumers' Council, the annual reports and accounts of the various regional boards, the *Handbook of Electricity Supply Statistics*, and a variety of reports from the House of Commons Select Committee on Energy and the Monopolies and Mergers Commission.

These sources, some of which are also relevant to studies of the gas industry, are supplemented by a number of national and international publications which indicate the performance of the UK energy industries in a world-wide context. An especially important source of statistical information and expert commentary is the IEA's annual review, *Energy Policies and Programmes of IEA Countries*. Information on energy production and consumption, at national and regional levels within the EC, can be obtained from Eurostat.

At national, regional and local levels a variety of other services of information on energy production and consumption are available. The *Family Expenditure Survey* (*FES*) provides information on expenditure on energy, while the *Annual Abstract of Statistics* and the *UK National Accounts* provide general information on the operation of the energy industries. Two important additional sources of local information on the characteristics and anticipated operation of energy production processing facilities, are the statements of evidence and reports of local planning inquiries, and the environmental impact assessments which now accompany many applications for planning permission (for example CEGB 1988).

Water supply in England and Wales was, until late 1989, the responsibility of 10 regional water authorities. These regional authorities were created by the Water Act of 1973, which brought together in the 10 authorities the water supply and sewage services previously provided by almost 1600 separate undertakings. The 1973 Act did not affect the operation of 28 statutory water companies which continued to supply water to a quarter of all domestic users. The Water Act of 1989 created 10 regional water companies and the National Rivers Authority, but did not incorporate into the 10 new PLCs the activities of the 28 statutory water companies. Water supply activities in Scotland are the responsibility of local government.

Existing sources of information on the supply of water and the disposal of sewage will not continue to be available in their present form. The 10 regional water authorities all published annual reports and accounts which provided details of water supply in their areas. A summary of the major items of information has been provided by the Water Authorities' Association in their annual *Waterfacts*. Other sources of information on water include the *Municipal Year Book*, the annual *Water Services Charges Statistics* (published by the Chartered Institute of Public Finance and Accountancy), the DoE's *Digest of Environmental Protection and Water Statistics*, Census of Production reports and textbook such as that by Parker and Penning–Rowsell (1980). The future availability and organization of information are, at present, uncertain. It is anticipated that the 10 new water companies will produce annual reports as will the National Rivers Authority and the Director-General of Water Services. A helpful commentary on the implications of water privatization has been provided by Rees (1988).

A common feature of many of the information sources discussed in this section of the chapter is the absence of a common spatial framework. This reflects the structure of the industries, the distribution of the resources which are used or supplied, and the nature of the pre-existing supply patterns which the industries have inherited from their predecessors.

15.6 Conclusions

This chapter commenced by making reference to the extensive and disparate nature of the sources of information which are available on the mining, utility and construction industries. From the detailed discussion of the major sources of information which has been presented in this chapter, three broad themes emerge.

First, the industries portrayed demonstrate the complete range of economic activities identifiable in the economic history of the UK; from early stage resource-based industries such as coal mining (Spooner 1981), to late stage consumer-centred industries such as housebuilding. Second, within the evolution of the industries there can be observed a historic process of concentration of ownership and control; this has caused many adjustments to be made to the operational areas and boundaries of these industries. Third, the quantity and quality of the information which is available on these industries vary considerably, and even when information does exist it is frequently difficult or impossible to gain access to it. Each of these themes is now elaborated.

Mining is an activity which commenced in a pre-industrial era. As such, information (often of an unreliable and dubious nature) is available on the operation of mining for a period of two centuries or more. It is only since the late nineteenth century that information on mining, especially coal mining, can generally be considered to be reliable. Even this generalization is, of course, only fully justified when considering the major producers (since nationalization a single major producer). Other sections of the mining industry are almost entirely in private ownership and it is an industry with a wide range and continual supply of new entrant firms. These firms vary from very small local producers, (small coal-mines and quarries), to major international conglomerates. Information on mining production is, by definition, spatially-specific, but is often difficult to obtain due to the constraints of commercial confidentiality. In summary, the mining industries, and to a lesser extent the utilities, are resource-based activities concerned with providing inputs to a national and global system of production. Information on construction activities is also spatially-specific, at least in terms of the visible outputs. Firms, government at all levels and many other organizations make information freely available on the number and location of houses built, the kilometres of roads and motorways constructed and other project achievements. What is more difficult to deduce is the size and initial and final destination of profits and other intermediate transfers.

A second theme is the historic process of concentration of ownership and control. This has tended to recast and redefine the operational and spatial parameters within which the industries function. Local coal-mining utility and construction firms have, through time, amalgamated and have been taken over; localized patterns of production and utilization frequently reflect all these operational changes. Until the 1980s the operational units and boundaries of coal-mining and the utilities industries had remained constant for some considerable time (since the post-1945 nationalizations). During the 1980s however, some of these industries have been subject to privatization and although area or regional boundaries have often been retained, in future it may prove difficult to compare information on the distribution and structure of production. The construction industry is mainly in private ownership and it is often difficult to trace the full details of the origins and operational characteristics of firms, although generally most firms started by serving a local market (or market segment) and then grew both through self-generated growth and by the acquisition of other (often competitor) firms. As noted above, although it is a relatively simple matter to trace spatially-specific outputs, it is more difficult to identify the intermediate transfers within and between construction firms.

The third theme relates to the enormous variations which exist in the quality and availability of information on mining, utility and the construction industries. Sources such as the CoE are nationally standardized and at a regional and local level provide generally consistent information through time. Other official sources of information have been noted elsewhere in this chapter and sources such as *UK Mineral Statistics*, the Census of Production, *Regional Trends* and the *Annual Abstract of Statistics* all provide useful and reliable information. Although the form and content of the reports of organizations such as the National Coal Board (now the British Coal Corporation), British Gas and the CEGB were for many years both consistent and directly comparable on a year-by-year basis, privatization has now resulted in an increasing diversity of information. Windfall events for researchers have occurred; the sale of the gas and water industries resulted in the general release of a vast amount of very useful information which hitherto was not easily accessible.

As public organizations move into the private sector, the consistency over time of the information which is available is likely to vary considerably. In one sense these industries (gas, water, electricity and possibly coal) will become more akin to the construction industry; an area of economic activity which, as has already been observed, is often difficult to analyse and research through time due to its changing pattern of ownership, control and organization. While information on housing starts and completions is available on a consistent spatial basis over time, it is far more difficult to discover which firms built the houses, how many persons were directly or indirectly employed and at which site, and what the levels of turnover and profitability were. Company reports often provide detailed total or operational unit statistics, but even a brief review of the annual reports of a single major company over a 20-year period will reveal many pitfalls for the unwary researcher. Even profit-related information can be misleading, for many major projects are undertaken by contractually binding but otherwise ephemeral consortia of firms and it is often difficult to determine which firms made what input, where and when. None of the aforementioned difficulties should be taken to suggest that there is a dearth of information on construction sector activities; it is simply the case that such firms have never regarded the needs of spatial analysts as superior to their own immediate requirements and the legal obligations placed upon companies to provide a statement of accounts.

Finally, it should be emphasized that in almost all cases known to the authors, a reasonable and specific request for information, within the preset conditions of commercial confidentiality and sensitivity, will be met, especially if the results of the research are made available to the providers of the information. Considerable value can often be added through the careful analysis of raw or seemingly inconsistent information; the providers of information can sometimes be persuaded to become the sponsors of research. The mining, utility and construction industries provide a fascinating and ever-changing spatial mosaic of economic activities, the study of which demonstrates the complexity and richness of the UK's economic landscape.

Note: changes to the organizations responsible for electricity generation are referred to in the Editor's preface (p ix).

References

Blunden J 1975 *The Mineral Resources of Britain: A Study in Exploitation and Planning* Hutchinson, London

British Coal Opencast Executive 1988 *Opencast Coal Mining in Great Britain* British Coal Corporation, London

CEGB 1988 *Proposed West Burton 'B' Coal-fired Power Station: Environmental Statement* CEGB London

DoE 1989a *Minerals Planning Guidance Note 6* DoE, London

DoE 1989b *The Future of Development Plans* HMSO, London

Fernie J 1980 *A Geography of Energy in the United Kingdom* Longman, London

Harris D J 1980 Coal, gas and electricity. In Maunder W F (ed.) *Reviews of United Kingdom Statistical Sources* vol 11, Pergamon Press, Oxford

Hills P 1984 Planning for coal: issues and responses. In Cope D, Hills P, James P (eds) *Energy Policy and Land Use Planning* Pergamon Press, Oxford, pp 21–68

Manners G 1981 *Coal in Britain: An Uncertain Future* Allen & Unwin, London

Parker D, Penning-Rowsell E 1980 *Water Planning in Britain* Allen & Unwin, London

Rees J 1988 *Water Privatisation and the Environment: An Overview of the Issues* Friends of the Earth, London

Roberts P, Shaw T 1982 *Minerals Resources in Regional and Strategic Planning* Gower, Aldershot

Roberts P, Shaw T 1984 *Planning for Gas in the United Kingdom.* In Cope D, Hills P, James P (eds) *Energy Policy and Land-use Planning* Pergamon Press, Oxford pp 101–22

Spooner D 1981 *Mining and Regional Development* Oxford University Press, Oxford

Sources

British Coal Corporation annual *Annual Report and Accounts* British Coal Corporation, London

British Gas PLC annual *Annual Report* British Gas, London

British Gas PLC annual *Facts and Figures* British Gas, London

CEGB annual *Statistical Yearbook* CEGB, London

CIPFA annual *Water Service Charges Statistics* CIPFA, London

Colliery Guardian annual *Guide to the Coalfields* Colliery Guardian, Redhill, Surrey

County Planning Officers Society annual *Opencast Coalmining Statistics* County Planning Officers Society, Durham

CSO annual *Annual Abstract of Statistics* HMSO, London

CSO periodic *The Business Monitor* HMSO, London

CSO annual *Family Expenditure Survey* HMSO, London

CSO monthly *Monthly Digest of Statistics* HMSO, London

CSO annual *Regional Trends* HMSO, London

DE monthly *Employment Gazette* DE, London

Department of Energy annual *Development of the Oil and Gas Resources of the United Kingdom* HMSO, London

Department of Energy annual *Digest of UK Energy Statistics* HMSO, London

DoE annual *Digest of Environmental Protection and Water Statistics* HMSO, London

DoE annual *Housing and Construction Statistics* HMSO, London

DoE annual *Local Housing Statistics* HMSO, London

Federation of Small Mines of Great Britain annual *Annual Report* Federation of Small Mines of Great Britain, Newcastle, Staffordshire

Financial Times monthly *International Coal Report* Financial Times, London

HM Treasury annual *UK National Accounts* HMSO, London

IEA annual *Energy Policies and Programmes of IEA Countries* OECD, Paris

Mineral Resources Consultative Committee periodic *Mineral Dossiers* HMSO, London

National House Building Council quarterly *Private house building statistics* NHBC, London

NERC annual *UK Mineral Statistics* HMSO, London

North of Scotland Hydro-Electric Board annual *Reports and Accounts* North of Scotland Hydro-Electric Board, Edinburgh

Northern Ireland Electricity Service annual *Annual Report and Accounts* Northern Ireland Electricity Service, Belfast

Northern Ireland Office annual *Northern Ireland Annual Abstract of Statistics* Northern Ireland Office, Belfast

Scottish Office annual *Scottish Abstract of Statistics* Scottish Office, Edinburgh

South of Scotland Electricity Board annual *Annual Report* South of Scotland Electricity Board, Glasgow

UK Atomic Energy Authority annual *Annual Report* UK Atomic Energy Authority, London

Water Authorities' Association annual *Waterfacts* Water Authorities' Association, London

Welsh Office annual *Digest of Welsh Statistics* Welsh Office, Cardiff

16

Manufacturing

Michael Healey

16.1 Main themes

Of all the sectors of the British economy manufacturing has received the most attention from local and regional research workers. Although the number of people employed in manufacturing in the UK was estimated in 1990 to be 5.1 million, a fall of 3.4 million or 40 per cent from its peak in 1966, the sector remains paramount as a generator of wealth and a source of technological innovation. Moreover, despite the growing recognition that shifts in the location of several service industries may influence spatial trends in manufacturing and population as well as respond to them (Daniels 1983; see also Ch. 18), variations from place to place in the changing level of manufacturing activity are still the most important contributor to spatial variations in economic change. As Fothergill and Gudgin (1982: 47) noted, even during a period when British manufacturing was in crisis at the end of the 1970s and early 1980s, 'the pattern of urban and regional growth depends more than ever on what happens to manufacturing employment'.

Unfortunately, during a period in which British manufacturing industry has undergone some of the most dramatic changes in history, the ability of research workers and policy-makers to monitor and analyse these changes and the differential impacts from one local economy to the next has been severely impeded by inadequacies in the information base and the deterioration in its quality. Not only is the amount of information about the nature of manufacturing activity in different localities and regions limited, and its quality in many instances declining, but the access to what does exist is often restricted by confidentiality and cost constraints. These problems are magnified by theoretical shifts which have increased the emphasis placed on locality studies (Cooke 1989).

Official statistics in the UK define manufacturing as those activities included in Divisions 2, 3 and 4 of the 1980 Standard Industrial Classification

(SIC) (Central Statistical Office (CSO) 1979). Statistics published for the 1970s mainly use the earlier 1968 SIC, in which manufacturing was taken as activities classified in Orders III–XIX (CSO 1968). The two classifications are slightly different (see Ch. 1). For example, mineral oil processing is included as manufacturing in the 1968 SIC but not the 1980 SIC, while the extraction of metalliferous ores (which for the purposes of this book is covered in Ch. 15) was reclassified to manufacturing in the 1980 SIC. The net effect of the changes was that in the 1981 Census of Employment (CoE) 76 000 more people were classified as employed in manufacturing industries in the UK using the 1980 SIC than with the 1968 SIC. For some purposes other groupings of primarily manufacturing activities may be appropriate, such as high-technology industries (Butchart 1987).

The main aim of this chapter is to review the principal sources of information of use in local and regional studies of manufacturing activity in the UK and the way in which these sources have changed over the last two decades or so. The discussion is structured around a scale hierarchy. The chapter begins with some brief comments on sources of information to provide an international and national context for local and regional studies of manufacturing activity; section 16.3 examines the main regional and subregional sources; while section 16.4 discusses the nature of company and establishment data. The division is not, however, mutually exclusive, as some data sources are appropriate at more than one scale.

16.2 International and national context

It is important in examining the nature and changes in manufacturing activities at the local and regional level to have access to sources of information at the international and national levels for at least three reasons:

1. It is often helpful to identify how the local and regional industrial characteristics and trends compare with the country as a whole and with other countries.
2. With the development of structuralist and realist critiques of neo-classical and behavioural approaches to industrial location in the 1970s and 1980s the processes operating at the international and national level have been increasingly emphasized.
3. Many of the topics on which research workers require information are only available at the national level.

Space limitations, however, allow only a few of the more important general international and national sources to be mentioned. For some purposes the more detailed industry-specific international and national statistical series may be more appropriate. For example, two useful annual publications on the iron and steel industry are the *International Steel Statistics* (UK Iron and Steel Statistics Bureau, London) and *World Steel in Figures* (International Iron and Steel Institute, Brussels). Much valuable information may also be gleaned from the numerous books, reports and articles on particular manufacturing industries. They provide insights into important issues facing different industries and often contain tables and figures derived from statistical publications. They should thus be consulted at an early stage in a research project, usually *before* turning to the more detailed statistical sources.

At the international level there are various useful compilations of statistics, which include information relevant to studies of manufacturing activity in a range of nations. Caution is required in using such international collections of statistics though, because definitions, methods of collection, and dates of collection vary between countries and over time. One of the most comprehensive compilations is the *Industrial Statistics Yearbook*. This comes in two volumes. Volume I, *General Industrial Statistics*, is divided into two parts: the first part contains the basic data for each country or area in the form of separate chapters, and the second part gives a selection of indicators showing global and regional trends in industrial activity over the last 5 years. Volume II, *Commodity Production Statistics*, contains detailed information on world production of industrial commodities for the previous decade. The 1986 edition (published in 1988) includes data on 94 countries or areas in volume I and information on 530 commodities for 200 countries or areas in volume II. The country-by-country coverage in volume I, however, makes international comparisons difficult. More useful in this respect is the *Handbook of Industrial Statistics* which provides statistical indicators relevant to the drawing of international comparisons of the process of industrialization. While relying largely on existing data the *Handbook* is original in two ways. First, the basic data accumulated have been combined to form indicators such as ratios, growth rates and indices that facilitate comparison. Second, many of the indicators presented are derived from estimates made by the United Nations Industrial Development Office to supplement data missing from traditional sources. The 1988 edition has comparable data for about 150 countries, 28 industries and more than 100 manufactured commodities. The most comprehensive international comparisons of labour statistics is provided by the *Yearbook of Labour Statistics*. The 1989 edition includes data on paid employment, hours of work, wages, and labour costs in manufacturing as a whole and by major industry groups for 181 countries, areas and territories.

Useful background information on international economic trends is provided each year by the United Nations, which publishes a *World Economic Survey*, and the Economic Commission for Europe, which prepares an *Economic Survey of Europe*. Eurostat annually produces *Basic Statistics of the Community*, an invaluable source of comparative information on some European countries, Canada, USA, Japan and the USSR, although it does not contain much information specifically on manufacturing industries. Trade statistics are well covered in the *International Trade Statistics Yearbook*, while a useful collection of statistical data relevant to the analysis of world trade and development is contained in the *Handbook of International Trade and Development Statistics*. Wherever possible in this publication data are presented in an analytical way through the use of rank orderings, growth rates, shares and other special calculations, so as to facilitate their interpretation. Caution is required, however, in making international comparisons because of the variety of base years, time periods, classifications, definitions and data collection procedures used. A number of these statistical publications are available on disc for use with a personal computer.

British Industrial Performance provides a link between the national and international level statistics as it includes a comparison of the performance of 10 British manufacturing industries with their main overseas competitors. The main source for national statistics on manufacturing industry is the annual **Census of Production**. It covers establishments in the UK engaged in production, energy and construction. In general, forms are mailed to all establishments employing 100 or more employees and to a sample of 1 in 2 of those in the 50–99 employment size band and 1 in 4 for those employing 20–49. Information is available on the following topics: employment, wages and salaries, purchases, total sales and work done, gross output, net output, capital expenditure, total stocks and work in progress at end of year, and operating ratios. Information from each census is published in the form of *Business Monitors* in the PA series. In general there is a separate monitor for each production industry group (three digit) of the 1980 SIC. Summary tables are published in PA 1002. There is a lag of about 3 years between the census being taken and the last of the findings being published and the figures are subject to revision for a further 2 years. The Department of Economic Development for Northern Ireland publishes separate reports for the Province similar to that for the UK as a whole, but includes additional information on wages, salaries and energy purchases.

In the last decade a series of changes have been made to the annual censuses of production as a means of reducing the burden of form filling on the business community. In 1980 the first of a series of 'slimline' censuses was introduced. These are interspersed with more substantial 5-yearly 'bench-mark' censuses, the first of which occurred in 1984. Further cuts in the amount of information collected were proposed in the Armstrong-Rees review, including a reduction in the current sample size from 16 000 (4 years out of 5) to 13 000 and reductions in some of the detail requested (Department of Trade and Industry (DTI) 1989). However, some of these changes would not permit the UK to meet European Community (EC) directives fully and 'will not be implemented until the requirements of the directives can be changed' (Norton 1989: 17).

The annual censuses of production are the most important source of data enabling the local and regional research worker to monitor industrial sectors at the national level. For example, North *et al.* (1983), in their study of industrial change in London boroughs, used the censuses as the source of data for characterizing the nature and structure of different industries and their medium- to long-term dynamics. They also used the information to investigate the variables which were correlated with employment trends in the sectors. However, discrepancies in the employment data with that found in the CoEs led them to conclude that as far as the dynamics of sectors are concerned 'the

Census of Production is likely to give a better representation of the *relation between* variables (e.g. such as between employment and gross value added in measuring productivity) than of their *absolute* trends' (p. 116).

Other useful national sources for providing a context for local and regional studies of manufacturing include:

1. *Employment Gazette.* This publishes data and articles on current labour topics and surveys, notes on forthcoming changes in labour statistics, and commentary analysing recent employment trends. The results of the CoE and the *Labour Force Survey (LFS)* are published here.
2. *Economic Trends.* This contains a monthly commentary and a selection of tables and charts on broad trends in the UK economy. The *Economic Trends Annual Supplement* provides long runs of 30–40 years and includes a useful section on notes and definitions.
3. *British Business.* This weekly publication from the DTI provided, until it ceased publication in September 1989, many useful statistics and features, such as analyses of the births and deaths of firms based on VAT registrations. Fortunately *Business Briefing*, its successor from the British Chambers of Commerce, is continuing to publish most of the business trend statistics and much of the DTI news.
4. *UK National Accounts ('CSO Blue Book').* This includes data on gross domestic product (GDP) and index numbers of production on 10 groups of manufacturing industries.
5. *Input–output Tables for the UK, 1984.* This enables the economic interdependence of various industrial sectors to be analysed.
6. *The Engineering Industry and Training Board.* This body publishes a range of sector and occupational profiles which provide detailed analyses of employment and training trends within the engineering sector.
7. *Industrial Performance Analysis* (published by ICC Business Publications Ltd). This provides a useful guide to profitability, productivity and growth in 27 major British industries.
8. *Annual Abstract of Statistics.* This contains several tables which summarize data on manufacturing industries from other sources, in most instances for an 11-year period.

A wealth of industry-specific data also exists, much of which may be traced through Key Note Publications (1990) and Mort (1990).

16.3 Regional and subregional information

Regional and subregional information is essential for studying spatial patterns of manufacturing and net changes in the amount and nature of manufacturing activity in different places. Central government statistics are the most important type of regional and subregional data, because most of the information cannot be obtained elsewhere. Only the state has the economic resources and political mandate to undertake the collection of large quantities of data on a national scale. The government, moreover, is in the unique position of being able to demand that individuals and organizations provide the required information. Only official statistics are discussed in this section. A basic distinction can be made between those government statistics which are published and those which are not.

Published central government statistics

Nature and uses

The principal advantages for the research worker using government-published statistics are their ready availability and the wide coverage they provide. The latter is true in three main ways. First, most series cover the whole population, or at least a very large sample of individuals or firms. Second, a national coverage is usually provided, which allows comparability between different areas within the country, though the level of spatial disaggregation varies and the data available for Northern Ireland, and to a lesser extent Scotland, sometimes differ from those collected for England and Wales. Third, most series are published at regular intervals and this gives the potential for analysing changes over time, though changes in the data collected, the boundaries of the areas used, and the definitions employed can limit this potential.

The most important central government aggregate statistics for local and regional research into manufacturing activity are those provided by the censuses of population, employment and production. The **Census of Population** (CoP) is taken every 10 years; the latest published census results are for 1981. The decennial censuses cover the whole population living in the UK on census night, though some tabulations, for example, workplace and journey-to-work statistics, are based on a 10 per cent sample of the returns because of the large amount of clerical work involved in their compilation. The census publishes data for local authority districts. In using the census data it is important to distinguish between tabulations based on places of work and those compiled for places of residence. The workplace data are generally more useful for examining the location of economic activity than the residential data, although the latter are an important source of data on labour supply characteristics (see Chs 2 and 3).

A more frequently used source for employment data is the CoE (Ch. 2). Before 1971, estimates of the number of employees were based on counts of National Insurance cards. Since then CoEs have provided detailed statistics of employees (not the self-employed). An annual census was held each June from 1971 to 1978. These were followed by censuses in September 1981, 1984, 1987 and 1989. The next census is due in 1991. Data are collected from the pay points listed in the Inland Revenue's register of employers' PAYE schemes, which means that while people with two jobs are counted twice, 'part-year workers' not in employment during census week are excluded. Up until 1981 full censuses were taken, although pay points with less than three employees were included only in 1973, 1976 and 1981. Since then sample surveys have been used in Great Britain. For example, the 1987 Census surveyed all pay points with 25 or more employees, and a sample averaging one in seven taken from the remainder. The latter were stratified according to size, location and industry. The Northern Ireland census has continued to survey all units.

The results of the censuses are published in the *Employment Gazette*, but the only spatial disaggregation presented is of employees in employment in each standard economic region of Great Britain by four-digit activity headings (see Table 2.3). 'The main purpose of the Census of Employment is to provide accurate national and regional "benchmark" figures with which to realign the employment estimates obtained from quarterly and monthly sample inquiries among employers and the labour force' (*Employment Gazette* Oct 1989: 542). While the estimates provide a good guide to trends in employment in existing businesses they are unable to provide a comprehensive measure of firms going out of business or of new businesses becoming established. Since the census has become less frequent the annual *LFS* has been used increasingly to update trends (see Ch. 2).

The main features of the third of the censuses, the annual Census of Production, were described in section 16.2. Until the 1979 Census each industry report also included a table giving the regional distribution of employment, net capital expenditure and gross value added at factor cost. From the 1980 Census onwards regional results are shown at the more aggregate class level (two digits) in the *Business Monitor* PA 1002. These statistics have been used to examine regional variations in output, some costs and investment in manufacturing (Tyler *et al.* 1988; see Ch. 4). PA 1002 also includes analyses of foreign manufacturing enterprises by country, standard region and assisted area. Some of the most interesting analyses from the Census of Production are those of manufacturing (local) units by employment size (PA 1003) which present information about UK manufacturing units recorded in the register of businesses maintained by the CSO. The numbers of manufacturing units in various size groups and the total number of persons employed in each category are shown by industrial classification and by county, region and country. This monitor has been published annually from 1971 onwards, with the exception of 1974. The data have recently been analysed to show the changing spatial structure of manufacturing plants in Great Britain, 1976–87 (Tomkins and Twomey 1990). Since 1985, with the integration of the two main CSO (formerly Business Statistics Office) registers (Perry 1985, 1987), local manufacturing units with employment below 20 have been added to the tables.

The current register is based on all businesses trading in the UK which pay VAT. Business VAT registrations, deregistrations and stocks have been analysed in a series of articles published in *British Business* (e.g. 25 August 1989). Some of the early articles have been compiled into a book (Ganguly 1985). With the demise of *British Business* the series is now published in *Employment Gazette*. VAT registration statistics are the most widely used source for examining the geographical distribution of new and small firms (e.g. Moyes and Westhead 1990). However, the data are not ideal. They include not only new start-ups but also businesses which have been trading for some time without previously being liable to pay VAT (e.g. because their turnover was below the VAT threshold), and some wholly-owned subsidiaries as well as genuine new firms. In addition, some industries are zero-rated or exempt from paying VAT (e.g. children's clothes, food, printed matter), although many firms in these sectors voluntarily register and so are included in the statistics (Mason 1987).

Useful summaries of regional and subregional data on manufacturing industry from the three main censuses and other sources are contained in *Regional Trends*, the *Scottish Abstract of Statistics*, the *Digest of Welsh Statistics* and the *Northern Ireland Annual Abstract of Statistics*. Other useful official sources include: the reports of various investment agencies, such as Invest in Britain Bureau, Scottish Development Agency and Welsh Development International (WINvest); and the Engineering Industry (previously) Training Board regional profiles of trends in engineering employment and training. Various House of Commons Committee reports also provide useful information, such as the reports *Inward Investment into Wales* (Welsh Affairs Committee 1988) and *Locate in Scotland* (National Audit Office Report by the Comptroller and Auditor-General 1989).

Limitations

Clearly official published statistics are indispensable for the study of many aspects of the industrial geography of the UK. However, they have a number of limitations which restrict their usefulness, some of which have already been outlined. These problems mainly concern the purpose for which the data are collected; the level of spatial disaggregation provided; accuracy; the frequency and temporal lag of publication; and the definitions and classifications used.

1. *Purpose of data collection* The Government Statistical Service (GSS) exists first to serve the needs of government. A secondary function is the provision of an information service for commerce and business. Academic research workers and other users of published statistics are, according to the Government Statisticians' Collective (1979), assigned a lower priority. Consequently, research workers often have to make the best use they can of data collected for other purposes (see Ch. 1).

2. *Level of spatial disaggregation* Perhaps the most serious limitation in using published official statistics for examining manufacturing industry is the general lack of data for small areas. Indeed, as has already been noted, many statistical series are published only at a national scale and where data are spatially disaggregated it is usually only for standard economic regions. The CoP is one of the few official sources to publish detailed data on the character of industry for each county and district council in the country, though in

1971 it was restricted to larger urban districts. Little official information on manufacturing industry is published for smaller areas. Even where suitable data are available, a major problem in examining temporal change is the frequency with which the boundaries of the spatial units alter. A dramatic change occurred in 1974 when local government reorganization affected not only the areas of many local authorities, particularly the district councils, but also the boundaries of several of the standard economic regions. In some cases the official figures published after such reorganizations are given for both the old and new boundaries. In the case of the 1971 CoP, for example, about 70 additional volumes were produced in which data were resorted to the new local authority areas. More often though alterations to spatial units occur in a piecemeal fashion and comparable statistics are not published. A variant of this theme is the way in which firms and individuals are allocated to areas. Postcode sectors are increasingly being used to allocate both establishments and unemployed persons to areas. For example, postcode sectors were used to allocate the CoE data units to jobcentres in 1978 and 1981 and to wards since 1984. The best-fit areas can, however, vary by up to 50 per cent from the areas they are intended to approximate (Goddard and Coombes 1983). Wards are now the basic building block used to aggregate CoE data to larger areas such as local authority districts and travel-to-work areas (TTWAs).

3. *Accuracy* The increased use of sampling and difficulties over verifying the register of businesses used for the sampling frames inevitably increase the problem of inaccuracies. The first problem is most apparent when dealing with small areas and the latter has most effect on new and small firms which may escape inclusion in the sampling frames. Problems over data accuracy is one of the reasons for the dearth of statistics published below the regional scale. For example, the Department of Employment (DE) expressed concern over data reliability from the 1984 CoE for industry detail below divisional level for counties, travel-to-work and local authority areas (*Employment Gazette* Aug. 1987: 409). Access is nevertheless possible to the small area unpublished data (see below). Townsend (this volume, Ch. 2) notes that improvements in sampling procedures in the 1987 CoE mean that 1987 local data are more safely compared with 1981 than 1984 data.

4. *Definitions and classifications* A major problem with official statistics is that definitions and classifications often vary between series and, over time, within a series. This makes it difficult to match data from different sources and to analyse temporal trends. The effect of changing spatial classifications has already been noted. A further example is the different definitions of employment used. The CoE, for instance, uses employees in employment, while the CoP and the LFS usually also include employers and the self-employed. Further differences arise from the counting method used. For instance, the CoE, as noted above, counts filled jobs in businesses during the census week rather than people, while the CoP includes persons in employment at any time during the week before the census. The annual Census of Production, in contrast, asks for the average number of persons on

Table 16.1 Numbers employed in manufacturing in Great Britain, 1981: three sources. (*Sources:* OPCS 1984; *Employment Gazette* December 1983; BSO 1984.)

	000s
Census of Population	6194
Census of Employment	6057
Census of Production	5662

the payroll of establishments during the year of return. The effects of these different definitions and methods of counting are illustrated in Table 16.1 where the numbers employed in manufacturing in 1981 are given from three different sources. The highest employment figure for Great Britain (CoP) is over half-a-million or 9 per cent greater than the lowest (annual Census of Production). Much higher differences occur the more disaggregated the data. For example, the gap between employment estimates in 1984 produced by the BSO and the DE for two-digit classes of the 1980 SIC is as large as 50 per cent in some industries (Table 16.2). Such differences may be reduced in the future if the proposal to integrate the different registers of employers used by the two departments are accepted (DTI 1989).

The changes in industrial classification in 1958, 1968 and 1980 also cause problems in analysing changes over time, though Fothergill and Gudgin (1978) have compiled regional employment statistics on a comparable basis from 1952 to 1975, using a composite industrial classification of 116 industries. The DE published regional employment data on a continuous basis from 1965 to 1975, but at the more aggregate order level (*Department of Employment Gazette* Aug. 1976: 839–49).

Unpublished official regional and subregional statistics

The majority of statistics collected by government departments are not published, because of the cost and the need to ensure the confidentiality of the respondents. A number are, however, available to some users on request; for instance, research workers and consultants may apply to the DE for a notice under the Statistics of Trade Act 1947 and the Employment and Training Act 1973 for access to unpublished employment statistics, usually through the **National On-line Manpower Information System** (NOMIS) (see Chs 1, 2 and 8). The maps in Fig. 16.1 showing changes in subregional manufacturing employment were constructed from data obtained in this way. The Engineering Industry Training Board also maintains a local labour market information system based on 1973 employment exchange areas which relates to engineering employment and training (Green and Owen 1989). A further useful source is the VAT data showing the total number of registrations and

Table 16.2 Comparison of CoE and annual Census of Production estimates of employment in manufacturing classes of the 1980 SIC, 1984 (thousands). (*Source:* Cabinet Office 1989 : 145.)

	Census of Employment (GB)*	Annual Census of Production (UK)	Difference	%
Extraction and preparation of metalliferous ores: metal manufacturing	195.4	164.3	−31.1	(−15.9)
Extraction of other minerals, manufacture of non-metallic mineral products	254.1	217.4	−36.7	(−14.4)
Chemical industry; production of man-made fibres	347.2	299.9	−47.3	(−13.6)
Manufacture of metal goods not elsewhere specified	331.1	338.3	+7.2	(+2.2)
Mechanical engineering	750.4	646.8	−103.6	(−13.8)
Manufacture of office machinery and data processing equipment	85.7	43.9	−41.8	(−48.8)
Electrical and electronic engineering	599.7	560.0	−39.7	(−6.6)
Manufacture of motor vehicles and parts	273.2	288.2	+15.0	(+5.5)
Manufacture of other transport equipment	289.2	304.1	+14.9	(+5.2)
Instrument engineering	104.2	79.7	−24.5	(−23.5)
Food, drink and tobacco manufacturing industries	587.8	612.6	+24.8	(+4.2)

Table 16.2 Cont.

	Census of Employment (GB)*	Annual Census of Production (UK)	Difference	%
Textile industry	234.9	236.4	+1.5	(+0.6)
Manufacture of leather and leather goods	22.1	21.5	−0.6	(−2.7)
Footwear and clothing industries	292.4	311.8	+19.4	(+6.6)
Timber and wooden furniture industries	202.8	192.7	−10.1	(−5.0)
Manufacture of paper and paper products; printing and publishing	481.6	458.0	−23.6	(−4.9)
Processing of rubber and plastics	194.2	203.4	+9.2	(+4.7)
Other manufacturing industries	80.5	80.7	+0.2	(+0.2)
Total manufacturing	5326.7			
plus Northern Ireland	108.4			
Total UK	5435.1	5059.7	−375.4	(−6.9)

Excluding the self-employed.

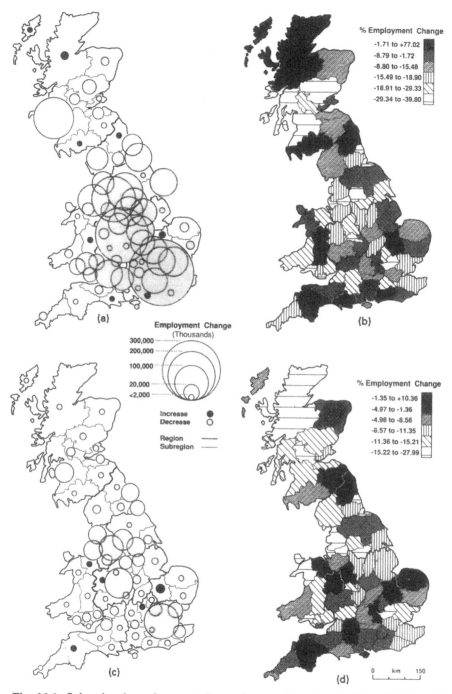

Fig. 16.1 Subregional employment change in manufacturing in Great Britain: (a) 1971–81 absolute change; (b) 1971–81 percentage change; (c) 1981–84 absolute change; (d) 1981–84 percentage change. (*Source:* Healey and Ilbery 1990: 323.)

deregistrations in each year in each local authority district in Great Britain, and broken down further by industry group for the counties in England and Wales and Scottish regions. The information is available on request from the Employment Department (Statistics C4, Room 405, Caxton House) for a fee of £75. Without this limited access much industrial research, particularly at the intra-regional and intra-urban scales, would be considerably more difficult than it is already.

Many data are collected by government departments for their own administrative use. For example, when the DTI set up the **Regional Data System** in the early 1980s it was intended to link information about the performance of companies and their operational units with details of the financial assistance they receive from the department (Nunn 1983). However, it also enables regional and subregional trends in manufacturing industry to be monitored (e.g. Macey 1982). The database holds information on manufacturing units whose employment has at some point exceeded 10. A key feature of the system is the inclusion of information from other computerized data systems within central government (i.e. CoE, DE quarterly employment survey, annual Census of Production, and Quarterly Production Inquiry). Aggregated data have been released, via DTI regional offices, to assist research projects (e.g. Dobson 1989) which 'illuminate areas of genuine policy interest' (Nunn 1983: 110).

Limitations

There are two main problems in making use of unpublished official statistics, in addition to those already noted concerning the published statistics. They are acquiring knowledge of their existence, and obtaining access to them.

1. *Knowledge* Many data exist in unpublished surveys of industry carried out by government departments. Sometimes the existence of these only become apparent through enquiries, personal contacts, conference papers and references to them by people who have previously managed to obtain access to the data. For example, the results of the survey by the Inquiry into Locational Attitudes Group (ILAG), established by the DTI, were discussed at conferences (e.g. Howard 1971) and made available to some academics (e.g. Keeble 1971), but would probably not have been published if they had not been given in evidence to a House of Commons Expenditure Committee Inquiry into *Regional Development Incentives* (DTI 1973). Sometimes statistics are only published infrequently; for example, data from the Record of Openings and Closures maintained by the DTI since 1966 were not published until 1980 (Nunn 1980), though analyses of statistics from this source were used extensively within central government and continue to be made available to some academics and local authorities (e.g. Taylor and Twomey 1988). Caution is required when interpreting some of the detailed data released for small areas, as already noted, because of inaccuracies.

2. *Access* Knowing that the required information exists is only one aspect of

the problem; obtaining access to it is another. The Government Statisticians' Collective (1979) suggests that, in general, the more 'acceptable' enquirers' political and academic credentials are, the more co-operative the department will be in releasing information to them. One of the constraints is that it should not be possible from released (or at least published) data to identify individual firms. The development of NOMIS has meant that, once permission has been received, obtaining much employment and unemployment information has become considerably easier (Ch. 8). Although some information is provided free by official bodies the trend is increasingly to charge for data (Ch. 1). The Tradeable Information Initiative has made government departments more aware of the commercial value of much of the unpublished information they hold. Special tabulations from the CoP and the Census of Production have been available for a long time, but for the academic research worker the cost can be exorbitant.

16.4 Company and establishment information

The aggregate statistics used in regional and subregional studies of manufacturing are derived from data referring to individual companies and establishments. It has become increasingly realized that, with the concentration of economic power in the hands of a relatively few multi-plant and multinational enterprises, it is crucial that the changing aggregate patterns are viewed within their corporate context (e.g. Gaffikin and Nickson 1984; Healey and Clark 1984; Healey and Watts 1987). There are several guides to sources of information on British companies published (e.g. Norkett 1986), but much of the information needed for a geographical analysis requires data at the establishment level, on which there is less information readily available. Hence this section focuses particularly on establishment-based data sources and ways of making their corporate links explicit.

Uses

Access to data at the company and establishment level is essential to enable the patterns and processes of manufacturing location to be analysed (e.g. Gudgin 1978; Healey and Clark 1985; Lloyd and Shutt 1985). Sometimes this is because the aggregations available are unsuitable for the task in hand, or because the tables contain more asterisks than figures, due to confidentiality constraints (Firn 1973). Data on individual units enable the research worker to combine them together into whatever size, industrial, ownership, spatial or other categories are desired. By linking data (e.g. employment and floorspace) interesting relationships may become apparent which are hidden by more aggregate correlations. This flexibility is particularly important when analysing changes over time (Thompson 1983). Aggregate statistics usually allow only *net* changes to be identified, for example, the net changes in manufacturing employment in a local authority area between two censuses. The availability of establishment data allows *components of change* to be calculated so that gross changes can be identified. For instance, in the last example, access to employment data at the level of the establishment would enable the net change in employment to be disaggregated into entries, exits and *in situ* expansion or contraction in surviving plants. Further investigation may allow greater disaggregation; for example, entries may be divided into births, transfer openings and branch openings, while exits may be divided into deaths, transfer closures and other closures. The ability to recognize components of change is important, because different components may have distinct explanations and may respond to different policies.

Company and establishment data are required not only to overcome some of the limitations of the available aggregate data but also to analyse the processes of change. For example, it is needed to test the many hypotheses in local and regional industrial research which relate to individual units. It is often suggested that different types of company and establishment respond differently to similar external conditions (Healey 1981, 1982). Such relationships cannot be adequately tested using aggregated data, even when they are available, because of the 'ecological fallacy' (Robinson 1950). In other words, it cannot be assumed that relationships discovered at one scale also apply at other scales. If, for instance, the correlation between large plants and the rate of industrial decline in towns and cities is found to be $+0.5$, it would not be justified, without more information, to extrapolate this relationship and say the same correlation of $+0.5$ between plant size and employment performance obtains for individual plants, or at the regional and national levels. Processes of change may also be identified for individual companies. Company reports can provide a useful source of information in the search for causal factors at this level, especially for large companies, although they are usually only a precursor to interviews with representatives of the companies (Ch. 11).

A further use of microdata is to provide a sampling frame. Ideally this should be a list of every member of a population, for example, all the manufacturing establishments in a particular area. Much effort is often expended in attempting to construct sampling frames and reduce the limitations inherent in the data sources used.

Company and establishment databanks

The need for a good microdata base for local and regional industrial research led, particularly in the 1970s and 1980s, to the construction of establishment databanks (Healey 1983a; Hobbs 1988; Swales 1976). An establishment databank may be defined as a file of data held on all establishments from a specified population in a particular area. There were at least 15 such databanks on manufacturing industry in academic institutions which were functioning in the 1980s, although lack of resources has prevented several of them being updated (Appendix 16.1). They vary in size from databanks covering 200 or so establishments in a small area to one based on Dun and Bradstreet's directory for the UK, which has data on over 460 000 companies. All of the databanks have information on the location, product group and employment size of the establishments and most have added other information such as ownership and status. The majority have data on more than one time period and this has allowed research workers to identify some or all of the main components of change. Several local authorities, for the most part county councils, have also constructed establishment databanks to help them monitor the nature of industry and industrial change in their areas. For example, Tayside have constructed a business establishment register (BERT) which is checked for accuracy annually with a postal questionnaire (McLeish 1987); while Tyne and Wear have established a county-wide business information service (COBIS) which uses several non-local authority sources for file creation and maintenance (Spicer 1988; see also Ch. 8).

Central government is also concerned with the construction of company and establishment databanks and, given the resources available to them, the databases are usually much larger and more comprehensive than those set up by academics. Several of the databases, such as the one operated by the CSO for its inquiries into production industries (Perry 1985, 1987), integrate information on establishments and the companies which operate them. However, the purpose of assembling the information has usually been different. Three main uses are made of establishment databanks by central government:

1. To provide sampling frames for the various government censuses and surveys;
2. To store the raw data for the construction of the aggregate statistics discussed earlier; and
3. To act as an information system to assist government departments in their administrative duties.

These uses sometimes overlap, as the same databanks are often used for all three purposes. For example, the company-based Regional Data System described earlier is primarily used by the DTI for administration, but it is also used to generate regional and subregional statistics and as a sampling frame for surveys.

Local authorities also have an interest in developing company and establishment databanks for their areas (Healey 1983a). Unfortunately, despite the access of local authorities to the pay point (ERI) data from the CoE, the introduction of sampling in the latter half of the 1980s has made the construction of such databanks increasingly difficult. Some authorities undertake their own surveys although these are sometimes only for industrial estates. Many of the Training and Enterprise Councils (TECs) and Local Enterprise Companies (LECs) are likely to set up their own employer databases in the next few years.

Sources

The establishment databanks set up outside local and central government have used a variety of official and non-official sources of information (Appendix 16.1). The most common official source has been the Factory Inspectorate records of the Health and Safety Executive. In the process of inspecting factories the inspectors collect information on the employment and products made at each establishment. Several databanks have used data from the CoE. Other official sources used include registers of redundancies, records of the industrial training boards, local authority and industrial development association censuses and surveys, and rating and valuation records. Local non-domestic rating lists associated with the introduction of the unified business rates in April 1990 are available for public inspection at the offices of the district authorities (Ch. 6). A further useful source is the *UK Directory of Manufacturing Businesses* (*Business Monitor* PO 1007). The list is based on larger local manufacturing units which contributors have agreed to be included. The 1989 edition included over 41 000 business addresses, about a quarter of those on the register, accounting for about 45 per cent of employment in manufacturing. Unfortunately, the published version is only available in activity heading order, though special geographical and alphabetical analyses in the form of computer print-outs may be purchased from the CSO. Respondents have to agree for their local manufacturing units to be included in the list. However, a helpful feature of PO 1007 is the inclusion for each industry of the estimated share of employment accounted for by the units listed. Financial and other details about specific enterprises may be obtained from one of the company registration offices or one of the commercial company databases such as Jordans or ICC Business Publications' *Regional Company Survey*.

The most commonly used of the unofficial sources is the Market Location (ML) – formerly Industrial Market Location – database which not only lists manufacturing establishments in a specific area with details of employment, products made, ownership and status, but also gives map references to enable the location of individual units to be pinpointed. In March 1990 their 'prime file' contained information on 75 000 manufacturing locations in Great Britain. Other sources include Dun and Bradstreet's (D&B) Market Facts File and British Telecom's (BT) *Yellow Pages* Business Database. Examples of the use of these databases may be found in Healey (1983b) (ML), Gallagher *et al.* (1990) (D&B) and Hakim (1989) (BT). Unfortunately accessing these databases can be very expensive. Much more widely available are the *Postcode Directories* which provide a comprehensive local lists of larger businesses. Larger users of the Post Office have their own postcode and are listed at the back of the directory. Telephone books are also sometimes used, particularly for checking dates of openings and closures and identifying local moves. Details of ownership and status are usually added from information in *Who owns Whom, Key British Enterprises, Kelly's Manufacturers' and Merchants' Directory* and *Kompass*. For further details about specific companies the *The International Stock Exchange Yearbook*, *Times 1000*, *Britain's Top 2000 Private Companies*, *Extel Cards* and company annual reports are particularly useful. Other sources which are sometimes used include local industrial directories, membership lists of chambers of commerce, trade directories, lists of new business registrations, and articles in newspapers and trade magazines. The McCarthy Information Service provides a very helpful abstracting service of press articles on companies and industries which is available on card, microfiche and on-line. Fortunately for the research worker many large business libraries subscribe to this service. These and other sources are reviewed in Owen (1988).

Limitations

There are several limitations to establishment databanks (Thompson 1983) and to the sources on which they are based (Foley 1983), which may be summarized under five headings – resources, access, coverage, accuracy and comparability.

1. *Resources*

Setting up an establishment databank is expensive in terms of both time and cost. Though much of the work is tedious, high-quality research assistance is needed for most tasks in order to identify and resolve anomalies and errors in the data sources consistently. The East Midlands Industrial Databank took between 5 and 6 full person-years' work to complete (Gudgin and Fothergill 1979). Merely adding ownership details to the North West Region Industrial Establishment Databank took 18 months' work. These examples are taken from some of the largest databanks in the country, but the smaller ones have taken proportionally as long to establish, or even longer where direct survey has been a major source of information. A considerable effort is required in 'cleaning' the data because much of it is derived from sources not designed for use in establishment databanks (Thompson 1983). The expense of constructing establishment databanks has led some to question their cost-effectiveness (Swales 1976). However, the number of establishment databanks set up outside government in the 1980s (Appendix 16.1), suggests that the consensus of opinion at the time was that the potential insights into patterns and processes of industrial location, as well as local and regional development, were worth the time and effort involved. Not surprisingly, the great majority of these establishment databanks are, or were to begin with, dependent on external funding.

2. *Access*

The difficulties of obtaining access to unpublished official aggregate statistics have already been discussed; the problem is much greater for establishment-

and company-level data. The Censuses of Employment and Production come under the terms of the Statistics of Trade Act 1947, though local authorities and some government consultants have been allowed access to the former. The guidance given by the DE to users of the CoE tapes is that:

> In accordance with section 9 of the 1947 Act, care must be taken to ensure that any subsequent report, summary or communication compiled from this information is so arranged as to prevent any particulars shown therein from being identified as particulars relating to any individual person or undertaking except with the previous consent in writing of that person or the person carrying on that undertaking, as the case may be (Sprott 1977: 13).

Access to other sources of official firm and establishment data (e.g. Factory Inspectorate records) is at the discretion of the provider, but may be dependent on the source not being named, as well as non-disclosure of information on individual establishments.

Most industrial directories are widely available in large libraries, though the cost of data from some organizations (e.g. ML), means that few libraries can afford to subscribe (see Ch. 1). A further problem is obtaining information for earlier periods. Many libraries dispose of their out-of-date directories, because of space problems; while organizations which maintain continuously updated computerized databases only produce hard copies on demand and hence do not have back copies. The latter applies to the Industrial Market Location data. However, before ML stopped producing individual directories the Department of Geography at Coventry Polytechnic obtained a set for virtually each county in Great Britain relating to the late 1970s and early 1980s.

3. Coverage

None of the data sources includes every manufacturing plant in an area or industry. Inevitably some, particularly the very small plants, are missed or do not respond to the request for information. The coverage can vary widely. Sometimes small plants are omitted deliberately, for example, the *UK Directory of Manufacturing Businesses* excludes all smaller local units. In some commercial directories small companies have to pay for an entry. There are also reports of the omission of records for some large firms from the CoE local unit tapes (Hubbard 1980). Unfortunately, it is not possible to identify manufacturing businesses separately in all sources. For example, New Business Locations prepares weekly lists of new registrations at Companies House, Cardiff which are sifted at source to exclude non 'bona-fide' companies, but contact has to be made with the companies to establish the exact nature of their business (Risk 1988).

Surprisingly, some of the non-official directories can give as comprehensive a coverage as the official ones (Aubrey *et al.* 1989). In a comparison of four data sources, Foley (1983) found that for the electrical engineering industry in West Yorkshire the Market Facts File and the Industrial Market Location Directory gave almost as good a coverage as the Factory Inspectorate data and better than that of the CoE, though the differences were not large. However, results vary widely. In Leicestershire Industrial Market Location found only about 65 per cent of the plants identified by the Factory Inspectorate in 1976. Using the Industrial Market Location Directory did increase the number of establishments in the databank by slightly over 10 per cent, but nearly all the additional plants were very small and added little to the total employment in the county (Gudgin and Fothergill 1979). In contrast, in Coventry in the area mapped by ML in 1978 relatively few of the plants which survived to 1982 were absent from the directory (Healey 1983b). A later study of large businesses in Coventry found that British Telecom's *Yellow Pages* Business Database omitted 34 per cent of the establishments identified by the city council mainly from the rating records (including two operated by BT!). However, most of the omissions were additional branches of large organizations and their coverage of establishments operated by small businesses was superior to that of the city council's database (Elias and Healey 1991).

Important changes, which affect the coverage of several of the establishment-based sources, have occurred in recent years. For example, the introduction of sampling to the CoE means that employment data are not available for many units. To enable sampling to occur the Inland Revenue provides the DE with a size indicator, based on the amount of tax collected, and a crude industry code for each unit. This information may be passed to local authorities under the 1987 Local Finance Act, but about a quarter of the units do not have a postcode and therefore the list of employment units available to a local authority is likely to be incomplete.

4. Accuracy

Even where the coverage is good the sources may include inaccuracies. Employment figures are probably most accurate from the CoE. In contrast, many of the Factory Inspectorate figures are rounded, while those from ML and D&B are probably the least helpful in that they are categorized into groups and are sometimes estimated. One of the weakest areas, as far as accuracy is concerned, lies in the classification of establishments into industrial groups. Some of the errors occur because only limited business activity information is available from some employers and some detailed descriptions are themselves difficult to classify.

Foley (1983) found that, taken together, the four sources he examined identified over a third of establishments incorrectly as electrical engineering plants. This may be a special case, because the majority of establishments wrongly classified were wholesalers, retailers and contractors, and in many industries this is unlikely to be a major problem; however, North *et al.* (1983) found a quarter of their panel of establishments turned out to be misclassified by minimum list heading (MLH) when the initial contact was made. Another study, by the DE, found that a sample of establishments coded to the 1980 SIC for the 1981 CoE revealed a 6 per cent error at the level of industry division, and a 17 per cent error at the full activity heading (AH) level (Healey 1983c). Several problems have also been noted concerning the 1984 Census. For example, some 16 000 employees in AH 2210 (the iron and steel industry) in the UK should have been classified elsewhere (*Employment Gazette* Oct 1989: 542).

Analysis of change over time can be markedly affected by these inaccuracies in industrial classification. Such errors can have a particularly serious effect on the interpretation of changes in the local economies where the industries involved are concentrated. A similar problem can occur from errors in the location coding of establishments. For example, at the local authority level severe distortions can occur in employment figures derived from the CoE, if these errors are not first corrected (Bradley 1989; Coppin 1980). The 1978 CoE figures suffer particularly from this problem because postcodes were used for the first time to allocate establishments to administrative areas (Elton 1983; Goddard and Coombes 1983). Further problems can occur because of the inclusion of out-of-date information. This sometimes happens because of the inclusion of unchecked data from one edition of a directory to the next.

5. *Comparability*

This may be a problem in two senses. First, there are difficulties in combining data from different sources. Second, there are problems with comparing information between databanks. A major reason for the difficulties in comparability is that the definitions, classifications and procedures used often vary between sources and databanks. The categorization of employment size into broad groups in the Market Facts File and ML directories has already been mentioned. Other examples are given by Foley (1983) and Thompson (1983). Different methods of estimating missing employment data, for example, can have a marked effect on the results, particularly on the size of the *in situ* employment change component (see e.g. Healey 1983d).

A related difficulty in comparing sources and databanks is the different time periods to which the data refer. For some sources, particularly Factory Inspectorate data and the most recent ML directories, the information is collected over several years. This lack of time specificity means that care needs to be taken in interpreting the results of studies using these sources, especially where comparison is made with findings based on other sources, or the analysis is of change over time.

The extent to which these limitations of establishment-based research are significant clearly varies from one database to another. The first two – resources and access to data – affect the viability of the enterprise and are influenced by the ability to obtain funding and the quality of the contacts made with those who provide the data. The effect of the next two limitations – the coverage and accuracy of the sources used – can, however, be reduced, principally by using as many different sources as possible, and by vigorous checking of the data used. Of course, the extent to which this strategy can be followed depends on access to data and resources. It also raises the last limitation – comparability – in the form of the difficulties of combining information from different sources.

16.5 Conclusion

Dramatic changes have occurred over the last 10–15 years to the manufacturing sector of the British economy both in the number of people employed and its character. Examining these changes is fundamental to understanding the operation of local and regional economies. Moreover, despite rapid deindustrialization manufacturing remains the prime sector in most local and regional economies because of its capacity for generating wealth and influencing the size of the other sectors. Unfortunately during a period in which many commentators are talking about a revolution occurring in the nature of manufacturing activity, the ability of research workers in local and central government, academic institutions and business to monitor and understand the changes is made considerably more difficult by a deterioration in the database. As many authorities suggest that the economy is at the beginning of the fifth Kondratieff long wave and a new regime of capitalist accumulation is commencing based on methods of flexible specialization (Healey and Ilbery 1990), the amount and quality of the information available, particularly from official sources, to explore the impact of these changes on local and regional manufacturing activity, have fallen.

Although some improvements have occurred in accessing what data are available through computerized information systems and the private sector has attempted to fill some of the gaps, these additions do not compensate for the losses. Furthermore, the cost of obtaining data from the private sector is often prohibitive for many research workers and government is becoming more aware of the commercial potential of the information it holds and is introducing charges, albeit sometimes differential for different client groups (see Chs 1 and 8). These changes in the database are forcing more research workers to collect their own information direct from business (Ch. 11). Although the construction of establishment databanks is not as popular as it was in the early 1980s, one-off surveys concerned with particular issues are becoming more common. Thus many of the savings envisaged by the Rayner cuts and later efficiency drives are largely illusory; the survey costs have simply been shifted from central government to other bodies. It is arguable that as more and more requests are made by a variety of organizations for information the burden on business has increased rather than fallen as was the intention.

Local and regional trends in manufacturing activities cannot be understood in isolation from changes occurring to other sectors of the economy. Producer services in particular are closely interrelated with manufacturing industries. Changes in finance, accountancy, marketing, transport and distribution, for example, have a critical effect on manufacturing industries. The next chapter examines the information base available for exploring local and regional trends in producer services.

Acknowledgements

The author is very grateful to Andy Pratt and Doug Watts for their encouragement and perceptive observations on an earlier draft of this chapter.

Appendix 16.1 Non-government manufacturing industry establishment databanks in the UK. (*Sources:* Mostly based on Healey 1983a, partly updated by Hobbs 1988.)

No.	Spatial coverage	Region	Industries	Units	Years	Sources	Status	Access	Institution	Contact
1.	South Hampshire	SE	M	2 000+	79, 85 (71)	FI, Sy	Op	Negot	Southampton Univ	Dr C Mason
2.	East Anglia	EA	M	4 000+	71, 81	FI, Of ML, Oth	Op	Yes	Cambridge Univ	Dr D Keeble
3.	Cornish estates	SW	M, S	900	84	Sy, Oth	Op	Negot	Coventry Poly	Dr A Pratt
4.	Devon and Cornwall	SW	M, S	150+	82	Sy	Op	Rest	Poly South West	Mr D King
5.	West Midlands	WM	M	10 000	48–88	FI, ED871 WMCC	Prog	Negot	Birmingham Univ	Mrs B Smith
6.	Coventry subregion	WM	M	3 000	74, 78 82/83/85	ML, Of Sy	Op	Negot	Coventry Poly	Dr M Healey
7.	East Midlands	EM	M	10 000	68, 75/ 81, 86	FI, ML	Op	Negot	NIERC Belfast	Dr G Gudgin
8.	East Midlands cities	EM	M	5 000	70, 75/ (86 Leicester	FI, ML Oth	Op	Negot	North London Poly	Dr J Fagg
9.	West Yorkshire	YH	M	6 500	79, 85 (Kirklees)	Of, Oth	Sh	Yes	Huddersfield Poly	Dr D Reeves
10.	Sheffield	YH	M, S	6 000	83 onwards	Sy, Of	Op	Negot	ETIC	Dr P Foley (Sheffield Univ)

Appendix 16.1 Cont.

No.	Spatial coverage	Region	Industries	Units	Years	Sources	Status	Access	Institution	Contact
11.	North West	NW	M	23 000	59, 66, 72, 75, 81	FI	Sh	Yes	Manchester Univ	Prof P Lloyd (Liverpool Univ)
12.	Lower Swansea Valley	Wales	M, S	280	81, 87	Sy	Op	Aggreg	Univ Coll Swansea	Dr R Bromley
13.	Northern Ireland	NI	M	4 500	73–86	FI, Sy, Oth	Op	Negot	NIERC Belfast	Dr G Gudgin
14.	UK	UK	M, S	460 000	71, 81–87	MFF, ML	Op	Limited	Newcastle Univ	Prof C Gallaher
15.	UK	UK	Worker co-ops	1 600	80 onwards	Sy, Oth	Op	Negot	Open Univ	Mr P Hobbs

Notes: M – manufacturing; S – services; FI – Factory Inspectorate; MFF – Market Facts File; ML – Market Location; Sy – survey; Of – official; Oth – other; Op – operational; Sh – shelved; Prog – in progress; Negot – negotiable; Rest – restricted; Aggreg – aggregated data only; Univ – university; Poly – polytechnic; Coll – college; NIERC – Northern Ireland Economic Research Council; ETIC – Employment and Training Information Consortium; ED871 – Employers' Register; WMCC – West Midlands County Council. Excludes single-industry databanks, e.g. footwear, foundry, electrical engineering, high technology – see Healey (1983a).

References

Aubrey J, Clarke G, Stillwell J 1989 *The Use of Establishment Data to Examine Employment Structure: Two Studies in West Yorkshire* Working Paper 529, School of Geography, University of Leeds

Bradley M 1989 Employment data and forecasting: problems and potential: a review of county councils in England and Wales. Unpublished MA Regional Planning Research Project, Coventry Polytechnic

BSO 1984 *Report on Census of Production 1981: Summary Tables* PA 1002, HMSO, London

Butchart R L 1987 A new UK definition of the high technology industries. *Economic Trends* **400**: 82–8

Cabinet Office 1989 *Government Economic Statistics: A Scrutiny Report* HMSO, London

Cooke P (ed.) 1989 *Localities: The Changing Face of Urban Britain* Unwin Hyman, London

Coppin P W 1980 The use of ACE and local surveys for employment monitoring. *BURISA Newsletter* **44**: 7–8

CSO 1968 *Standard Industrial Classification Revised 1968* HMSO, London

CSO 1979 *Standard Industrial Classification Revised 1980* HMSO, London

Daniels P W 1983 Service industries: supporting role or centre stage? *Area* **15**: 301–9

Dobson S M 1989 Jobs in space: some evidence on spatial uniformity in the job generation process. *Urban Studies* **26**: 611–25

DTI 1973 Memorandum on the inquiry into location attitudes and experience. In Expenditure Committee (Trade and Industry Sub-Committee) *Regional Development Incentives, Session 1973–74, Minutes of Evidence (from July 1973), Appendices and Index* The House of Commons, HMSO, London

DTI 1989 *Review of DTI Statistics* DTI, London

Elias P, Healey M J 1991 *People and Work in Coventry: A Survey of Employers* Coventry City Council, Coventry

Elton C J 1983 The impact of the Rayner review on unemployment and employment statistics. *Regional Studies* **17**: 143–6

Firn J R 1973 Memorandum. In Expenditure Committee (Trade and Industry Sub-Committee) *Regional Development Incentives, Session 1973–74, Minutes of Evidence (from July 1973, Appendices and Index* The House of Commons, HMSO, London: pp 694–712

Foley P 1983 A comparison of four data sources. In Healey M J (ed.) *Urban and Regional Industrial Research: The Changing UK Data Base* Geo Books, Norwich pp 91–104

Fothergill S, Gudgin G 1978 *Regional Employment Statistics on a Comparable Basis, 1952–75* Occasional Paper 5, Centre for Environmental Studies, London

Fothergill S, Gudgin G 1982 *Unequal Growth: Urban and Regional Employment Change in the UK* Heinemann Educational, London

Gaffikin F, Nickson A 1984 *Job Crisis and the Multinationals: Deindustrialisation in the West Midlands* Trade Union Resource Centre, Birmingham

Gallagher C, Daly M, Thomason J 1990 The growth of UK companies 1985–87 and their contribution to job generation. *Employment Gazette* Feb: 92–8

Ganguly P 1985 *UK Small Business Statistics and International Comparisons* Harper & Row, London

Goddard J, Coombes M 1983 Local employment and unemployment data. In Healey M J (ed.) *Urban and Regional Industrial Research: The Changing UK Data Base* Geo Books, Norwich, pp 31–8

Government Statisticians' Collective 1979 How official statistics are produced:

views from the inside. In Irvine J, Miles I, Evans J (eds) *Demystifying Social Statistics* Pluto Press, London pp 130–51

Green A E, Owen D W 1989 The changing geography of occupations in engineering in Britain, 1978–1987. *Regional Studies* **23**: 27–42

Gudgin G 1978 *Industrial Location Processes and Regional Employment Growth* Saxon House, Farnborough

Gudgin G, Fothergill S 1979 *The East Midlands Industrial Databank: A Guide to Sources, Methods and Definitions* Working Note 559, Centre for Environmental Studies

Hakim C 1989 Identifying fast growth small firms. *Employment Gazette* **97** Jan: 29–41

Healey M J 1981 Locational adjustment and the characteristics of manufacturing plants. *Transactions of the Institute of British Geographers New Series* **6**: 394–412

Healey M J 1982 Plant closures in multi-plant enterprises – the case of a declining industrial sector. *Regional Studies* **16**: 37–51

Healey M J 1983a Establishment databanks in the United Kingdom. In Healey M J (ed.) *Urban and Regional Industrial Research: The Changing UK Data Base* Geo Books, Norwich pp 131–72

Healey M J 1983b *The Coventry Region Industrial Establishment Databank: A Guide to Sources, Methods and Definitions* Industrial Location Working Paper 2, Department of Geography, Coventry Polytechnic

Healey M J 1983c The changing data base: an overview. In Healey M J (ed) *Urban and Regional Industrial Research: The Changing UK Data Base* Geo Books, Norwich pp 1–29

Healey M J 1983d Components of locational change in multi-plant enterprises. *Urban Studies* **20**: 327–41

Healey M J, Clark D 1984 Industrial decline and government response in the West Midlands: the case of Coventry. *Regional Studies* **18**: 303–18

Healey M J, Clark D 1985 Industrial decline in a local economy: the case of Coventry, 1974–1982. *Environment and Planning A* **17**: 1351–67

Healey M J, Ilbery B W 1990 *Location and Change: Perspectives on Economic Geography* Oxford University Press, Oxford

Healey, M J, Watts H D 1987 The multiplant enterprise. In: Lever W F (ed.) *Industrial Change in the United Kingdom* Longman, Harlow, Essex pp 149–66

Hobbs P 1988 *Survey of Industrial Databases: Summary Report* Co-operative Research Unit, Faculty of Technology, Open University

Howard R S 1971 The movement of industry: its role and generation. Paper presented to the Regional Studies Association Conference on the movement of industry, Cambridge

Hubbard R K S 1980 Monitoring employment trends in urban areas. In *Policy Analysis for Urban and Regional Planning: Proceedings of Seminar A held at the PTRC Summer Annual Meeting, University of Warwick* PTRC Education and Research Services, London pp 161–84

Keeble D E 1971 Employment mobility in Britain. In Chisholm M, Manners G (eds) *Spatial Policy Problems of the British Economy* Cambridge University Press, Cambridge pp 24–68

Key Note Publications Ltd 1990 *The Source Book* Key Note Publications, Hampton, Middlesex

Lloyd P, Shutt J 1985 Recession and restructuring in the North West region 1975–82: the implications of recent events. In Massey D, Meegan R (eds) *Politics and Method: Contrasting Studies in Industrial Geography* Methuen, London pp 16–60

Macey R D 1982 *Job Generation in British Manufacturing Industry: Employment Change by Size of Establishment and by Region* Government Economic Service Working Paper 55 London

McLeish J 1987 Business establishment register for Tayside: computerised company database. *BURISA Newsletter* **78**: 4–5

Mason C M 1987 The small firm sector. In Lever W F (ed.) *Industrial Change in the United Kingdom* Longman, Harlow, Essex pp 125–48

Mort D 1990 *Sources of Unofficial UK Statistics* The University of Warwick Business Information Service, Gower Press, Aldershot

Moyes A, Westhead P 1990 Environments for new firm formation in Great Britain. *Regional Studies* **24**: 123–36

National Audit Office Report by the Comptroller and Auditor-General 1989 *Locate in Scotland* HMSO, London

Norkett P 1986 *Guide to Company Information in Great Britain* Longman, Harlow, Essex

North D J, Leigh R, Gough J 1983 Monitoring industrial change at the local level: some comments on methods and data sources. In Healey M J (ed.) *Urban and Regional Industrial Research: The Changing UK Data Base* Geo Books, Norwich pp 111–29

Norton R 1989 Statistics: less of a burden. *British Business* **30** (14 July): 16–17

Nunn S 1980 *The Opening and Closure of Manufacturing Units in the United Kingdom 1966–75* Government Economic Service Working Paper 36 London

Nunn S 1983 Information systems in central government: the role and potential of the regional data system. In Healey M J (ed.) *Urban and Regional Industrial Research: The Changing UK Data-base* Geo Books, Norwich pp 105–10

OPCS 1984 *Census 1981: Economic Activity GB* HMSO, London

Owen T 1988 *Mind your Local Business* Eurofi, Newbury

Perry J A 1985 The development of a new register of businesses. *Statistical News* **70**: 13–16

Perry J 1987 Company reporting for the production industries. *Statistical News* **77**: 18–21

Risk J 1988 New business registrations. *BURISA Newsletter* **86**: 11–13

Robinson W S 1950 Ecological correlations and the behaviour of individuals. *American Sociological Review* **15**: 351–7

Spicer J 1988 Tyne and Wear business information (COBIS). *BURISA Newsletter* **83**: 15–17

Sprott T F 1977 Annual census of employment data. *BURISA Newsletter* **24**: 13

Swales J K (ed.) 1976 *Establishment Based Research: Conference Proceedings* Discussion Paper 22, Centre for Urban and Regional Research, University of Glasgow

Taylor J, Twomey J 1988 The movement of manufacturing industry in Great Britain: an inter-county analysis, 1972–81. *Urban Studies* **25**: 228–42

Thompson A 1983 The prospects for establishment-level databanks. In Healey M J (ed.) *Urban and Regional Industrial Research: The Changing UK Data Base* Geo Books, Norwich pp 65–89

Tomkins J, Twomey J 1990 The changing spatial structure of manufacturing plant in Great Britain, 1976 to 1987. *Environment and Planning A* **22**: 385–98

Tyler P, Moore B C, Rhodes J 1988 *Geographical Variations in Costs and Productivity* HMSO, London

Welsh Affairs Committee 1988 *Inward Investments into Wales and its Interaction with Regional and EEC Policies* 2 volumes HMSO, London

17

Producer services

Neill Marshall

17.1 Introduction

Data sources provide a lens through which researchers regard the economy. They shape views of its nature and of its dynamics. There is also a relationship between researchers' theorizations of economic change, the subject chosen for research, the research methods used in the enquiry and the data required to conduct the analysis. The data collected, the importance attached to them, the way they are presented and used, all reflect received views of the way the economy works (Massey and Meegan 1985).

The British economy has witnessed a major change during the last two decades, what Lloyd (1989) describes as a 'phase-shift'. Important components of this change include two major recessions, a growing interdependence of the British and other European economies, fierce international competition, a reorganization of business including the introduction of different labour practices, rapid technological innovation and the emergence of the state as a champion of market forces, enterprise culture, consumer sovereignty and technologically-aided restructuring.

It is clear that the growing prominence in the economy of service activities (that is activities relatively detached from material production), brought about by both an expansion of employment in service industries and contraction in manufacturing employment, is deeply involved in these structural changes. Services are dominant employers, employing approximately twice as many workers as manufacturing, and even in output terms they far exceed production industries. Equally important is the fact that services are becoming increasingly involved in trade, both directly through their own exports and indirectly because of their contribution to firm competitiveness. Many of the government's attempts to reduce state interference and enhance competition through regulatory change have also involved services such as finance, the media, public services and the legal profession. Finally, services are in the forefront of restructuring because of their increasing use of information and communication technologies, which is encouraging reorganization and bringing in train a host of new services and reshaping labour practices.

The growing prominence of services demands a reworking of our views of the economy, requiring that we shed our preoccupation with manufacturing activities and reorientate academic enquiry towards services. Services should be seen not as an independent entity separate from manufacturing, but as part of an economy where the final value of a good contains an ever more substantial input from the service sector; where the production of goods depends more and more on service provision, computer-aided design and production methods and advanced information systems; where the growth of the service sector is itself creating a strong demand for information technology (IT) goods; and where financial services are acting as a vanguard sector in a round of service innovation based on information and communication technology products (Barras 1986).

Investigating this 'new industrial economy' (Miles 1988) is making new demands of the official statistical database, demands which it is currently unable to satisfy. The statistical data available for service activities is weak in quality and limited in extent. Data collection has concentrated on measuring the goods sector because this has been regarded as the dominant part of the economy. This view is being modified, but it will be some time before this will be fully reflected in official government statistics.

The starting-point for any sophisticated analysis of services is a disaggregated set of statistics which identify different types of service activities. This chapter is concerned with producer or intermediate services, that is services that supply customers in business and commerce. These services include research, consultancy, physical distribution, transport and communications, advertising, marketing, legal, computer and (for some) financial services (see Marshall *et al*. 1988). Such functions are performed, not only in service industries, but also in-house by non-specialist producers elsewhere in the economy.

The analysis of these activities is limited by the aggregate nature of most official statistics on services which hinders attempts to identify producer, consumer, public and private services. This chapter also highlights the difficulties involved in measuring output and employment change in service industries using official statistics, and the implications of this for attempts to understand producer service growth. To appreciate the mechanisms behind the growth of producer services it is essential to monitor shifts in economic activities between manufacturing and service industries and between the service industries themselves, yet most official statistics present a rigidly sectoral view of the economy. The limited value of input–output (I–O) tables and the Census of Population (CoP) in monitoring the linkages between service and other industries is demonstrated.

As services increasingly participate in international trade the geographical impacts of export activity become significant, and are likely to be more so after the creation of a Single European Market. The treatment of services in the trade statistics and problems of measuring service transactions within companies are briefly outlined. The problems of international comparisons are also reviewed.

General themes throughout this chapter are the lack of continuity in the official statistical database, and the poor quality and reliability of the material produced. For these reasons researchers are facing new challenges to their ingenuity in data collection and analysis. A variety of new avenues of data collection are being pursued to supplement official secondary source statistics. In turn these source materials, such as industry sub-literature and ephemera, are posing new challenges to the analytical abilities of researchers.

17.2 The official statistical database

The current debate

> Current statistical systems emphasise older, mature or even declining economic activities and provide only little information on emerging new industries, products and services, new investment and employment (OECD 1986: 20).

Concern over the state of official statistics has reached fever pitch (House of Commons 1988a). At the time of writing (September 1989) hardly a week seems to go by without another report that the quality of official statistics is under threat (Hencke 1989a, b). In 1988 the Treasury and Civil Service Committee was moved to argue economic statistics were so faulty that it was 'very hard to understand where the economy has been or is, let alone where it is going' (House of Commons 1988b: XXIII). This concern goes back to the Rayner review in the early 1980s which instituted a major reorganization (some would call this a cut) in official statistics, and it has been sustained by an endless round of changes ever since. Changes and cuts in microstatistics are in turn threatening the whole superstructure of the **National Accounts**. So much so that government have conducted a scrutiny of their economic statistics, and a further reorganization of their statistical services has been implemented (Cabinet Office 1989; DTI 1989). Notwithstanding this change, some still argue that the role of government in the collection and dissemination of statistics is being systematically downgraded (Phillips 1989a) and call for the statistical service to be made independent of government to avoid improper pressures (Phillips 1989b).

The decline in the quality of a number of official statistics is very real, and concern about it is firmly grounded, but in the ensuing debate it has almost gone unnoticed that the quality of the official statistical database is deteriorating not just because of changes in the government's statistical service (GSS), but because government data collection is failing to keep pace with structural changes in the British economy and most notably the new demands being placed upon it by the growth of service activities. In one of the few discussions where the issue has been pursued it is considered that developments in the service sector make economic monitoring more difficult. Thus commenting on concern over the large discrepancy in the official balance of payments statistics the Central Statistical Office (CSO) relates this to:

> . . . the major changes which have taken place in financial markets over the last four years, notably deregulation.
>
> There has been an accompanying upsurge in competition and innovation, a move towards greater use of securities markets for channelling finance and hedging techniques have come into more widespread use.
>
> In these circumstances, it is possible that estimates of UK net credits from providing and receiving financial services . . . are harder to track (quoted in Coutts *et al.* 1990: 17).

However, as this paper argues, a rather more serious problem is the way in which official statistics restrict our ability successfully to monitor service activities and, therefore, present a partial and even distorted picture of the dynamics of that part of the economy.

Classifying economic activities

Official statistics present by and large a sectoral view of the economy, notwithstanding the deep links between individual sectors. Since the Second World War the main scheme for the presentation of industrial statistics has been the **Standard Industrial Classification** (SIC). Established in 1948, it identified, for the first time, a separate service sector (Table 17.1). Though it has become progressively more detailed, the overriding fact, which has remained constant over the whole of the last 30 years, is that the SIC provides much less information on service activities than it does for the manufacturing sector. By 1958, the time of the first revisions to the SIC, service employment exceeded manufacturing, yet employment data were provided for only 32 service divisions while there were 100 manufacturing industries, a ratio of three to one. In 1980, when the most recent (and major) revisions to the classification were concluded, service employment was more than double that of the manufacturing sector, and yet there were still only 102 service divisions compared to 210 for manufacturing, a ratio of two to one.

This caused *The Economist* to protest, 'Anybody analysing the restructuring of the British economy can still get far more information on the footwear industry (with 50,000 employees) than on business services not elsewhere specified (with 182,000)' (quoted in Daniels and Thrift 1987: 4). The constraint imposed by the lack of detail in the industrial classification used in official data sources runs very deep because it affects our very ability to identify meaningful types of service activities. The seminal works of Greenfield (1966) and Katouzian (1970), which distinguished between services which supply business and those which serve individual customers, and the development of this classification to distinguish within market production

Table 17.1   Classification of service industries

SIC 1948	SIC 1958	SIC 1968	SIC 1980
Transport and communication	*Transport and communication*	*Transport and communication*	*Distribution, hotels, catering, repairs*
Railways	Railways	Railways	*Wholesale distribution*
Tramways and omnibus	Road passenger transport	Road passenger transport	Wholesale distribution of agricultural supplies
Other road passenger transport	Road haulage	Road haulage contracting	Wholesale distribution of industrial materials
Goods transported by road	Sea transport	Other road haulage	Wholesale distribution of building materials
Sea transport	Ports and water transport	Sea transport	Wholesale distribution of motor vehicles
Ports and water transport	Air transport	Ports and water transport	Wholesale distribution of transport equipment
Harbour, etc.	Posts and telecommunications	Air transport	Wholesale distribution household goods
Air transport	Miscellaneous transport	Posts and telecommunications	Wholesale distribution of clothing, footware, etc.
Posts and communications		Miscellaneous transport	Wholesale distribution of food and drink
Other transport			Wholesale distribution of pharmaceuticals, etc.
Storage			Other wholesale distribution
Distributive trades	*Distributive trades*	*Distributive trades*	*Dealing*
Coal, builders' materials	Wholesale distribution	Wholesale distribution of food and drink	Dealing in scrap metals
Other industrial materials	Retail distribution	Wholesale distribution of petroleum	Dealing in other scrap
Wholesale food and drink	Dealing in coal, etc.	Other wholesale distribution	
Retail food and drink	Other industrial dealing	Retail distribution of food and drink	*Commission agents*
Wholesale non-food	Other retail distribution		
Retail non-food		Dealing in coal, etc.	*Retail distribution*
Retailing confectionery, etc.	Other industrial dealing		Food retailing
			Confectioners, etc.
Insurance, banking and finance	*Insurance, banking and finance*	*Insurance, banking, finance and business services*	Chemists
		Insurance	Retailers clothing
		Banking	Retailers footware, leather
		Other finance	Retailers fabrics, textiles
		Property	Retailers household goods
		Advertising and market research	Retailers motor vehicles
		Other business services	Filling stations
		Central offices	Retailers books, stationery
			Other retailers
			Mixed retailers

Table 17.1 Cont.

SIC 1948	SIC 1958	SIC 1968	SIC 1980
			Hotels catering
			Eating places
			Take-aways
			Public houses
			Clubs
			Canteens
			Hotel trade
			Other tourism
			Repair consumer goods
			Repairs to motor vehicles
			Repairs to footware and leather
			Other repairs
			Transport and communication
			Railways
			Other inland transport
			Road passenger transport, etc.
			Other road passenger transport
			Road haulage
			Transport
Professional services	*Professional services*	*Professional and scientific services*	
Accountancy	Accountancy	Accountancy	
Education	Education	Education	
Legal	Legal	Legal	
Medical	Medical	Medical	
Religious	Religious	Religious	
Other professional services	Other professional services	Research and development	
		Other professional services	
Miscellaneous services	*Miscellaneous services*	*Miscellaneous services*	
Theatre, cinema, etc.	Cinema, theatre, radio, etc.	Cinema, theatre, radio, etc.	
Sport and recreation	Sport and recreation	Sport and recreation	
Catering, hotels, etc.	Betting	Betting and gambling	
Laundries	Catering, hotels	Hotels, etc.	
Dry cleaning, etc.	Laundries	Restaurants, etc.	
Hairdressing	Dry cleaning, etc.	Public houses	
Domestic service (resident)	Motor repairers, distributors	Clubs	
Domestic service (non-resident)	Repair boots and shoes	Catering contractors	
	Laundries	Hairdressing	
		Dry cleaning, etc.	
		Motor repairers, distributors	
		Repair boots and shoes	
		Other services	
			Sea transport
			Air transport
Public administration	*Public administration*	*Public administration*	
National government	National government	National government	
Local government	Local government	Local government	

Table 17.1 Cont.

SIC 1948	SIC 1958	SIC 1968	SIC 1980
			Support services
			Support inland transport
			Support sea transport
			Support air transport
			Miscellaneous transport
			Postal services and telecommunications
			Postal services
			Telecommunications
			Banking, finance, insurance, business services
			Banking and finance
			Banking
			Other finance
			Insurance
			Business services
			Auxiliary services banking
			Auxiliary insurance
			Estate agents
			Legal
			Accountancy
			Professional technical nes
			Advertising
			Computer services
			Business services nes
			Central offices

Table 17.1 Cont.

SIC 1948	SIC 1958	SIC 1968	SIC 1980
			Renting
			Hiring agricultural equipment
			Hiring construction machinery
			Hiring office equipment
			Hiring consumer goods
			Hiring transport equipment
			Other hiring
			Owning, dealing real estate
			Other services
			Public administration
			National government nes
			Local government nes
			Justice
			Police
			Fire
			National defence
			Social security
			Sanitary services
			Refuse
			Sewage
			Cleaning
			Education
			Higher education
			Schools
			Education nes
			Driving, flying schools

Table 17.1 Cont.

SIC 1948	SIC 1958	SIC 1968	SIC 1980
			Research and development
			Medical
			Hospitals
			Other medical
			Medical practices
			Dental practices
			Private nurses, etc.
			Veterinary
			Other services
			Social welfare
			Trade, professional organizations
			Religious
			Tourist offices
			Recreation
			Films
			Radio, television, etc.
			Authors
			Libraries, etc.
			Sport, recreation
			Personal services
			Laundries
			Dry cleaning, etc.
			Hairdressing
			Personal nes

Table 17.2 Classification of private producer service industries. (*Source:* Marshall *et al.* 1988: 20–1)

MLH	NIEC (1982)	Greenfield (1966)	Gershuny*† and Miles (1983)	Marquand (1979)	Hubbard and Nutter (1982)‡
Transport and communications					
701 Railways				§	
702 Road and passenger transport				• §	•
703 Road haulage for hire	/			/	/
705 Sea transport	•			/	/
706 Port and inland water transport			/	/	/
707 Air transport					•
708 Postal services and telecommunications				/ §	/
709 Miscellaneous transport	/			•	/
Distributive trades					
810 Wholesale distribution, food/drink	•	/		/	/
811 Wholesale distribution, petroleum	•	/		/	/
812 Other wholesale distribution	•	/		/	/
820 Retail distribution, food/drink.					
821 Other retail distribution		/			
831 Dealers in builders' materials and agricultural supplies	•	/		/	/
832 Dealers in other industrial materials	/	/		/	/

Table 17.2 Cont.

MLH	NIEC (1982)	Greenfield (1966)	Gershuny*† and Miles (1983)	Marquand (1979)	Hubbard and Nutter (1982)‡
Insurance, banking, finance and business services					
860 Insurance	.	.	/	§	/
861 Banking	.	.	/	§	/
862 Other financial institutions	.	.	/	§	/
863 Property owning and managing	.	.	/	§	.
864 Advertising and market research	/	/	/	/	/
865 Other business services	/	/	/	/	/
866 Central offices	/	?	?		/
Professional and scientific services					
871 Accountancy	/	.	/	§	/
873 Legal services	.	.	/	§	/
875 Religious organizations	/	/	/	/	/
876 Research and development	/	.	/	§	/
879 Other professional and scientific services	/		/		

Table 17.2 Cont.

MLH	NIEC (1982)	Greenfield (1966)	Gershuny*[†] and Miles (1983)	Marquand (1979)	Hubbard and Nutter (1982)[‡]
Miscellaneous services					
881 Cinemas, theatres, radio, etc.					
882 Sport and recreation					
883 Betting and gambling					
884 Hotels and residential establishments					/
885 Restaurants, cafés, etc.					
886 Public houses					
887 Clubs					
888 Catering contractors					
889 Hairdressing					
891 Private domestic services					
892 Laundries					
893 Dry cleaning, dyeing, etc.					
894 Motor repairers, distributors, garages					
895 Repair of boots and shoes					
899 Other services					

*These definitions are not totally consistent with the UK SIC.
[†]Gershuny and Miles (1983) regard distributive services as a separate category.
[‡]Defined as 'basic' services, but excluding consumer activities.
[§]These services though classified as 'mixed' producer–consumer produced a higher r^2 with population than employment.
. 'Mixed' producer–consumer services.
/Producer services.

between office and blue-collar services (Stanback *et al.* 1981; Stanback and Noyelle 1982) are constrained by the lack of detail in government statistics on service activities and the integration of business, consumer, market and non-market activities. As the debates of the Producer Services Working Party show (Marshall *et al.* 1988), arbitrary decisions must be made concerning the boundary between individual types of service, and it follows that differences . in definition can easily cloud analysis (Table 17.2).

Measurement problems in output and employment statistics

The measurement problems for services are multiple and serious. They relate to a long list of economic performance indicators such as real output, productivity, prices, or to inputs such as employment, hours of work, skill and education, or to institutional factors such as the size of firms, the registering of new firms or to impeding obstacles such as regulations of service (Ochel and Wegner 1986: 27).

There are in general terms two reasons why producer service industries may be growing in a region:

1. The growing prominence of producer service industries may reflect slow growth of labour productivity, so that as output grows more employment will appear in service activities;
2. In contrast, employment change could reflect a strong underlying growth in demand for producer service output.

Real changes in gross domestic product (GDP) are the key to understanding the dynamics of producer services. The central issue is how to measure output change and in turn how to compare this with trends in labour usage to compute labour productivity. The National Accounts (CSO, Blue Book) contain three measures of GDP, current price GDP calculated by both income and expenditure, and indicators of output in constant prices. The latter may be used to deflate current price GDP data (see CSO 1981 for the conventions involved). Since the early 1970s the CSO has also produced regional accounts via the income definition of GDP, largely using information on employment and wages by industry in each region (CSO 1978). But significantly there is no officially produced measure of regional variations in prices, though *Reward Regional Surveys* (1972–91) do produce such information (biannually) for consumer, but not producer, prices.

Output measures

Furthermore, there are doubts concerning the veracity of the national indices of output for producer service industries, and many argue that they under-represent service growth compared with goods production (Barras 1983; Smith 1972; Hill 1971). In many areas of the intermediate service sector direct output indicators do not exist because there is no clear and unambiguous output. In these cases proxy measures such as direct employment are used. An employment series by definition excludes a productivity effect (other than due to skill mix). This proxy is used for 11 per cent of all private services.

In some services it is relatively easy to estimate the physical quantities produced and these in turn give a measure of output (e.g. the distance travelled in transport or distribution). Direct output measures account for 18 per cent of the output of private services. However, service output incorporates a substantial quality component and changes in this may be more important than changes in the quantity produced. Quality changes are ignored in direct output measures.

Price adjustment of wage bills or sales can also be used to calculate service output in constant prices. Deflated sales account for 43 per cent of private-sector service output and other proxies 28 per cent. However, finding adequate price deflators is particularly difficult in the computer and telecommunications industry, and the increasing role of these in the economy, especially where they cross international boundaries, is a problem. In addition, the deflation of values by price indices implies an ability to separate quality improvements from inflation. This may be difficult to achieve where product differentiation is high (Nankivell 1985).

Employment measures

Labour inputs are poorly measured in many intermediate services, casting further doubt on measures of labour productivity. Though employees in employment, derived from the Census of Employment (CoE) and published in the *Employment Gazette* (see Ch. 2), are measured reasonably well, there is doubt about the coverage of part-time (usually female), temporary and more especially self-employed workers. Bench-mark estimates for self-employment are derived from the **Census of Population** (CoP). Between these dates estimates are obtained from the *Labour Force Survey (LFS)*. So not surprisingly the figures are invariably available at a less detailed level of disaggregation than in employees in employment (Graham *et al.* 1989). They are also frequently out of date, and unlikely to pick up recent changes.

Small firm problem

One of the most difficult problems in intermediate services is the number of small firms and the relatively short-lived and changeable nature of their operations. This creates significant problems in creating and maintaining a sample register for government survey purposes. This is less of a problem with sources such as the CoE which are based on Inland Revenue pay points but it can make estimates between censuses unrealistic.

Accepting the conventional statistics at 'face value' Table 17.3 presents percentage output and employment change in various services and manufacturing as available for the regions of the UK between 1974 and 1985 (see

Table 17.3 Percentage change in service and manufacturing employment (E) and output (O) (1980 prices), 1974–85. *Source:* Marshall 1989a – based on *Regional Trends* and CoE data.

	South East		East Anglia		South West		West Midlands		East Midlands		Yorkshire and Humberside		North West		North		Wales		Scotland	
	E	O	E	O	E	O	E	O	E	O	E	O	E	O	E	O	E	O	E	O
Man	-28.5	6.4	-6.3	36.9	-18.7	8.7	-34.9	-17.2	-21.1	4.0	-38.3	-12.4	-39.4	-9.3	-40.8	-2.4	-38.1	-12.9	-36.9	-6.4
Dist	10.3	6.9	25.8	22.2	9.2	8.3	8.8	15.2	19.7	20.5	-4.3	17.4	+5.8	5.3	-3.4	6.3	13.1	20.9	3.3	20.8
T&C	-7.9	13.7	37.5	69.6	-4.6	30.2	-13.1	15.7	2.8	34.5	-3.7	9.3	22.5	2.4	-13.8	28.8	-23.3	4.2	-17.3	26.0
F & BS	34.8	59.9	65.7	90.4	72.6	103.9	61.2	84.0	50.0	84.2	40.0	89.5	27.0	69.4	45.8	89.7	42.5	82.3	32.4	80.5
OD		19.3		72.6		18.0		19.3		25.5		23.8		14.6		7.5		17.0		4.8
Pub*	1.0	-15.2	6.1	27.6	15.3	-0.5	4.6	8.0	4.9	0.9	0.8	-0.1	2.4	7.8	-3.3	6.2	4.3	24.4	3.7	10.9
Ed/Health		5.1		19.3		48.9		11.0		36.1		27.5		18.9		47.6		41.0		33.5
Other*	17.9	51.3	27.4	75.6	22.1	17.3	25.2	21.2	59.8	30.4	15.2	23.1	5.9	50.6	33.5	66.3	22.4	45.5	24.6	38.7
Total Services	12.2	12.9	28.5	40.2	18.6	22.2	15.9	17.5	28.4	24.2	13.1	19.3	4.1	14.0	10.8	23.7	10.4	24.9	10.2	23.4

* Public administration includes compulsory social security in the regional GDP data which *Employees in Employment* allocates to other services.

Notes: Man – manufacturing; Dist – distribution; T & C – transport and communications; F & BS – finance and business services; OD – ownership of dwellings; Pub – public administration; Ed/Health – education and health; Other – other services.

Marshall 1989a for details). Current price GDP for each region has been adjusted using the appropriate price deflator for individual service industries implicit in the National Accounts. The results must, therefore, be examined with considerable caution and for this reason the precise labour productivity change has not been calculated.

Service-sector employment and output changes look similar. Even in those regions where service-sector output growth has been greater than the corresponding employment increase (e.g. East Anglia, the North, Wales and Scotland) services do not match the performance of manufacturing, where the recovery of the latter from recession has been characterized by a major improvement in labour productivity. This suggests that slow growth in labour productivity must be an important explanation for the growing prominence of services in regional economies.

It is exceedingly difficult to examine service industries in any detail. However, closer inspection suggests that while employment grows in parallel with output in the public sector and distribution (where consumer demand is important), in contrast in services where business markets are more important, such as finance and business services and transport and communications, more rapid growth in labour productivity is implied in the figures.

The weight attached to this conclusion hinges on the attitude taken to the quality of the official statistics. None the less, there is evidence to support the view that there has been a significant increase in the labour productivity performance of some intermediate services during the latter 1970s and 1980s, as a consequence of a growing capital intensity (Green 1985). This suggests that for these services the labour productivity explanation for their growth is insufficient. Indeed if one accepts that the problems with the constant price series for service-sector output and the difficulty of measuring changes in small firms probably means that service growth is underestimated (Barras 1983), this supports the view that a growth in demand for service output is an important factor in the growing prominence of intermediate service industries in regional economies. This reinforces the need to examine the mechanisms which underlie the changing demand for intermediate services in the economy, which requires an understanding of the linkages between service industries and other activities.

Examining service linkages

Input–output (I–O) tables

These are produced every 5 years on the basis of a full purchasing enquiry, the most recent survey having been carried out in 1984. They provide in principle a neat and convenient description of the economy, illustrating how national income is built up, and how it is related to national expenditure and output. They display the intermediate current account transactions (capital purchases are part of final demand) which are not apparent in other tables in the National Accounts. Estimates are provided for industry groups of their purchases of goods and services from other industries, and the sales of that industry which supply intermediate output to each of the other industries, as well as final output to persons, exports, the public sector and capital formation (see CSO 1975, 1988 for the best reviews of sources and methods).

Thus, it is possible, in principle, to calculate the output of service industries which supply intermediate and final demand (critical for identifying producer services) to begin to attach weights to the various factors which influence service demand, and to answer questions about the way in which services might respond to changes in the economy as a whole (Robertson *et al.* 1982).

Work by Wood (1988) and Siniscalco (1989) demonstrates the value of I–O tables. The authors grapple with the obvious problem of how (if at all) to deflate current price I–O data on services, and are forced to assume that output can be equated fairly readily with employment for service industries; in other words, that there is homogeneous labour productivity within broad service sectors. Nevertheless, Siniscalco demonstrates that in Italy the growth of service-sector employment is critically related to the fast-growing demand for intermediate services from the goods-producing industries, rather than directly by final consumer demand, and this analysis highlights the growing integration between manufacturing and services. Wood confirms the importance of intermediate demand in service-growth, but argues in a Canadian context that a growing indirectness of service production within the service industries themselves (including that supplying intermediate demand) is also important.

Marshall *et al.* (1988) find evidence of both trends in the UK I–O tables. There are, however, a number of problems which restrict the analysis. Most obviously official statistics are not produced below the national scale, though a number of unofficial estimates have been produced at the level of the standard region and below, most notably for Scotland (Industry Department Scotland, 1979). Leaving aside this difficulty, the national tables do not identify individual services in the 1950s, while in the 1960s the bulk of the service sector is lumped into the miscellaneous category. During the 1970s and 1980s the classification of services improves and finance, business services, distribution, transport and communications, all vital to the monitoring of producer services, are separately identified.

However, the value of this information is limited by the character of the tables and the methods of data collection employed. The I–O tables rather unhelpfully classify the public sector to final demand, making it difficult to explore the intermediate links between the public sector and other parts of the economy. The **Census of Production**, the main source of information for the I–O tables, provides reasonable information on the purchases of services made by manufacturing, though there is some concern about the coverage of head offices, where many service purchases take place. But there is no census for many intermediate service industries (the enquiries into wholesale, retail and catering trades are of some help – see Ch. 18). All too frequently statisticians rely upon company reports, one-off surveys and anecdotal evidence, or alternatively service data emerge as residual items in the tables (CSO 1988: 11). Random errors which bedevil some tables and apocryphal stories about data collection lend credence to the view that I–O tables in the UK tell us more about the methods used to collect the data than about trends in the service sector of the economy (Marshall *et al.* 1988). Finally, the latest scrutiny of official statistics was somewhat ambiguous concerning the value of I–O tables, and while the tables will continue to be produced it will be on the basis of a restricted purchasing enquiry (Cabinet Office 1989; DTI 1989).

Census of Population (CoP)

Another approach towards examining the interdependence of service and manufacturing industries is to examine the cross-classification of industry by occupation produced at the standard region level via the 10 per cent sample of returns from the CoP. Using these data Crum and Gudgin (1977) were able to show that in 1971 there was a marked under-representation of non-production workers in manufacturing in the northern regions of Great Britain, and to suggest that this was only partly explained by industrial structure. More recently the debate concerning the importance of the externalization of service functions by manufacturing in the growth of service industries has encouraged researchers to compare employment trends in non-production occupations within manufacturing and service industries to see how many jobs have been simply transferred from manufacturing to services (Perry 1990).

In Great Britain the CoP has a more limited disaggregation by industry than the CoE, and this significantly affects the division between consumer and producer services (see Marshall *et al.* 1988). Research is also limited by the reorganization of the industrial and occupational classification used in 1980, which restricts the comparison of census returns from 1971 and 1981 (Office of Population Censuses and Surveys (OPCS) 1980). This problem will also materialize in 1991 because the occupational classification will be different again (Department of Employment (DE) 1988).

Work carried out by the Producer Services Working Party and more detailed analysis conducted in the West Midlands Economic Planning Region (Marshall 1988) indicates that it is impossible to compare Crum and Gudgin's earlier results with non-production employment data for 1981. Green and Owen (1984) and Marshall and Green (1990), however, have used a collapsed version of the socio-economic group (SEG) classification, which remains comparable between 1971 and 1981, to identify white-collar non-production employees in manufacturing. This, of course, excludes blue-collar service workers in physical distribution from in-house service work in manufacturing. More importantly, their analysis, which is conducted at the level of the local labour market, uses data classified by place of residence (place of work data are not available below the level of the standard region) and this means that trends are distorted by commuting (The LFS also provide data on occupation by industry for counties, but the small size restricts the disaggregation).

It is also possible at the national scale to update the CoP evidence by using the annual statistics of the DE on administrative technical and clerical staff in manufacturing, which are published in the *Employment Gazette*. These statistics go as far back as 1924 and provide employment disaggregated by industry (Marshall *et al.* 1988). These data show an interesting decline in the share of service workers in manufacturing employment during the 1980s, probably indicating significant rationalization of non-production staff and contracting out (Marshall 1989a). However, this classification is not comparable with the census classification of occupations or the SEG classification.

International perspectives

While this volume is primarily concerned with UK sources, in services as in many other fields it is necessary at the very least to view developments in the UK in an international perspective. Indeed, the growing interdependence of national economies demands an investigation of service trade in its own right (see Riddle 1986 and Nusbaumer 1987 for good reviews). It is impossible to do justice to the vast number of relevant international sources here, but international organizations such as the Commission of the European Communities (CEC), the World Bank, the International Labour Organization (ILO) and the United Nations are valuable sources of employment, output and trade statistics on services. The OECD (Organization for Economic Co-operation and Development) historical series on employment are of particular value (OECD 1987a). However, as with the national statistics, those provided by international organizations are characterized by limited disaggregation, poor measurement and under-reporting. In addition, the different national nomenclatures for industrial statistics, and differences in the way National Accounts are compiled, make international comparisons difficult. Of particular note in the latter regard are different attitudes to the boundary between goods and services. A recent critical evaluation of the statistics for the industrialized economies shows just how few international comparisons are possible (OECD 1987c). Nevertheless, attempts have been made to harmonize the International Monetary Fund (IMF) balance of payments statistics (see UNCTAD 1988: 149).

Such data on international trade in services deserve a special mention given the attempts to open up service trade in GATT (General Agreement on Tariffs and Trade), and through the legislation to establish the Single European Market. The IMF balance of payments statistics differentiate between transportation, shipping, travel and 'other services'. However, this categorization leaves 'unpacked' the most interesting category. So national balance of payments statistics have to be used to supplement these data. The CSO's *Balance of Payments* (Pink Book) contains the appropriate domestic statistics (see Liston and Reeves 1988 for an up-to-date review of these), and the OECD (1987b) provides a useful summary source for the major industrial countries. However, different conventions as to what constitutes service trade and when it actually occurs are adopted within national statistics. Indeed the issue of defining service trade is central to the GATT negotiations on services. The United Nations (UNCTAD 1988) argue that for an international transaction to take place either goods, capital, persons or information must cross a frontier to receive or obtain a service. But this locational approach ignores questions of ownership and control which may be significant when domestic service transactions are in the hands of large foreign-based multinationals.

There is also a view reported by Riddle (1986) that many business service transactions are truly 'invisible', because they are hidden in the balance of payments statistics on services. As far as service trade is concerned, little activity takes place at the border, and many service transactions are difficult to identify because they are transferred within companies. It is also significant that the data on service exports and imports at a world scale do not match, and that the discrepancy has been increasing, especially in the growth area of 'other services'. This is partly because virtually every country reports items as being traded on the basis of the primary function of the parent company involved, and it is believed this results in the reporting of service transactions by manufacturing firms as trade in goods.

Finally, central to the concerns of this chapter, statistics on service trade do not provide much information simultaneously on both the origin and destination of transactions which is central to the analysis of geographical questions (see Ch. 19 for a useful discussion of the information available).

17.3 Alternative sources of data

Research into intermediate services frequently wishes to explore the economy in more detail than is permitted by official statistics. It deals with issues where up-to-date information is absolutely essential, and considers research questions which link together services and other activities. Key research issues include:

(a) the changing manufacturing demand for services;
(b) the impact of technological changes on service activities;
(c) the financial sector and regional development;
(d) corporate structure and service location; and
(e) changes in the quality and character of service employment.

There are a plethora of data sources available to explore these questions in addition to official statistics, including other official sources of information such as reports from the committees of the Houses of Commons and Lords, as well as legislation and published studies. The press, industry journals, reports produced by a variety of commercial and non-commercial bodies, as well as business information systems provide other sources of data (Table 17.4). As the official statistical series have become increasingly unsatisfactory researchers have turned to these sources.

Table 17.4   Main sources of data on producer services

Secondary sources

1. Official statistics (produced by national government and international organizations):
   employment
   output
   input–output data
   earnings
   financial and trade data

2. Other official publications:
   House of Commons and Lords debates, legislation,
     select committees, etc.
   official research reports and one-off studies (e.g. see
     NEDO)
   official journals, e.g. *Employment Gazette, British
     Business*
   sectoral reviews and reports by EC, OECD, ILO, UNCTAD

3. Unofficial publications:
   company reports and accounts
   firm directories
   newspapers
   library and documentation centre information
   ephemera
   consultancy documents
   academic research in other disciplines
   business computer information systems

Primary data

Discussions with professional and trade associations
Discussions with trade unions
Questionnaire surveys of firms
Case studies, qualitative and 'anthropological' studies

Secondary sources

The secondary sources listed in Table 17.4, especially documents published by professional and trade associations, industry journals and ephemera available to practitioners, can be used to provide good background on any service industry being studied. This can include evidence on trends in the market, competition, the activities of major organizations, acquisitions and topical issues such as new legislation. Such material can shed light on the processes behind crude changes in employment and output, and perhaps alert researchers to problems in the official statistical series.

Table 17.5 provides a list of 'alternative' publications used in research on producer services. While not a complete guide, this should act as a useful starting-point for future research. These publications can be supplemented

Table 17.5 Industry publications used by research on
producer services

General
Accountancy
Accountants' Weekly
Advertising Age
Banker
Banker's World
Building
Business
Campaign
Euromoney
Futures
International Accounting Bulletin
International Journal of Marketing
International Financial Law Review
Investors' Chronicle
Services Industries Journal
Telecommunications Policy
The Accountant
The Economist
The Financial Times
UK Accounting Bulletin
World Property

Directories
Advertisers' Annual
Bankers' Almanac
British Banking Directory
British Research and Data Advertisers and Agency List
Building Society Year Book
Chamber of Commerce Directories
Computer Services Association Directory
Computer Users' Year Book
Crawford's City Directory
Insurance Year Book
Institute of Management Consultants Year Book
International Directory of Software
Kelly's Directory
Key British Enterprises
Kompass Directory
Marketing Services
Marketing Yearbook
Property Industry Year Book
RICS Yearbook
Times 1000
Yellow Pages

by reports produced by market research organizations such as Key Note, Mintel, International Data Communication Services and Logica where these are available. A number of researchers have been able to use such literature to highlight the factors encouraging an increase in the demand for intermediate service output. These factors include the growing internationalization and differentiation of markets, changing legislation and regulation concerning trade and the growing technological sophistication of production which is requiring in turn new workforce skills and the increasing complexity of organizations (Marshall 1985, 1988, 1989a; Ochel and Wegner 1986; Petit 1986; Miles 1985). These analyses add somewhat to the conclusions, based on official statistics, that labour productivity explanations for producer service growth are insufficient.

Commission of the European Communities sector studies and reports from international organizations such as OECD, UNCTAD, GATT and the ILO provide an invaluable source of data on trends in service development and trade. Howells (1988) and Howells and Green (1988) show how the literature listed in Table 17.5 can be used to address the issue of corporate reorganization, a topic which official statistics are singularly unsuited to address. They document the diversification of major companies in Europe into new markets, and examine in detail the origins of companies operating in computer services. The research demonstrates the way in which growth industries in the service sector spin-off from other parts of the economy and in turn attract entrants from other businesses, so that their spatial dynamics are shaped by existing business locations.

Annual reports and accounts can also provide more detailed information on the changing policies of 'major players' in the business, including new subsidiaries acquired, new markets developed, new offices established and closed down, as well as a list of existing sites. It must always be remembered, however, that these reports are to an extent public relations exercises. But they can be supplemented and cross-checked by the use of business directories which list the main activities of companies and often their offices.

Using such data is not less scientific than using official sources, which, as has been shown above, are seriously flawed as far as services are concerned. But analysing the materials raises different problems from those posed by the official statistics. Using any information source which is serving a particular sectional interest requires special care and a sceptical attitude on the part of the researcher (and this issue continually recurs when using 'alternative' data sources (Sayer and Morgan 1985)). A strong theoretical framework is necessary to avoid taking as given a particular view of the world, and to avoid 'getting lost' in masses of information.

Coverage of the subject under investigation is a major problem with research using 'alternative' sources. Data tend to be fragmentary and have to be put together from a variety of different and incompatible sources. Like the official statistics the coverage of smaller firms is patchy, but professional organizations which provide lists of individual as well as corporate members may help to identify these.

It is also difficult to obtain fine-grained information on the geography of services. All too often the research is thrown back on official statistics at this point. One-off publications, though, can be of some help here. Howells (1988) and Gillespie *et al.* (1984), for example, use a variety of sources to demonstrate the impact of technological changes on the geography of services, including government information on existing value added telecommunications networks and data from consultancy companies on network termination points. The surveys of office decentralization from London conducted by Jones Lang Wootton have also proved helpful. Leyshon and Thrift (1989) show how this information and the regional reviews of the *Financial Times* can usefully add to the aggregate and somewhat outdated information provided in government employment statistics. The local press often provides more detailed information, and a lot of hard data collection can be avoided where a local library has a good cuttings service.

Key organizations collecting relevant data can also be approached. Research into information and communication technology can, for example, seek to obtain information from the telecommunications network operators, and Gillespie *et al.* (1984) obtained regional data on telex, telephone and data modems partly from this source. Increasingly, though, in this area, as in many others, information is being commodified, and researchers find that data which were once provided free and without question, become commercially sensitive, or require a large fee to a consultancy company. As a consequence, for more detailed insights survey research is frequently required.

Survey research

During the 1960s, 1970s and early 1980s large-scale surveys of firms (often by post) tended to be the most popular approach. The emphasis was on improving the information available on a little researched sector (Crum and Gudgin 1977; Daniels 1975, 1983; Goddard 1978; Marshall 1979, 1982, 1983; Polese 1982). This research added substantially to the evidence gained from official statistics. It demonstrated the importance of information and business communications in the operation of service organizations and the implications of this for their location. The diverse market linkages of producer services were examined, and this supported the view that services were not simply dependent on manufacturing or final demand. Surveys of company organization showed that managerial, administrative and support staff in manufacturing and service organizations were spatially concentrated in head offices in southern and eastern UK, while production units in the provinces possessed few senior managerial personnel.

The last few years have witnessed a change in approach to survey research, reflecting the search for a deeper understanding of economic change, and an appreciation that service activities are more complex than originally perceived. The emphasis is on small carefully chosen case-study interviews usually with 'prime movers' in a business, or with firms supplying a specific market (Thrift *et al*. 1988; Marshall 1989b). Such a shift has been encouraged by the growth of interest in realist philosophy and the emphasis on intensive case studies that goes with it (Sayer and Morgan 1985, see also Ch. 11).

Considerable detail has been provided on the role of producer services in the growth of city centres during the 1980s. This research has also added to our understanding of service organizations, and indicated that stereotype corporate structures based on manufacturing conglomerates are inappropriate. Corporate structures seem not as heavily concentrated on the capital as earlier work showed and there seem to be possibilities for local development in the provinces.

Interviews with professional and trade associations are used as a complementary stage in this type of research, to supplement the knowledge obtained from secondary sources, highlight additional sources of information, or check information gained from other sources. Table 17.6 provides a list of the main professional bodies and trade associations used in research on producer services. This can be supplemented by consulting the DE's *Directory of Employers Associations, Trade Unions and Joint Organisations*. A browse through the reference section of a well-stocked local library can also be worth while.

Interviews with professional and trade associations give a good overview of the business being studied and provide some interesting examples of reorganization which may merit further scrutiny, but they tend not to give detailed insights into the operations of individual companies. They are heavily reliant (as are all interview studies) on the knowledge of the individual interviewed, and some associations may be dominated by older professionals who have been shunted to one side by proactive business management. Employment issues also tend to be neglected by such interviews, but these can be addressed by discussions with trade union representatives. Here, though, trends in the business of the companies concerned may often only be vaguely appreciated.

Approaching firms in producer services provides information on the way abstract processes identified in secondary-source analysis work out in practice. More detailed information on location is also obtained. Nevertheless, interview studies in producer services present a number of difficulties common to other sectors and some special problems (see Ch. 11). Problems of over-survey leading to a reluctance to be interviewed are beginning to build up. The large number of small firms in the sector can mean low response rates because small business entrepreneurs are notoriously difficult to survey. Individual producer services are also quite different so that standard questionnaires do not work very well. Some industries such as the financial sector present special problems. Here the sheer complexity of the business and the variety of global, national and local markets served makes sample design and interviewing unusually problematic. Multiple interviews at various levels and in different parts of the institutions are essential. Even so

Table 17.6 Professional bodies, trade associations and trade unions in producer services

---

Finance*

General
Association for Payment Clearing Services
Association of British Insurers
Association of Futures and Bond Dealers
Association of Investment Trust Companies
Banking Information Service
British Merchant Banks and Security Houses' Association
British Venture Capital Association
Building Societies' Association
Institute of Actuaries
Institute of Bankers
Unit Trust Association

Regulatory organizations
Bank of England
Brokers' Regulatory Association
Financial Intermediaries, Managers and
   Investment Management Regulatory Organization
Securities and Investment Board
The Life Assurance and Unit Trust Organization
The Securities Association

Business services†
Advertising Association
British Ports Association
British Institute of Management
Computer Services Association
European Computer Services Association
Institute of Chartered Accountants
Institute of Chartered Surveyors
Institute of Freight Forwarders
Institute of Physical Distribution Management
Institute of Public Relations
Institute of Practitioners in Advertising
Institute of Management Consultants
Institute of Marketing
Incorporated Society of British Advertisers
International Bar Association
Law Society
Management Consultants' Association
Market Research Society
Office of Telecommunications
Road Haulage Association
The National Association of Estate Agents
The British Property Federation
The Incorporate Society of Valuers and Auctioneers
The National Association of Warehouse Keepers
The Association of Certified Accountants

Table 17.6 Cont.

Trade unions
Association of Professional, Executive, Clerical and Computer Staff
Banking, Insurance and Finance Union
Manufacturing, Science and Finance Union
National Union of Insurance Workers
Technical, Administrative and Supervisory Section of the Amalgamated Union of Engineering Workers
Transport and General Workers' Union
Union of Shop, Distributive and Allied Workers

*Chris Gentle provided much of the information on the financial sector.
†This section draws on Thrift *et al.* (1988).

the paranoia of some parts of the sector concerning confidentiality can lead to vague answers to questions.

Confidentiality in fact is a recurring theme in detailed case-study interview research, and the way respondents wish the researcher to treat the information obtained can have a marked influence on the presentation of results. There are also ethical questions for the researcher to deal with where information may be of value to various interest groups in business. However, such difficulties need not be insurmountable. Much of the information of importance to the research will not be sensitive. Furthermore, it is surprising what firms are prepared to have published about themselves (see Marshall 1989b). Indeed, going back to consult participants about potentially confidential information can act as an additional second stage of data collection when participants comment on the data assembled so far and check its veracity.

17.4 Conclusions

The last decade has seen a substantial growth of research into producer services. This is making new demands of the official statistical database, demands which it is not at present able to meet. The outline of official statistics in this chapter has concentrated on measurement problems and cast doubt on our ability to monitor adequately trends in producer services. This does not suggest that official statistics should be ignored; they still provide a useful background or a sensible starting-point for research. But they are insufficient to satisfy the increasingly incisive questions being posed by research on producer services.

There needs to be a substantial effort to improve the quality of official statistics on producer services in particular and the service sector in general. The SIC used for measuring economic activities needs to be overhauled to provide substantially more detail on services. This detail needs to be provided across the range of government sources. Furthermore, government statistics need to be more aware of the interdependent nature of services and other activities, and to provide better facilities to monitor shifts in economic activities across the sectors. Data also need to be disaggregated more fully to at least the regional level.

There is, of course, little likelihood that this will happen, even though the recent *Scrutiny Report* (Cabinet Office 1989) has committed government to review ways of making the statistical series more sensitive to the changing nature of the economy. More hopefully, the CEC has recently reviewed the changing statistical demands of its directorates in the light of the growing dominance of services in the economy. This includes the need to monitor changes in the quality of service jobs, develop ways of improving and standardizing existing statistics and obtain new sources for official information. Such developments may ultimately affect the UK because the needs of the EC can encourage national governments to produce new data, as witnessed by the production of regional GDP figures when the UK entered the EC (CSO 1975). Indeed a major survey to provide additional data on business services is shortly to be conducted in the UK by the CEC.

None the less, researchers wishing to investigate producer services will have to be increasingly innovative in their use of unofficial sources of data. This chapter has identified a wide variety of alternatives to official statistics, and has shown the ways in which these have been used. Some approaches to data collection such as interview surveys are the tried and trusted tools of industrial researchers. But using other sources of data, such as industry sub-literature and ephemera, are making new demands and requiring new skills. Data collection is becoming an increasingly arduous business. Research methodologies are needed which facilitate the detailed investigation of particular cases while allowing broader generalization to the level of the economy.

References

Barras R 1983 *Growth and Technical Change in the UK Service Sector* Technical Change Centre, London

Barras R 1986 New technology and the new services. *Futures* **18**: 748–72

Cabinet Office 1989 *Government Economic Statistics: a Scrutiny Report* HMSO London

CSO 1975 *Input–output Tables for the United Kingdom 1968* Studies in Official Statistics No 22, HMSO, London

CSO 1978 *Regional Accounts* Studies in Official Statistics No 31, HMSO, London

CSO 1981 *The National Accounts: A Short Guide* Studies in Official Statistics No 36, HMSO, London

CSO 1988 *Input–output Tables for the United Kingdom 1984* HMSO, London

Coutts K, Godley W, Zezza G 1990 Is Britain in credit with the world? *The Guardian* 26 Jan

Crum R E, Gudgin G 1977 *Non-production Activities in UK Manufacturing Industry* Regional Policy Series 3, CEC, Brussels

Daniels P W 1975 *Office Location: An Urban and Regional Study* Bell, London

Daniels P W 1983 Business service offices in British provincial cities: location and control. *Environmental and Planning A* **15**: 1101–20.

Daniels P W, Thrift N 1987 *The Geographies of the UK Service Sector: A Survey* Working Papers on Producer Services No 6, Portsmouth Polytechnic and University of Bristol, Bristol

DE 1988 Standard occupational classification: a proposed classification for the 1980s. *Employment Gazette* **96**: 214–21

DTI 1989 *Review of DTI Statistics* DTI, London

Gershuny J, Miles I 1983 *The New Service Economy* Frances Pinter, London

Gillespie A E, Goddard J B, Thwaites A T, Smith I J, Robinson F 1984 *The Effects of New Information Technology on the Less-favoured Regions of the Community* Regional Policy Studies Collection 23, CEC, Brussels

Goddard J B 1978 Office location in urban and regional development. In Daniels P W (ed.) *Spatial Patterns of Office Growth and Location* Wiley, London pp 37–62

Graham N, Bentson M, Wells W 1989 1977–1987: A decade of service. *Employment Gazette* **97**(1): 45–54

Green A, Owen D 1984 *The Spatial Manifestation of the Changing Socio-economic Composition of Employment in Manufacturing 1971–1981* Discussion Paper 56, Centre for Urban and Regional Development Studies, University of Newcastle upon Tyne

Green M 1985 The development of market services in the European Community, the United States and Japan. *European Economy* **25**: 69–96

Greenfield H 1966 *Manpower and the Growth of Producer Services* Columbia University Press, New York

Hencke D 1989a Young in row over figures. *The Guardian* 12 March

Hencke D 1989b Ministry bans Scottish new job figures. *The Guardian* 13 April

Hill T P 1971 *The Measurement of Real Product* OECD, Paris

House of Commons 1988a *The 1988 Budget* Treasury and Civil Service Committee Fourth Report, HC 400, HMSO, London

House of Commons 1988b *Autumn Statement 1988* Treasury and Civil Service Committee Fifth Report, HC 89, HMSO, London

Howells J 1988 *Economic, Technological and Locational Trends in European Services* Gower, Aldershot

Howells J, Green A E 1988 *Technological Innovation, Structural Change and Location in UK Services* Gower, Aldershot

Hubbard R K B, Nutter D S 1982 Service employment in Merseyside. *Geoforum* **13**: 209–35

Industry Department Scotland 1979 *The Scottish Input–output Tables* Vols 1–5 Scottish Office, Edinburgh

Katouzian M A 1970 The development of the service sector: a new approach *Oxford Economic Papers* **22**: 362–82

Leyshon A, Thrift N 1989 South goes north? The rise of the provincial British financial centre. In: Lewis J, Townsend A (eds) *The North–South Divide* Chapman, London pp 114–56

Liston D, Reeves N 1988 *The Invisible Economy* Pitman, London

Lloyd P 1989 Research policy and review 28: fragmenting markets and the dynamic restructuring of production: issues for spatial policy. *Environment and Planning A* **21**: 429–44

Marquand J 1979 *The Service Sector and Regional Policy in the UK* Research Series 29, Centre for Environmental Studies, London

Marshall J N 1979 Ownership, organisation and industrial linkage: a case study in the Northern region of England. *Regional Studies* **13**: 531–57

Marshall J N 1982 Linkages between manufacturing industry and business services. *Environment and Planning A* **14**: 1523–40

Marshall J N 1983 Business service activities in British provincial conurbations. *Environment and Planning A* **15**: 1343–60

Marshall J N 1985 Business services, the regions and regional policy. *Regional Studies* **19**: 533–63

Marshall J N 1988 *Producer Services and the Manufacturing Sector in the West Midlands* West Midlands Enterprise Board, Birmingham

Marshall J N 1989a Private services in an era of change. *Geoforum* **20**: 365–79

Marshall J N 1989b Corporate reorganisation and the geography of services. *Regional Studies* **23**: 139–50

Marshall J N, Green A 1990 Business reorganisation and the uneven development of corporate services in the British urban and regional system. *Transactions of the Institute of British Geographers* new series **15**: 162–76

Marshall J N, Wood P A, Daniels P W, McKinnon A, Bachtler J, Damesick P, Thrift N, Gillespie A, Green A, Leyshon A 1988 *Services and Uneven Development* Oxford University Press, Oxford

Massey D, Meegan R (eds) 1985 *Politics and Method* Methuen, London

Miles I 1985 *The Service Economy and Socio-economic Development* Science Policy Research Unit, University of Sussex

Miles I 1988 *Services and the New Industrial Economy* Science Policy Research Unit, University of Sussex

Nankivell O 1985 Productivity in services. Paper presented to the Long Range Studies Group, NEDO, London

Northern Ireland Economic Council (NIEC) 1982 *Private Services in Economic Development* Report No. 30, NIEC, Belfast

Nusbaumer J 1987 *Services in the Global Market* Kluwer Academic Publishers, Boston

Ochel W, Wegner M 1986 *The Role and Determinants of Services in Europe* FAST Occasional Paper No 132, CEC, Brussels

OECD 1986 *Trends in the Information Economy* Information Computer Communications Policy Series 11, OECD, Paris

OECD 1987a *Historical Statistics, 1960–85* OECD, Paris

OECD 1987b *National Accounts Statistics 1960–85* OECD, Paris

OECD 1987c *Production and Employment of Member Service Industries* Working Party of the Trade Committee, OECD, Paris

OPCS 1980 *Classification of Occupations* HMSO, London

Perry M 1990 Business service specialisation and regional economic change. *Regional Studies* **24**: 195–209

Petit P 1986 *Slow Growth and the Service Economy* Frances Pinter, London

Phillips M 1989a Lies, damned lies and Nigel Lawson. *The Guardian* 11 March

Phillips M 1989b Standing up to be counted. *The Guardian* 8 Dec

Polese M 1982 Regional demand for business services and inter-regional service flows in small Canadian region. *Papers of the Regional Science Association* **50**: 151–63

Reward Regional Surveys 1972–91 *Salary and Cost of Living Reports* Reward Regional Surveys, Stone, Staffs

Riddle D I 1986 *Service-led Growth* Praeger, New York

Robertson J A S, Briggs J M, Goodchild A 1982 *Structure and Employment Prospects of the Service Industries* Research Paper No 3, DE, London

Sayer A, Morgan K 1985 A modern industry in a declining region: links between method, theory and policy. In Massey D, Meegan R (eds) *Politics and Method* Methuen, London pp 147–81

Siniscalco D 1989 *Structural Change and the Integration between Industry and Services* El Paper dels Services a les Empreses en el Desenvolupament Regional i Urba, Ajuntament de Barcelona, Deputacio de Barcelona, Mancomunitat de Municipis de l'Area metropolitana de Barcelona

Smith A D 1972 *The Measurement and Interpretation of Service Output Changes* NEDO, London

Stanback T M, Noyelle T J 1982 *Cities in Transition* Totowa, Rowan and Allanheld, Osmun, New Jersey

Stanback T M *et al.* 1981 *Services: The New Economy* Totowa, Allanheld, Osmun, New Jersey

Thrift N, Daniels P, Leyshon A 1988 *Growth and Location of Professional Producer Service Firms in Britain* Final report to Economic and Social Research Council, London (Grant D00232194)

UNCTAD 1988 *Trade and Development Report* United Nations, New York

Wood P A 1988 The economic role of producer services: some Canadian evidence. In Marshall J N *et al. Services and Uneven Development* Oxford University Press, Oxford pp 268–78

18

Consumer services

John Dawson

18.1 Introduction

The provision of consumer services in the British economy constitutes a large and diverse sector. Even conservative definitions of the scope of consumer services point to the sector employing over 21 per cent of the labour force and contributing over 17 per cent of gross domestic product (GDP) in the UK. The sector is large and diverse across several dimensions. It is large in number of employees, in number of organizations involved and in the number of establishments. In the part of consumer services concerned with retailing products through shops over 2.1 million people are employed in 240 000 businesses which operate 345 000 shops. To appreciate the full scale of the consumer service sector various other activities selling a mixture of products and services must be added to these already substantial numbers. The list of additional consumer services is long but includes:

1. *Leisure services* such as restaurant and food services, entertainment activities;
2. *Financial services* such as banks, building societies, accountancy, insurance and assurance agencies;
3. *Household services* such as estate agents, house and durable good maintenance undertakings, household design agencies;
4. *Personal services* such as beauticians, hairdressers and opticians;
5. *Social services* such as private-sector advisory and legal services and also consumer social services provided by public-sector and quasi-public-sector agencies such as employment services, consumer advice and housing advisory facilities.

If the definitions are spread even wider then even personal transport, health and education services can be included in a broadly based approach to consumer services (Marshall *et al.* 1988). The sector can become very large in this manner with the inclusion of the final distribution of any goods and services on which consumer expenditure is made or which are provided by public agencies for personal consumption. This chapter excludes transport (see Ch. 19), health and education.

Because of its size the sector is also very diverse. Dimensions of diversity include:

1. *Character of employment* – hours worked, skill level, status (proprietor, employee, etc.), etc.;
2. *Character of the business* – size, product/service sold, trading method, legal status, organizational structure (head office, branch, outlet etc.), etc.;
3. *Location of service provision* – direct to the residence, in a town centre, etc.;
4. *Method of delivery of the service* – level of technological sophistication, use of personal service, etc.

With such diversity, interactions and processes in the sector are complex because they involve relationships among very different firms and institutions which may have quite different aims and objectives in providing a consumer service. The consumer obtains goods and services from a wide spectrum of types of provider. Choice is large for the consumer and it is becoming wider. A feature of the increasing level of affluence of society is the widening of choice in consumer service provision. A significant feature of all consumer services is the need to deliver them in a spatially disaggregated fashion, simply because they are required where the consumer can gain best access. While the organization of the business may be centralized the pattern of provision of establishments has to be spatially dispersed.

Allied to the large scale and the diversity of the sector is its dynamism. Many of the processes, which operate and define levels of activity in the sector, are characterized by rapid rates of change. One of the reasons for this dynamism is the closeness of these services to final consumption by consumers who constantly change their pattern of demand in response to income changes, personal circumstances, information, particularly media, stimuli and so on (Gardner and Sheppard 1989).

The size, diversity and dynamism of the sector result in considerable difficulties of definition and of measurement of consumer services. The 1980 Standard Industrial Classification (SIC) includes consumer services mainly within Divisions 6, 8 and 9. Within all the other SIC divisions, however, consumer services are present to some extent. In Division 0 shops attached to agricultural and horticultural holdings and nursery gardens are included; gas and electricity showrooms are included in Division 1; the preparation of ornamental and funerary stonework is included in Division 2; the preparation of spectacles and lenses is assigned to Division 3; bakehouses associated with retail shops and print-shops are both included in Division 4; plumbers, electricians and other household service agents are part of Division 5; post offices are included in Division 7. While the SIC is the most widely used framework for classification in the economy it is not well suited to analyses of consumer services. The determination of what exactly is a consumer service is difficult. In many cases firms supply goods and services to several types of end users, not only those involved in final consumption but also those involved in intermediate stages. The petrol filling station illustrates this problem in that it sells its product not only to personal consumers but also to business users. As product and service diversification have increased in the economy in recent years so it has become more difficult to establish the distinctions among consumer services, producer services and production itself.

Difficulties of measurement occur in trying to relate concepts such as output and productivity to somewhat ephemeral consumer services. As a simple example, a discount retailer selling shirts may be compared with a retailer providing a bespoke shirt retail service. Even if quantitative measures of sales values and profits are the same in the two cases the output of the two is different in qualitative terms with a significantly different service being provided in each case. How are such qualitative differences measured? If difficulties occur in a simple case involving a product, consider the difficulties involved in measuring output and productivity in consumer services where no product is present, such as are provided by a solicitor or a branch bank. The remainder of this chapter reviews the various sources, both official and unofficial, of statistics and of other types of information. In the pattern of provision of this information there has been a considerable expansion, in recent years, of information from private-sector sources and a relative, sometimes absolute, decline in information provision by official government agencies. There has also been a notable reduction in detailed spatial referencing of information in official statistical sources. This point is made strongly by Jones (1984) in his review of retail sources. This chapter does not attempt to be a comprehensive review of sources along the lines of the Economic and Social Research Council (ESRC)-sponsored series of monographs related to sources of statistics by sector. The aim here is to provide a brief review of major current sources with no attempt made either to consider the historical development of these sources or to provide a complete review of all sources of information across consumer services. Legislation relevant to consumer services is also excluded from this review which is aimed at being of practical use to policy-makers who are faced with trying to gain an understanding of the consumer service sector and the spatial patterns within it.

18.2 Sources of official statistics

Consumer expenditure and consumption

Major sources of information on consumer activity, which is at the core of consumer service activity, are the Family Expenditure Survey (FES), National Food Survey (NFS) and the National Accounts (Blue Book). The three sources cover different elements of consumer expenditure and have been widely used by private-sector firms to generate estimates of locality-specific expenditure patterns.

The Blue Book provides broad-scale data on consumers' expenditure against several headings including food, alcoholic drink, tobacco, clothing, recreation, etc. Data are reported in *Economic Trends* as well as in the Blue Book. The detailed account of methods of collection is provided by a Central Statistical Office (CSO) publication, *United Kingdom National Accounts, Sources and Methods*, which has run to several editions as the data collection procedures have evolved since they were first undertaken in 1946.

The estimates of food expenditure in the Blue Book are derived from the *National Food Survey (NFS)*. This provides detailed information on what foods are bought and the relationship of food expenditure to income and price. Both quality and value of food sales are reported. Results are reported in an *Annual Report* with selected data being reproduced in the *Annual Abstract of Statistics, Social Trends* and *Regional Trends*, which bring out some of the spatial variations in patterns of food expenditure. Detailed statistics including locality-specific material are lodged with the ESRC Data Archive at the University of Essex. The data in the *NFS*, and also in consequence in the Blue Book, refer to food consumption in the household. Consumption out of the house is excluded but a table in the Annex to the 1985 Blue Book provides estimates of this expenditure on out-of-home consumption.

The procedures of the *NFS* are described in detail by Lund (1983) and in appendices to the *Annual Report*. In summary the survey is a continuous sampling enquiry based on a detailed log of the description, quantity and cost of food entering a sample of households over a 7-day period. The sample is drawn on a three-stage stratified random sampling scheme. The first stage is the selection of 44 Parliamentary constituencies; the second stage is the selection of polling districts; the third stage involves the selection of addresses. The sample size is approximately 8000 but varies due to aspects of non-response and non-contact. The survey data are classified into a total of about 180 codes which detail food items. These are aggregated to the 45–50 main food groups which are reported in the *Annual Report*. Some regional figures are provided, unlike the Blue Book data.

The *Family Expenditure Survey (FES)* collects data under approximately 250 headings, from a structured sample of households. It is conducted each year and reported annually with excerpts also published in the *Annual Abstract of Statistics* Economic Trends and *Employment Gazette*. The *FES* provides some regionally disaggregated information but for less frequently purchased products the standard error in some expenditure estimates is high and figures have to be treated with caution. The Department of Employment (DE), Statistics Division, will provide more detailed information on request (see also Ch. 19).

The *NFS* and the *FES* provide the basis for a number of non-official statistical series on local expenditure patterns. These are discussed below in section 18.5.

Sectoral statistics

Consumer services transcend several sectors of the economy but they are most in evidence in retailing and it is in this sector that the official statistical base is strongest. The main source of statistics for the retail sector in the UK is the **Retail Inquiry**. This has superseded the Census of Distribution which was last taken in 1971 in Great Britain and in 1975 for Northern Ireland. The first Retail Inquiry was taken in respect of 1976 and was conducted annually until 1980. The pattern changed to a full inquiry every other year, namely 1982, 1984 and 1986, with a slimline inquiry providing aggregate data in the intervening years. Since then the pattern has changed again with a full Retail Inquiry now being conducting every 5 years and smaller surveys in the intermediate years. Although the Retail Inquiry provides much valuable information it provides no spatially referenced data other than gross figures for England, Scotland and Wales. The reason for this is that the survey method is based on firms, using the VAT register as the sample frame, not on establishments. The earlier 1950, 1961 and 1971 full Censuses of Distribution and the 1965 and 1975 censuses in Northern Ireland used the establishment as the sample frame and so were able to produce location-specific data, at very considerable detail in the case of Northern Ireland. The Retail Inquiry is published, typically 18 months after the end of the year to which it relates, in the *Business Monitor* series.

The Retail Inquiry provides estimates for the number of businesses, outlets, persons engaged, sales volume, purchases, stocks, gross margin and capital expenditure. Businesses are classified to one of 7 kinds, for example food retailers, household goods retailers, mixed retail businesses, etc., and are further subdivided into 33 detailed kinds of business in years when enhanced surveys are undertaken. Data are also provided for firms classified into size groups in respect of number of outlets operated, sales volume of the firm and number of employees in the firm. Data on floorspace are not provided.

The Retail Inquiry is supplemented by major official statistical series on retail sales, investment and employment and also *ad hoc* studies on a variety of topics by different official agencies.

The **Retail Sales Index (RSI)**, produced monthly, provides indications of short-term changes in retail sales. The index in effect updates the bench-mark figures produced by the Retail Inquiry. The RSI shows the monthly change in average weekly sales since a base year and is published provisionally in a press release, reproduced in *Business Briefing* and then published in final, often revised form, in *Business Monitor* SDM 28. The information shown is the volume and value of retail sales for all retailers and for five kinds of business. The **Retail Prices Index (RPI)** is also produced monthly and records movements in the price of a basket of products and services. The

DE (1987) have produced a non-technical description of the RPI and some additional aspects of methodology are reported in a study by the National Audit Office (1990). No regional breakdown is provided in the RPI. None the less the index is extremely useful as a measure of the general buoyancy, or otherwise, of the retail sector.

Some data on capital investment are shown in the Retail Inquiry but this is supplemented by data collected quarterly for use in the generation of National Accounts. The data are reported in the *Monthly Digest of Statistics*. A qualification in the use of these data rests on their showing ownership, not use, of capital assets. Investment by property companies in shops will not appear as retail investment but will be allocated to the property or financial sector. The increasing amount of sale and leaseback arrangements in retailing causes additional difficulties in interpreting these data which are disaggregated into investment in land and buildings, plant, machinery and capital equipment and, thirdly, vehicles. Again, regional data are not provided.

Employment in retailing is also reported in the Retail Inquiry, but more detailed official information is available from the Census of Employment (CoE) and the *Labour Force Survey (LFS)* both of which are reported in *Employment Gazette*. The New Earnings Survey (NES) provides specific data on earnings and wages. The DE (1985) provides a good general description of these statistical materials and others which they collect on employment topics. These three major sources provide data by economic sector. Therefore, as well as providing data on retailing, they also provide information on various other consumer service sectors. The CoE is the largest and most comprehensive survey of employment and includes information on employees but not the self-employed who are a significant group in many consumer services. Regional data are published and small area data are available through the Training Agency/Durham University database, NOMIS (see Chs 2 and 8). Information on retail employment is provided according to 12 types of retailing which broadly conform to the kind of business classification in the Retail Inquiry, but the classifications are not directly comparable. Through the 1980s the census was carried out every 3 years after a period in the late 1970s when considerable delays in processing called into question the utility of an annual survey. The *LFS* is carried out annually and contains useful supplementary data on employee gender, youth workers and the self-employed, with additional cross-tabulations on size of firm, qualification level and training. Detailed data are held at the ESRC Data Archive and results of the survey are published in *Employment Gazette* (see also Chs 2 and 3). The *New Earnings Survey (NES)* provides data on levels, distribution and changes in average hourly and weekly earnings of employees in the retail industry and in several other consumer services sectors. Data are produced annually according to SIC classifications (Ch. 4).

Ad hoc official surveys fill in various gaps left by the major series. Thus, for example, the reports of the Monopolies and Mergers Commission address particular topics: various cost and profit issues in retailing were addressed in their report *Discounts to Retailers*, while the competitive processes in the on-sale of alcohol have also been studied recently. Reports on specific proposals for mergers involving firms in the consumer services sector provide additional useful, often detailed, information. The Office of Fair Trading also undertakes one-off studies, for example into car sales and competition in retailing. The Auld Report into a possible revision of the law on Sunday trading is a further example of the *ad hoc* official study which addresses specific issues in consumer services. Various reports of the now closed Price Commission, National Board on Prices and Incomes and the Commission for Industrial Relations addressed specific issues in consumer services but for the most part may now be considered as providing useful background information rather than material relevant to contemporary decision-making. Lists of the reports of these various official commissions and offices are widely available and easily accessible.

Discussion in this section has been concerned with official statistics and has focused on the retail sector. Other sectors of consumer services have surveys broadly comparable to the Retail Inquiry but in less detailed form. Again, however, a spatial dimension is absent. Catering and allied trades, motor trades and a number of leisure services are the subject of annual reviews which give information on numbers of businesses, turnover, capital expenditure and in some cases stock values. The results of these various inquiries were formerly published in *British Business* but, since it ceased to be published, now appear as an annual CSO *Business Bulletin* on distributive and service trades with preliminary results appearing in *Business Briefing*, which is produced weekly by the Association of British Chambers of Commerce. The various information sources on employment which use the SIC as a base for sample design and data collection provide data across a wide range of consumer services. The *National Travel Survey (NTS)* contains information on travel patterns to a variety of consumer service providers (see also Ch. 19).

The system of official surveys of the consumer service industries was severely disrupted in the early 1980s by a wide-ranging review of statistical provision requested by the Prime Minister. This followed closely on the Rayner review initiated in 1979, of the Department of Trade and Industry's (DTI) statistical service. The detailed results of these reviews and the recommendations to change the various procedures for collection are available in reports from the Business Statistics Office (BSO) which published the several reviews in 1986 and 1987. For example, the *Review of the Annual Service Traders Inquiry* was produced in 1987 (BSO 1987). The effects of the reviews was to reduce the amount of information collected and limit official statistics to only those directly necessary for governmental purposes. Thus a number of generally useful statistical series were curtailed in the interest of parsimony. A major casualty was spatially referenced data with a feeling in government that its needs could be met adequately by nationally based information. This curmudgeonly attitude has characterized the collection of official statistics throughout the 1980s.

18.3 Quasi-official sources

A number of agencies who are supported by government produce statistics and information on the consumer services sector but such data are not part of the body of official statistics. Various consumer groups, government research stations, the Scottish Development Agency, the Development Commission and the National Economic Development Office (NEDO) fall into this group of producers of quasi-official information.

The National Consumer Council, and its Scottish and Welsh counterparts, have on occasion addressed various issues associated with consumer services. These reports, in general, review the relative effectiveness and efficiency of the interaction between customer and service provider. Thus, for example, the Welsh Consumer Council and Scottish Consumer Council have sponsored and published reports on rural retail provision and the problems of consumer access to stores in rural areas, and the latter agency published the results of a substantial study of the operation of the used car market and the effects of licensing of used car traders. The National Consumer Council has a substantial list of publications which report the results of research and make suggestions for policy changes in the area of consumer services. The reports on banking and credit have been particularly influential in debates on the relationships between consumers and business and on the need for governmental intervention to adjust the balance of this relationship.

Government research stations, including the Building Research Station and Transport and Road Research Laboratory (TRRL), produce periodic reports on subjects in the consumer service sector. The undertaking of the large-scale shopping diary survey in 1969 and its publcation as *Shopping in Watford* remains an important reference document despite its age, and the series of studies produced through the 1970s by TRRL on traffic generation at various consumer service land uses form the basis from which extrapolation has occurred to give many present-day estimates of traffic generation.

Economic development agencies all generate project-specific information which can be a major source of regional study. Examples are the Development Commission with its work on rural retail provision, the Highlands and Islands Development Board with its sponsorship of a price index specific to its area and the Scottish Development Agency with its considerable commissioning of research projects into various aspects of sectoral change in the consumer service sector. As with the consumer councils and research stations, lists of available publications can be obtained from these various development agencies. The Training and Enterprise Councils (TECs) and Local Enterprise Companies (LECs) which are coming into existence in the early 1990s have a survey of resources as part of their remit, and it is likely that projects on consumer services will be undertaken and reported. In some cases these new bodies will extend the work presently undertaken by enterprise trusts, several of which have reviewed consumer service sector activity in their areas of operation. The specific reports from quasi-government agencies are wide-ranging and numerous, usually with a rigorous research and survey methodology making their findings robust and for the most part impartial.

Also in this mould is the considerable output from NEDO and in particular the publications from its, now closed, Distributive Trades Economic Development Committee (EDC) – the only EDC in the service sector until the recent establishment of one for leisure industries. A substantial number of influential reports have been produced, some statistical in nature such as the *Structural Model of British Retail Trade* (1979), several concerned with employment and training issues and others related to the strategic structural changes under way in British retailing. In this latter group are reports on the *Future of the High Street* (1989), *Technology: Issues for the Distributive Trades* (1982), *Retailing in Inner Cities* (1981) and *Future Pattern of Shopping* (1971). NEDO have also provided more specific material such as its advice to local planning authorities on the undertaking of retail and related surveys (1988). A number of NEDO publications reworked official statistical material into more consistent and user-friendly forms. *Retailing Prospects* (1983) and *Employment Perspectives and the Distributive Trades* (1985) are of this genre. The demise of the Distributive Trades EDC and the general reduction in activities in NEDO have removed an important source of information on consumer services, particularly but not only the distributive trades.

These various quasi-official sources have been notable providers of information and, importantly, providers of regional and local information. They are important because they provide, for the most part, independent commentary and interpretation of the official statistical sources. The result of the various reductions has been an increase in the commercial provision of information whose providers rework, represent and supplement with commentary, official and quasi-official statistics.

18.4 Local government agencies

Local agencies responsible for economic development initiatives and for land use planning are again sources of information on consumer service provision and use. The studies produced by these agencies invariably relate to part or the whole of the local authority area and generally involve definitions and/or methodologies which make comparisons among studies very difficult. Even on issues which are widely applicable in several areas, such as the estimation of the impact on existing firms of new superstores or shopping centres and the subsequent monitoring of changes in trading patterns and consumer response, a great variety of methodologies are used with each study having unique elements in it. Some such studies have been undertaken within the local government group by its own officers and in other cases the studies have been carried out by consultants. Studies are often part of the structure plan making process. Examples of this type of information source are the studies of *Tesco, Neasden* undertaken by Brent London Borough (1986) and of *Sainsbury's at Nine Elms* by Hammersmith and Fulham London Borough (1986). A report *Out-of-Centre Superstores* was produced by Greater London Council (1984) and the *Future of Local Shopping* was published by Newcastle City Council (1985). Examples of the many studies of consumers are the survey in 1981 of shoppers at *Duke Street Shopping Centre* undertaken by Glasgow District Council (1984) and *Trading Patterns of Retail Warehouses* by Leicester City Planning Department (1988). Many dozen such studies exist. Good sources which report and in some cases review these studies are the Newsletter of the Data Consultancy (formerly Unit for Retail Planning Information), *Planning*, and the guide to retail information sources by Kirby (1988).

There are some types of information on consumer services which official local agencies have a statutory duty to collect. More consistency from area to area is present with this information. A number of consumer services require licensing and information is collected, usually in register form, of licences granted, renewed and revoked. Consumer services involved with the sale of alcohol, gaming clubs, cinemas, taxis, street and market vendors, used car dealers in parts of Scotland, etc., are all of this type with basic information available on registers. The valuation sections of local authorities have collected data from many years on land and property values, including the premises used by consumer services of all types. Practices differ on the availability of these data and the level of aggregation at which they are available but they represent an important and valuable source of information. For several years they were aggregated to the level of local authorities and made available by broad land use types, in a series prepared by the Department of Environment (DoE), and its forerunner the Ministry of Housing and Local Government. The cuts in the official statistics mentioned earlier also affected this series, but Hillier Parker May and Rowden were instrumental in bringing together for publication the material for England for 1986 (see Ch. 5). Planning applications are a further body of information which are collected, in generally consistent form, by certain levels of local government. These data can provide useful primary source of information as Davies and Sparks (1989) have shown in the compilation of a database of grocery superstore applications in the UK. More detailed data may be collected on specific programmes of urban renewal such as are reported in the evaluation of urban aid programmes in the West Midlands (DoE 1988).

18.5 Non-governmental providers

Increased provision of information

During the 1980s there has been a substantial increase in the number of non-governmental providers of information on consumer services, the range of services about which information is available and the mechanisms of access to the information. Although the expansion has been relatively recent there are a number of long-established information sources in the consumer service sector. The series of annual statistical reports and conference proceedings of the Consumer Co-operative movement, for example, extends over more than

100 years and commercial directories are a source with an even longer history (Alexander 1970). By nature these sources provide locally based information. Market research agencies have increased their output in recent years but again, reports by these groups have been available for over 50 years. The recent rapid expansion of output by commercial providers can be explained by several factors:

1. The reduction in the base of official statistics;
2. The expansion in the range, variety and complexity of consumer services;
3. Demands by consumer service providers for more information on the consumers, market and competition in the sector;
4. Realization by consumer service providers that information is a resource to be used to obtain competitive advantage;
5. Enhanced demands for information coming from investors in the consumer service sector and the analysts of consumer service providers;
6. A change in attitude by information providers towards a greater appreciation of the realistic value of information and of the consequential high level of potential profit yielded by these providers;
7. The adoption of information technology (IT) and information management systems which can utilize and integrate information of several types and from a variety of sources.

The large increase in the amount of information being demanded and supplied results in information of very varied quality being produced. Not all producers of commercial data work to the same levels of rigour in respect of design of methods of collection or of checking information collected from non-primary sources. Estimates, or even informed guesses, may be made by some providers when secondary sources are unavailable and primary survey is deemed too expensive to undertake. The number of such estimates has increased in recent years because many commercial providers of information depend on official statistical or other sources for their base information and as these have been reduced so estimation procedures have had to be more widely adopted. To compensate for the many providers who rework each others' data and information there are some instances where commercial information providers such as Audits of Gt Britain (AGB), Textile Marketing Services (TMS) and Mintel, have instigated new primary data survey and information collection, with rigorous research methodologies.

Information on consumers

Detailed official statistics on consumer expenditure are available. They are supplemented in two ways. First, some providers publish more detailed material related to different product groupings. Secondly, several organizations provide spatially referenced small area statistics on expenditure.

In the first group , Economist Publications Ltd produce *Retail Business: Market Report* (monthly) and *Retail Business: Retail Trade Review* (quarterly); these contain updated and projected figures of consumer expenditure by sector. The statistics are based on official statistics, market research and company data. Euromonitor produce two reports each month, *Market Research Great Britain* and *Retail Monitor*, which include expenditure estimates by products and product groups. Corporate Intelligence Group produce 10 issues per year of *Retail Research Report* which has similar coverage. Official statistics are used as a base which is supplemented by market research. Major stockbrokers such as County Natwest Wood Mackenzie produce, through their research teams, gross figures and forecasts for consumer spending by broad groups and ally these to commentary documents on a regular basis. Mintel produce *Retail Intelligence*, six times per year, which includes information on expenditure by sectors and also includes reports of behavioural and attitudinal surveys of consumers. Most of these data do not provide regional or local information. In the second group there are three major providers of small area data: CACI, Pinpoint and the Data Consultancy. All use estimation procedures based on the FES and NFS. Although methods of calculation differ in detail, in essence all three providers offer similar packages which generate estimates of consumer expenditure by broad product group in the defined catchment areas of retail developments or in specified wards and local authority areas. The Data Consultancy also provide useful summary statistics of spending in product groups related to shop types rather than the rather generalized figures obtainable in the FES.

Consumer market measurement panels also provide considerable information on consumer expenditure. Audits of Great Britain produce information from six main panels covering food and convenience products, including cosmetics, consumer durables and a variety of specialist products including baby products. Typically data are available on a 4-weekly basis, and AGB are the leading provider of this type of panel data (Kent 1989). Textile Marketing Services provide data on consumer expenditure and market shares in the clothing and footwear, home textiles, ceramics and luggage and products groups. Data are available on various demographic dimensions and these are now linked to a geo-coded database by Sales Performance Analysis (SPA) to provide consumer spending patterns by postcode areas.

Sectoral studies

Private-sector commercial providers of information have become increasingly important generators of reports, studies and data on the consumer services sector. The major market research groups who provide regular information on consumers' expenditure also generate substantial periodic reports on various sectors involved in consumer services. Euromonitor, the Economist Intelligence Unit and Mintel all produce sectoral analyses of a substantial nature. Key Note produce a range of these studies across the total spectrum of consumer services. In late 1989 they had available nine product sector reports associated with retailing, for example fish retailing, six reports by type of trading, for example convenience stores, seven reports of travel and leisure sectors, for example cinemas and theatres, and seven market reviews of the major subsectors of consumer services, for example catering and leisure and recreation. There are several other providers of major sectoral reviews and commentaries. The Centre for Business Research at Manchester Business School, Staniland Hall, Henley Centre for Forecasting, and Cambridge Econometrics, all have produced substantial reviews of the retail sector since 1987. Jordans has produced comparable reviews of the catering sector and Lafferty has published several reports on the banking sector. These reports extend across particular industries within the consumer service sector and consider strategies of firms in the sector, innovations in marketing and technology, consumer attitudes and behaviour, market shares of companies, etc. Many of them work from a common base of knowledge and there can be considerable duplication in the material across the different publications.

A unique source is the Confederation of British Industry (CBI) monthly survey of perceptions of trade levels. This is published in the *Financial Times* and in report form from the CBI and shows traders' reported views on turn-over in the current month and expectations for the forthcoming month. The results are qualitative indications of trends but provide useful and consistent indicators of broad-scale movements in the sector (Burton 1990).

The only substantial source of consistent locational information are Goad Plans (see also Ch. 5) (Rowley 1984; 1987). Goad provide two major infor-mation sources. First, in map form there is a listing of building occupants in selected town centres. Secondly, in annotated directory form is a listing of large retail units trading from outside town centres. The major benefit of this source is the consistency with which it has been collected across many towns and over many years. The raw information which is provided can be aggregated and analysed in a variety of ways but care must be taken in the use of the information. Measurement from the map, for example, only provides information on ground-floor building area. Extrapolation of this into a measure of floorspace is difficult and can be misleading. Sales and Marketing Information (SAMI) have undertaken some reworking of the Goad material to provide a computer-based package to enable easy interrogation of information by town. SPA have linked this information on store location to a geo-demographic and local expenditure database enabling production of integrated studies of building use and customer profiling.

A second major group of information providers are various types of trade associations. The Institute for Grocery Distribution (IGD) is a substantial producer of quantitative and qualitative information monthly through its *Food Industry Digest* and *Grocery Market Bulletin*, annually via its various directories of supermarkets, superstores, freezer centres, etc. and a review of food retailing and on a more occasional basis through special project studies which have included *Product and Fixture Measurement Guidelines*, *Food Franchising* and *Retail Software Directory*. Other consumer goods sectors lack a comparable research unit associated with their respective trade associations. Associations in the building society and insurance sectors produce information but it lacks the authority of that produced by the IGD. Within the retail sector again the Retail Consortium produces a regular bulletin which provides a useful, free, updating service on key issues, often of a political nature, in the retail sector. The Article Numbering Association (ANA) is typical of various specialist associations, each of which produce newsletters and annual statistical reviews of activity. The ANA produces material about the new information technologies being used in the distribution sector and the growth of paperless trade whereby distributors and suppliers are linked electronically with resulting increase in labour productivity and a reduction in some traditional clerical occupations. Data provided in these lists can be used for regional and local analysis as the primary information frequently has spatial reference.

The magazines and journals of professional associations can serve as a specialist source of information. *Shopping Centre Horizons*, the journal of the British Council of Shopping Centres, is a useful source of ideas and information on this particular element of consumer services infrastructure. The more traditional trade magazines such as *Grocer*, *Retail Week*, *Supermarketing*, *Superstore Management*, etc. provide valuable qualitative information on issues in consumer service but like any ephemeral newspaper-style source the material needs careful checking before it can be used to justify arguments or test hypotheses. These sources are perhaps most useful for the interviews they carry with key decision-makers, allowing insights into corporate strategy.

A final group of sectoral studies produced by private-sector agencies are thematic studies which explore issues or subsectors in depth. Hillier Parker May and Rowden produce statistics and commentary on the number of new shopping developments, their size and ownership. They also produce indices of rental levels in town centres grouped by region. Property issues are the subject of reports by other agents including Healey and Baker, Investment Property Databank (IPD), Bernard Thorpe, Richard Ellis, and Debenham Tewson and Chinnocks. These studies tend to take a rather optimistic view of the topic area reviewed as they are intimately involved in its continuance as a buoyant economic sector. This rather rosy view of conditions can also be seen in other topics addressed by various interest groups. The Retail Management Development Programme produce studies and surveys of computer use and IT applications in consumer service sectors. ICL similarly commission and publish specialist reports on IT. Perhaps rather more impartial are surveys of a single topic such as the synopses of planning appeals prepared and published by Lee Donaldson Associates or the annotated directories of pedestrianization schemes produced by Transport and Environmental Studies (TEST). Reports from the major retail research groups based in Templeton College, Oxford (Oxford Institute of Retail Management), University of Stirling (Institute for Retail Studies) and University of Reading (Department of Geography) fall into this general group. Research from these groups has been substantial in recent years. There are a large number of these thematic studies which are produced and sold by the newly emergent commercial research agencies. Many contain useful information but may require careful study of their research design and methodologies before being used to support policy decisions or academic argument.

Company information

The consumer service sector consists of many different and individual companies. The numerous providers of company information include within their databases statistics and other information on firms involved in consumer services. Datastream and Extel provide specific corporate information to supplement the freely accessible and very useful material available in annual reports of companies. Various data which are more specific to consumer services are produced by Management Horizons, Inter-Company Comparisons (ICC), Corporate Intelligence Group (CIG) and financial analysts linked to stockbrokers.

The Management Horizons Yearbook is highly regarded as an authoritative compilation of financial and marketing information on companies in the retail sector. Brief corporate histories are provided together with reports on current activities. ICC have an on-line service and a series of annual reports which provide in a compatible and comparable form statistics on the operations of firms in several sectors of consumer services. Although they have a substantial coverage of retailers, information is provided on other service groups – insurance brokers, travel agents, hotels and leisure complexes, automatic vending, launderers and dry cleaners, and many others. Data are provided on individual firms and also for the activity group to allow comparison of individual firms against the subsector. Because of the highly disaggregated nature of these data they are useful for regional and subregional analyses. The CIG provide several reports which analyse and represent financial information on companies largely obtained from annual reports of these companies. The most comprehensive are *Retail Rankings* and *Retail Ratios* whose titles accurately describe the contents. The research sections of major brokers write company analyses which are produced in individual form for clients and on subscription. These analyses were given wider distribution in journals such as *Retail* and *Market Place* and they are used to underpin sectoral studies published by the financial advisory groups. These studies provide a wealth of substantive factual material on companies and can be very useful in understanding current and future locational decisions by the large firms in consumer services. Most of these reports draw on the financial information provided by individual companies and it is possible to access this directly through the annual reports of companies. These reports often provide information on new and proposed developments which, over several years, allows analysis of spatial patterns. The study of the spatial spread of Kwik Save by Sparks (1990) and of Asda–MFI by Davies and Sparks (1986) used several types of company information to produce valuable case histories.

More passive in nature are directory-based sources of information. These generally provide contact names, addresses and limited information about the firm. A large number of these sources of list-type information exist, for example the *Scottish Financial Yearbook* and *Retail Directory*. In parallel with directories of firms there are also directories of establishments, for example the lists by Data Consultancy and by the IGD of food superstores and supermarkets and the *Markets Yearbook* which lists stall and street markets in Great Britain.

A final group of sources of company-specific information are the sectoral news service publications. A number of information providers produce regular, usually monthly, publications which report news items and so act as an updating service. The Co-operative Wholesale Services (CWS) provide a *Retail Review*, the IGD produce *Grocery Market Bulletin*, Mintel provide a series, *Monthly Digest*, each of which reviews newspaper and trade press reports for a consumer product or service sector. Catering, financial services and retailing are covered in this way by Mintel. In similar vein is *Scottish Retail Newsletter* produced by the University of Stirling. This monthly digest of news items from newspapers and trade magazines provides material specifically relevant to Scotland.

The range and variety of information provided by commercial agencies are large and this chapter can only be indicative. The Business Information Service of the Science Reference and Information Division of the British Library provides an on-line database ACOMPLINE which, while not fully comprehensive, is an excellent entry point into commercially provided sources of information on consumer services.

18.6 Conclusion

Information on the consumer service sector in aggregate is plentiful but of greatly varying quality and value. The amount of explicity spatially reference information, however, is small. Across the various sectors which comprise consumer services the range and quality of information vary considerably. The retail sector has the best, although still imperfect, information base. Personal and leisure services probably have the least well developed sources of information. The general reduction in the quantity of official statistical information has resulted in an increase in value of these data as the basis for added value services by commercial providers. There has been considerable expansion of these private-sector information providers and continued growth seems likely in the coming years. Information is available but at a price. The already large research agencies seem likely to become larger and the structure of the sector will become more concentrated. The larger firms are able to provide a full range of integrated information, often using on-line databases, which are beyond the reach of the smaller specialist agencies. They are also more able to address the problems of providing spatially referenced data than the smaller firms.

The growth of information providers in the private sector means that information is generated primarily for use by corporate groups. The public-sector research community which has thrived on official data now has to look either to these other providers or to become more involved themselves in primary data collection. Either route tends to increase the cost of research for researchers who have become used to using the heavily subsidized official statistical sources. There have been relatively few substantial exercises undertaken by individual researchers aimed at primary data collection in consumer services. The surveys which have been carried out have been small, specialist and usually highly focused, which renders very difficult their use by other research workers. The size and variety of consumer services, suggested in the introduction to this chapter, may well be the cause of the absence of substantial, broadly based, research-orientated databases. Some of the difficulties of a DIY approach to information collection are considered in Chapter 11.

The main thrust of this chapter has been to review descriptive information sources. Information sources of a more analytical nature have not been reviewed but they provide a very valuable source for policy-maker, manager and researcher. It is not possible in this chapter to review the large number of books, journals and conference proceedings containing analyses of the sector. The *Service Industries Journal* comprises articles, reviews and references to a substantial number of these research publications. Abstracting services such as Anbar allow entry further into this Aladdin's cave of research.

References

Alexander D 1970 *Retailing in England during the Industrial Revolution* Athlone Press, London

BSO 1987 *Review of the Annual Service Trades Inquiry* BSO, Newport

Burton C 1990 Launching the CBI/FT survey of distributive trades. In Moir C M, Dawson J A (eds) *Competition and Markets* Macmillan, London pp 207–18

Davies B K, Sparks L 1986 Asda–MFI: the superstore and the flat-pack. *International Journal of Retailing* 1: 55–78

Davies B K, Sparks L 1989 Superstore retailing in Great Britain 1960–1986: results from a new database. *Transactions of the Institute of British Geographers* new series 14: 75–89

DE 1985 *Employment Statistics: Sources and Definitions* DE, London

DE 1987 *A Short Guide to the Retail Prices Index* HMSO, London

DoE 1988 *Improving Innercity Shopping Centres: An Evaluation of Urban Programme Funded Schemes in West Midlands* HMSO, London

Gardner C, Sheppard J 1989 *Consuming Passion* Unwin Hyman, London

Jones P M 1984 General sources of information. In Davies R L, Rogers D S (eds) *Store Location and Assessment Research* Wiley, Chichester pp 139–62

Kent R 1989 *Continuous Consumer Market Measurement* Edward Arnold, London

Kirby D A 1988 *Shopping in the Eighties: A Guide to Sources of Information* British Library, London

Lund P J 1983 The national food survey. In *Statistics Users' Conference on the Distributive Trades* NEDO, London pp 42–50

Marshall J N *et al.* 1988 *Services and Uneven Development* Oxford University Press, Oxford

National Audit Office 1990 *The Retail Prices Index* HMSO, London

Rowley G 1984 Data bases and their integration for retail geography: a British example. *Transactions of the Institute of British Geographers New Series* **9**: 460–76

Rowley G 1987 The Goad shopping centre reports: a new data set for town centre research in Britain. *Area* **19**: 277–8

Sparks L 1990 Spatial–structural relationships in retail corporate growth: a case study of Kwik Save Group plc. *Service Industries Journal* **10**: 25–84

19

Transport, trade and tourism

Tony Hoare

19.1 Introduction

Interaction and movement of goods and people are fundamental to the very existence of a 'space economy' and a 'space society'. Non-subsistence economies produce goods and services above and beyond local needs, and exchange the surpluses in the market-place for those available elsewhere. Whether the customers or the items they purchase move, interaction is an inevitable concomitant of economic progress. Equally, the separation of workplace and home generates major daily pendulum flows of commuting traffic to and fro between them, while elsewhere in advanced societies personal movement is driven by *social* desiderata of a particular life style and a widely cast network of friends and relations to visit.

While transport demand is mostly 'derived' demand, emanating from the need for what or who is thus made accessible, it may sometimes be desired for its own sake – as with tourism or other recreational tripping. Either way, the British transport industry has become a major employer, consumer of public invest-ment, occupier of land use and generator of foreign earnings in its own right.

Even so, the attention that research has devoted to transport and related themes in the UK has sometimes fallen short of their importance to the nation. Within the academic field of geography, for example, transport studies enjoyed a golden age in the 1960s with the rise of the quantitative/positivist perspective (Rimmer 1985, 1988a), while a plethora of policy initiatives in the UK on transport issues – such as Beeching on railways, Jack on rural buses, Roskill on airports, Buchanan on urban roads and Richmond on ports – provided both a further stimulus for, and a reflection of concern with, transport matters more widely within academe, as within the polity. But academic fashions and political priorities change, and today fewer academics would claim an interest in the topics covered by this chapter than in other aspects of the space economy, such as industrial or agricultural activity.

One cannot easily link this relative decline with changes in the availability of statistical sources in these same sectors. Since 1974 the valuable, if increasingly dated SSRC/ESRC/RSS *Reviews of United Kingdom Statistical Sources* series for example, has (so far) devoted 7 of its 36 published sections to transport and tourism themes, in each of which the list of referenced sources is never less than extensive (see Aldcroft 1981; Baxter 1979; Lickorish 1975; Mort 1981; Munby 1978; Phillips 1979; Watson 1978). In part, of course, this wealth of statistical material reflects the multi-media nature of the modern transport industry, and its consequent administrative segmentation, with each mode and agency generating its own data series. In addition, tourism and overseas trade are the responsibility of yet more public bodies which spawn their own statistics.

However, many of these data are severely constrained for contemporary regional and subregional research in the UK in two respects. First, they inevitably concentrate upon the bare numerical bones of transport systems – passengers carried, tonnes shipped, motorway kilometres opened, articulated lorries registered – rather than the social relevance of transport to the UK, to which transport researchers have increasingly turned (Hay 1981; Williams 1981). For good examples of this work see that of Cloke (1984), Halsall (1982), Moseley *et al.* (1977), Nutley (1980, 1982, 1983) and Smith and Gant (1982). So, for instance, published data are better at revealing levels of car ownership in East Anglia than how 'no-car' householders in small Suffolk villages are handicapped in finding and keeping a job.

Second, the spatial formatting of transport data series is often disappointing. Many are not disaggregated subnationally at all, or if so, only to a regional level. And such spatial detail as *is* provided is often a minor, vulnerable part of a larger enterprise: the 'car ownership' question in the 1981 Census of Population (CoP) was only reprieved at the eleventh hour following protest at earlier plans to discard it (Rhind 1983).

Taken in conjunction with the inevitable hazards to which *all* data series are subject (changing terms of reference and definitions, belated publication, confidentiality limits on disaggregations and cross-tabulations) this may seem a depressing picture. But things are improving in important ways (see section 19.6), and geographical research has shown an ability to glean valuable insights and information from such sources as are to hand. These can sometimes prove sufficient in their own right (e.g. Chisholm and O'Sullivan 1973, Hoare 1977) or as used alongside other information-gathering pathways (see below). At this latter level, publicly available sources paint the background scenery or provide the first act of a larger research production. The adequacy of rural bus services can be investigated some way, but not all the way, through studying company timetables, for example.

As supplementation, two options are commonplace. First, as with the previous example, researchers may conduct a 'DIY' survey, tailor-made to the specifics of their inquiry (e.g. Cook 1967; Hayter 1986; Hoare 1975). Guidance on surveying of this sort in the transport field is offered by Bruton (1970), Lane *et al.* (1971) and Powell (1985). Second, commercial companies and authorities involved in transport, trade and tourism usually hold their own stocks of in-house statistical information. Many are very ready to make this available for outside researchers to work through, once assured of academic credentials and honourable intentions, as over matters of confidentiality.

Any chapter-length overview of the vast range of transport–trade–tourism data sources in the UK must work within stringent boundaries. First, attention focuses on sources published, or otherwise readily available to 'outside' researchers. Second, with the exception of a few major international-scale sources, those covered here contain at least regional-level disaggregations within the UK. Third, they relate to the contemporary scene of the last two decades. Fourth, sources focusing on just a single town, locality or county are usually excluded, valuable though these can be for those interested in a particular area (local libraries, local government offices and commercial associations are useful ways of identifying them). Finally, transport as one type of *employment* or *land use* among many is left to other chapters (see Chs 2 and 6).

Within these somewhat fuzzy house rules, research sources are reviewed under four headings: first, those dealing with media-specific transport systems, their operators or customers; second, household-based sources, covering a range of media for many or specific purposes; third, tourism is given separate treatment; and the chapter concludes with a short review of British urban/subregional transportation studies. Two appendices are included; the first (19.1) lists in alphabetical order the independently published statistical data series referred to in the text, the second (19.2) includes details of three libraries specializing in aspects of transport and tourism.

19.2 Media-specific sources

Most media generate their own separate statistical series, and much transport research is similarly segmented. Even so, some conurbation-wide transport agencies have responsibility across such modal divides. The Passenger Transport Executives (PTEs) of major conurbations (London, Manchester, Merseyside, Strathclyde, South and West Yorkshire, Tyne and Wear and West Midlands) are responsible for between two and four separate media. Their annual reports and accounts and other publications (such as *Trends and Statistics 1975–1989* from Greater Manchester) provide useful statistical material on yearly operations, as well as new initiatives taken and the local and national context of policy formulation. These sources apart, for individual modes a useful distinction can be drawn between information on network *capacity* (potential to transport) on the one hand and its *performance* (interaction in practice) on the other.

Roads

Internal transport in the UK is currently dominated by road to the tune of some 83 per cent of passenger-kilometres and 58 per cent of the freight equivalent (86% if fuels and petroleum are excluded). This is mirrored in the profile of surveys conducted by or for the Department of Transport (DTp), and of tables in the major national digest of transport statistics, the annual *Transport Statistics Great Britain (TSGB)* series. *British Road Statistics* is another useful compendium of some road network and road-user statistics at regional and county levels, through drawing on sources discussed below and in section 19.3.

Capacity

One obvious way to identify changes in the modern **road network** is through the handbooks of the Automobile Association (AA) and Royal Automobile Club (RAC), which have an accuracy matching their being updated annually for a road-using, subscription-paying clientele. Current issues are readily available, while for advice on back issues consult the Archivist at the AA head office, the Librarian of the RAC club or the office of the Divisional Director at RAC House (see Annex 1 at the end of this book for address).

A number of computer-based systems have also been developed by the DTp, storing *very* detailed local-scale information on road networks to which additional data can be 'bolted on', such as accident locations. At present their use seems confined to the county highway authorities, but interested readers will find more in White (1985).

Easier to access is the record of road length by the standard and DTp regions (i.e. two different spatial frameworks), county and metropolitan districts begun in 1985. *Road Lengths in Great Britain* specifies, annually, 12 types of road 'maintainable at public expense' as well as 'other public passageways' (bridleways, footpaths, etc.), while another annual, *National Road Maintenance Conditions Survey*, indicates the state of repair of eight types of road (based on visual inspection) at the same county and district scales.

On road financing, *Local Road Maintenance, England* identifies yearly current account spending by local highway authority (LHA) – counties and London boroughs – again under several heads. This expenditure is partly underpinned by Whitehall through the rate support grant. New-road expenditure by these same LHAs, also centrally subsidized (at 50% in England and Wales through the 'transport supplementary grant'), is specified in *TSGB* and indicates the differing priorities of these authorities to this call upon their budgets. These grants are agreed with Whitehall, based on the highway authorities' transport plans, which can themselves be a valuable source of local transport information (see, for example, Moyes n.d.). *Highways and Transportation Statistics – Actuals* details expenditure on roads and transport more generally, though at similar spatial scales, based on independently conducted annual surveys. On new capital expenditures, periodic ministerial statements in the House of Commons identify new commitments for trunk roads and motorways beyond such local responsibility, illustrated at the time of writing by the DTp's earmarking an extra £6bn for network upgrading, as specified in Fig. 19.1.

Turning to **operators**, many annual national business directories such as *Kelly's Business Directory* and *Kompass* provide relevant listings with names and head office addresses of road freight hauliers. Spatial differences in service availability can thus be traced. Individual companies commonly provide details of freight rates and collection/delivery territories (see Hoare

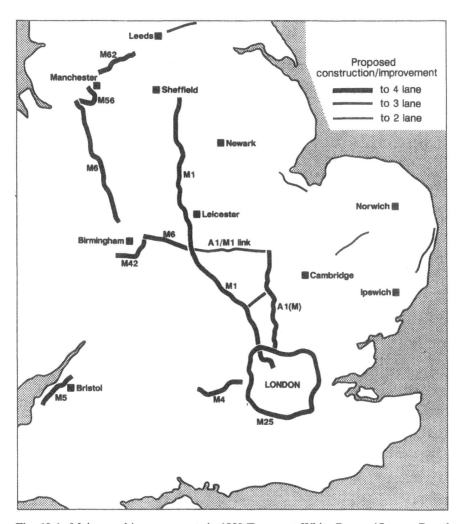

Fig. 19.1 Major road improvements in 1989 Transport White Paper. (*Source:* Based on Transport White Paper.)

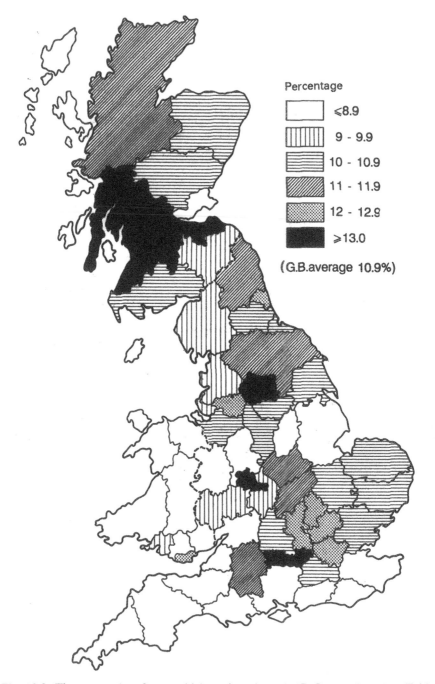

Fig. 19.2 The geography of new vehicle registrations, 1987. *Source:* Based on Table 2.26 in *TSGB* 1977–87.)

1983 for examples). Equivalent listings for road passenger operators are found in two annuals, *Who's Who in the Bus Industry* and *The Little Red Book*. Direct approaches to operators are again necessary to investigate timetabled services and fare structures, although some socially significant routes may be omitted from published schedules (such as free shopper services to supermarkets, school- and works-bus runs). Nutley's work (1983, 1984) illustrates and extends this discussion.

Finally, a detailed tabulation of motor vehicle **registration** by British region and county appears in *TSGB* based on Driver Vehicle Licensing Centre records. Current and earlier registration levels are shown in absolute and relative-to-population terms along with average vehicle age and new-car registrations. Figure 19.2 illustrates the geography of first-year registrations, which is probably influenced by cheap-rate purchases by car plant workers and by the centralized registration of company fleets. Appropriate details of registration year for 11 unique (i.e. less than helpful) 'traffic areas' appear in *Heavy Goods Vehicles in Great Britain*.

Performance

The first and most extensive subdivision here is **traffic flows** – volumes and speeds – each of which has a long history of collection in the UK, notably by the Transport and Road Research Laboratory (TRRL). A series of *TSGB* articles describes this endeavour (e.g. Channing 1984; Jones 1987), wherein the emphasis has been less on a detailed mapping of volumes and speeds than on sampling their changes over time in relation to particular types of roads. However, surveying densities have become sufficient to produce all-traffic and heavy goods flow maps of motorway and major roads on nation-wide and selective regional scales in *TSGB*, starting in 1983. The 1988 version for the infamous M25 appears in Fig. 19.3.

Published road traffic **origin–destination** (O/D) surveys are currently confined to freight flows, and then only at the regional scale. Two relevant annual series are *International Road Haulage by United Kingdom Registered Vehicles* and *The Transport of Goods by Road in Great Britain*. Both are sample surveys. The former, of 25 roll-on roll-off ferry routes to the continent, provides overseas country × UK region loading/unloading matrices for UK-registered vehicles. The second, more comprehensive, survey is stratified by vehicle weight and traffic area. Both are subject to sampling caveats and neither guarantees identification of ultimate start and end locations.

For road **accidents** the relevant annual series since 1982 has been *Road Accident Statistics: English Regions* (although accident records stretch back to 1909). These are DTp 'regions', and are supplemented with some county-level data, as applies also in the parallel *Road Accidents* volumes for Scotland and Wales. Northern Ireland's accident series, like its other major transport data, is summarized in the *Northern Ireland Annual Abstract of Statistics*, supplemented by a finer published disaggregation by 11 police divisions in *Road Traffic Accident Statistics* (from the Royal Ulster Constabulary,

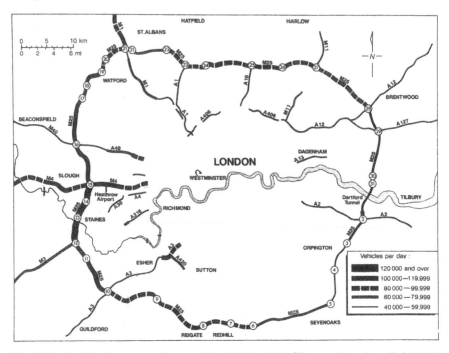

Fig. 19.3 Traffic flows on and around the M25, 1987. (*Source:* Based on Table 2.26 in *TSGB* 1977–87.)

Belfast). Police activity also features in the regional-scale tables on motor-vehicle related prosecutions published irregularly in the annual *Regional Trends*.

Other aspects of carrier performance are sparsely reported, not surprisingly so given the commercial sensitivity of such data in a very competitive industry. A very crude spatial disaggregation (London, Scotland, Wales, metropolitan English counties, and the rest) of passenger- and vehicle-kilometres for bus and coach operators is provided in *Bus and Coach Statistics: Great Britain*. For anything better, recourse will again often be necessary to individual operators with no guarantee of success. However, until its disbandment in 1986, annual reports of the National Bus Company recorded the operating details of its subsidiaries, many of them now independent again. The same date saw the end of the annual *Operating Costs of Bus Undertakings*, a mixed blessing as its tabulations of the income and expenditure heads of some 50 local enterprises are hedged about with confidentiality and other technical qualifications.

Rail

In comparison to road, rail is poorly served with statistical information. Not only is its economic role less but, as a near monopoly, much rail activity can be planned, administered and monitored in-house. The annual reports from

British Rail (BR) provide an easily accessible reference for the past year's developments, and various aggregate tabulations covering the rail system as a whole.

Capacity

Information on the BR network is available through various railway atlases (especially Baker 1988; but see also Freeman and Aldcroft 1985; Wignall 1983; Jowett 1989). Separate rail systems are administered by London Regional Transport, and the Strathclyde and Tyne and Wear PTEs, for whom the *TSGB* provides variable details of track, staffing, rolling stock, etc. The PTEs can be approached for further details of their services. Major public libraries will stock current nation-wide BR timetables, but are unlikely to hold past issues, and no comprehensive source of these can be suggested for consultation by researchers.

Performance

The Serpell Report's (DTp 1983) maps of BR freight and passenger flows nationally are the only modern equivalent of those provided by Beeching (British Railways Board 1963), prior to his savaging the post-war network. Serpell also shows infrastructure costs (traffic ratios per network segment), although much more performance information must be available within BR (Rubra 1975). *TSGB* records financial and passenger journey details for the main rail operators (BR and the PTEs) and also BR passengers separately for the Inter-City, London/South East and provincial routes. Two other more detailed sets of geographical data exist. First, *Regional Trends* occasionally publishes a (Standard) region-by-region matrix of rail freight flows, derived from BR records and produced in turn to comply with European Community (EC) statistical requirements. Second, O/D surveys of London commuter traffic were undertaken in 1964, 1971 and 1981, the last two to marry with census-derived commuting information (see section 19.2). In 1981 these were analysed geographically by the South East's conventional 'annular ring' structure (Sly 1986), although post-survey checks revealed unfortunate response rate biases in the self-completion traveller questionnaire. An alternative longitudinal (1973–79) London commuting analysis based on BR ticket sales is described in *TSGB* (see in Overson 1982 and Brown 1981). Its county- and ring-formatted analyses highlight recent trends and their susceptibility, *inter alia*, to house prices and competing fares on public transport. Other reports on railways provide occasional data of interest geographically, as in the DTp's (1986) published report *Crime on the London Underground* by stations of incidence.

Air transport

Capacity

Licensed public services, both domestic and international, are simplest to trace through two monthly sources. *World Airways Guide* records timetables, airlines and other passenger details of flights from individual British (and world) airports, while the *Air Cargo Guide* provides equivalent details, including freight rates. The half-yearly *ABC Air Travel Atlas* maps passenger services but without specifying carrier or frequency. Private-owned business aircraft and courier air services are not covered by these sources.

Performance

Details of aircraft movements, air passenger use and air cargo handled at each commercial airport in the country can be found in *UK Airports* (annual, since 1973). The British Airports Authority's annual reports provide similar data supplemented by employment totals and runway availability for its seven airport locations, as well as the customary review of the year's happenings. A very rich O/D airport tabulation for UK overseas passengers, shown separately by scheduled and charter flights, is also available in *UK Airports* for the current and previous year, while *UK Airlines* (annual and monthly) shows similar airport-paired data for internal services. On overseas flights the *International Passenger Survey* is an important complement (see section 19.3), while some sources for the UK's airborne foreign visible trade by cargo types and overseas trading partner are covered under the next heading (and Table 19.1). Finally, a very detailed *Traffic by Flight Stage* series records the aircraft type, carrier airlines, carrying capacity and passengers and cargoes moved each year for every route world-wide. Its availability is restricted to specialist libraries: the Civil Aviation Authority (CAA) Library in London is one such.

Internal surface travel by terminating air passengers to/from major UK airports has been surveyed irregularly since about 1970. These have been brought together for the 18 major airports in *UK Airports* (see Fig. 19.4), while more detailed, subregional information (e.g. to the county scale and below), and further disaggregated by trip purpose, resides in the various regional airport survey reports (e.g. CAA 1980, 1984). The potential of such data for tracing the shifting catchments of rival airports is considerable, but underused.

Water transport

Such is the global nature of world shipping and the position of British commerce within this (e.g. the role of Lloyd's of London) that a number of statistical series on shipping published in the UK fall outside the present remit (Mort 1981).

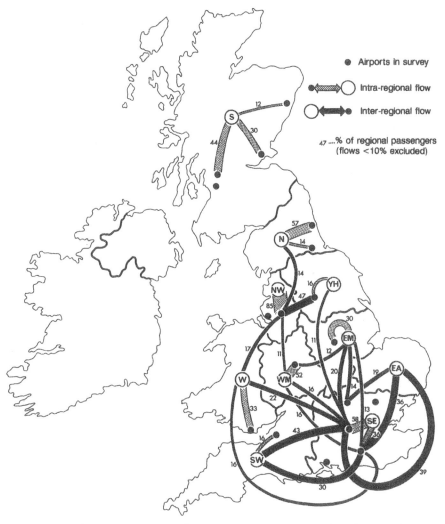

Fig. 19.4 Regional passenger surface flows to and from major airports. (*Source:* Based on *UK Airports*, 1987 (CAP 536, Table 16).)

Many of the remainder change their titles, coverage and availability with infuriating regularity and for no immediately obvious reasons. The series now called *Port Statistics*, whose ancestry dates from 1966, forms the main annual source on port **capacity**, measured as manpower and capital investment respectively for (in 1987) 15 port groups and 21 individual ports in the UK. (Before 1987 Northern Ireland data were produced separately by the NI Department of Economic Development.) The definitive source on seaborne trade sailings is the *Lloyd's Loading Lists*, produced weekly but covering a month's sailings on all port-to-port routings world-wide. Ship size and nationality are specified, but its availability is limited to a few libraries or to subscribers. Much care is also necessary in interpretation: many shipping companies may block-book space on any one sailing so the number of entries overstates the actual movement of ships. Finally, details of companies supplying services associated with sea (and air) freight and passenger transport (agents, carriers, etc.) can be found in the commercial directories detailed earlier.

Port Statistics also details the financial **performance** of each port, but most geographers interested in seaborne traffic have concentrated on the triad of **trading port**, **traded commodity** and **overseas trading partner** alone or in combination, as measured by tonnage and/or value (Hoare 1986; Schrach-Szmigiel 1983). Here the range of suitable sources is particularly confusing, not least when tabulations disappear from one source only to reappear, slightly modified, in another.

Table 19.1 attempts to summarize the main features on offer, in addition to which *Port Statistics* includes a breakdown of traffic by transport mode (containers, roll-on roll-off, etc.). Some of the sources in Table 19.1 also cover air freighting. For a three-way – port × trade partner × commodity – breakdown of visible trade, by sea or by air, reference (and payment) has to be made to HM Customs and Excise for unpublished tabulations. Bassett and Hoare (1987) offer an example of their use.

Excluded from Table 19.1 but of likely interest to some readers are the occasional reports on the O/D regions within the UK engaging in overseas trade, either of particular commodities or with specified trading countries or country groups. By now three such reports are completed, but detailed, unpublished results from the first (1964) survey (Martech Consultants Ltd 1964) were pulped along ago. The last (for 1986) was published 3 years later at an extortionate price (DTp 1989), but opens up possibilities for comparing temporal changes in trade geography, in conjunction with the similarly formatted report for 1978 (DTp/National Ports Council (NPC) 1980). Earlier (1964–78) comparisons have already been published (Hoare 1985), as has an examination of Scottish trade for 1978 alone (Anon 1981). More generally, a classification of the research problems appropriate to these overseas trade data has been suggested by Hoare (1988), even though the accompanying account of sources themselves is now out of date, while Chisholm (1985) offers an interesting regional analysis of the 1978 external trade flow patterns in conjunction with internal road and rail movement information.

Academic interest in inland, coastal and one-port (port-to-offshore) water-borne freight has been very limited. The traffic volumes involved, other than for coal and oil/petroleum, are insignificant but at least the range and reliability of such data have been much enhanced in the 1980s (DTp/Maritime and Distribution Systems (MDS) 1988; Sowerbutts 1987). A set of tables for 12 major rivers and 14 port areas, including some O/D matrices, now appears in *TSGB*. The British Waterways Board produce an *Annual Report* of their yearly activities, but lacking much statistical content.

Finally, *TSGB* also shows car and coach numbers on roll-on roll-off ferries by overseas-port country (not usually the ultimate O/D, of course) and

Table 19.1 Visible trade statistics

	Series (current or final title)	Frequency	Produced by	Dates	Data types	Weight (W) Value (V)	Medium
1.	*Annual Statement of Overseas Trade of the UK*	A	HMCE	1944–75	UK port × commodity, country × commodity	W + V	Air and sea
2.	*Direction of Trade Statistics*	M and A	IMF	1950–	Country × country (135 countries: M 165 countries: A)	V	Unspecified
3.	*Overseas Trade Statistics of the UK*	M	DTI	1965–	Commodity × country	W + V	Unspecified
4.	*Port Statistics*	A	BPF	1966–	UK port × commodity (from 1985), UK port × country	W	Sea
5.	*Quarterly Statistical Abstract of the UK Ports Industry*	Q	BPA	1980–86	UK port × commodity, UK port × mode × trading area	W + V W	Sea
6.	*Statistics of Trade Through UK Ports*	Q	HMCE	1976–80	UK port × commodity, UK port × country	W + V	Air and sea
7.	*UK Trade Data* (successor to 5)	Q	BPF	1987–	UK port × commodity	W + V	Sea

Notes: At times 'country' may be groups of countries. Commodity detail specified varies widely among sources. M – monthly; Q – quarterly; A – annual; BPA – British Ports Association; BPF – British Ports Federation; DTI – Department of Trade and Industry; HMCE – Her Majesty's Customs & Excise; IMF – International Monetary Fund.

by ports (individually or grouped) for all routeings, whether domestic or international. A coarse-grained 'port/port group × overseas region' table appears in *Port Statistics*. These are of strictly limited value, and international passenger traffic is better served in some respects by sources reviewed later (see section 19.4). Information on the traffic handled by particular carrier companies, and on their operations in general, can be found in the relevant annual reports and in published timetables. Thus the yearly activities of the major internal carrier, Caledonian MacBrayne, are now available in the annual report of its parent, the Scottish Transport Group, although its timetables of ferry services to and among the Scottish islands are available separately.

19.3 Household-based surveys

Here four surveys are reviewed which focus on transport and travel from the standpoint of individuals and households, identifying personal mobility potential and experience, though without necessarily specifying transport media. Only one exclusively concerns personal movement, but all enable it to be seen in a wider social context than the sources in the previous section (see also Maultby 1984).

The Family Expenditure Survey (FES)

The *FES* represents a 'unique and reliable source of household data on expenditure, income and other aspects of household finances' (DE 1989: iii). In its modern format it dates from 1957, being now an annual sample survey of (in 1987) over 7000 households (which represents a response rate of slightly over 70%). A complex stratification design includes a spatial component (by Standard regions and four types of administrative area based on population density). Elaborate cross-checks and reliability estimates are made by its parent agency (the Office of Population Censuses and Surveys (OPCS)), and the survey mechanics are described in detail in the introductory notes to the published annual reports. Northern Ireland runs a separate but similar exercise through the Department of Finance and Personnel (NI) (see also Ch. 18).

One of the household weekly expenditure categories identified is 'transport and vehicles', within which five subheadings are further specified. Although the response rate among multi-car-owning households is particularly low, no evidence of under-reporting on 'transport' spending is noted (unlike some other heads). Only a fraction of the *FES*'s extensive tabular output is geographically structured, but significant interregional variations in absolute and relative spending levels can be traced, and over time, nevertheless (Fig. 19.5). Based on 1960s *FES* data Chisholm (1971) suggested that income-elasticity and urban–rural effects might be at work here, with higher transport spending

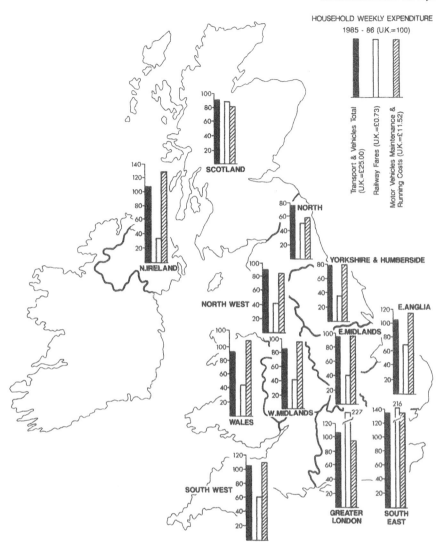

Fig. 19.5 Regional expenditure on transport-related items, 1985-86. (*Source:* Based on Table 29 in FES, 1986.)

in affluent and/or low-density localities. But with no subregional data to hand these remain as just suggestions: geographical interest in the *FES* as a whole has been very limited.

The Census of Population

This familiar source is discussed in detail elsewhere in this volume (e.g. Chs 2 and 16) and among its many social science research uses two are important

Table 19.2   Local area census data on car ownership

	No. of cars per household cross-tabulated by	For
1971		
England and Wales	1. No. of households and persons	LAs
	2. Size and age structure of households	Conurbations and regions
	3. Housing tenure	
Scotland	As (1)	LAs
	As (2)	Planning subregions
	As (3)	
Northern Ireland	4. No. of households, tenure and garaging facilities	LAs
	5. No. of households and No. of earners	
1981		
England and Wales	6. No. of households, tenure and other amenities	LAs
	7. No. of persons, tenure and other amenities	
	8. Household size and amenity*	
	9. New Commonwealth citizens, tenure and other amenities*	
Scotland	10. No. of households and persons	LAs
Northern Ireland	11. Housing tenure	LAs
	12. No. of earners	

*Car ownership data limited to 'No car' identifier.
Note: Subdivisions above do not necessarily correspond individually to specifically published tables.

Table 19.3 Local area census data on journey to work (JTW)

	JTW variable	cross-tabulated by	For
1971			
England and Wales (10% sample)	IOB	Gender Social class, Socio-economic class, Industry, Occupation Mode	LAs Conurbations, who were in- or outflow >1000 Conurbations
	Mode	No. of cars (by residence area) Mode	Conurbations
	O/D	Gender	Conurbations and LAs
Scotland (10% sample)	IOB	Gender Social class, Socio-economic class, Occupation, Industry	LAs Conurbations, where in- or outflow >1000
	O/D	Mode Mode, O/D Gender	LAs and NTs Conurbations LAs
Northern Ireland (100% census)	IOB	Gender Socio-economic group, Occupation, Industry Mode	LAs LAs where in- or outflow >2000 LAs
	Mode O/D	No. of cars (by residence area) Gender	LAs

Table 19.3 Cont.

	JTW variable	cross-tabulated by	For
1981			
England and Wales (10% sample)	IOB	Gender	LAs, NTs / City centres
		Social class, Socio-economic group, LAs / Industry, Occupation	LAs
	Mode	Gender, Mode	LAs, NTs Counties
	O/D	Gender, No. of cars (for resident population) / Gender	LAs
Scotland (10% sample)	IOB	Gender, Employment status, Age, Socio-economic group, Industry	LAs where in– or outflow > 1000 or > 106 residents
	O/D	Mode, Distance to work for resident population / Mode, Distance to work for working population	LAs
Northern Ireland (100% census)	IOB	Socio-economic group, Industry, Occupation	LAs
	Mode	Mode / No. of cars / Start time of JTW } for resident population	
	O/D	Gender	

Notes: LAs – local authority areas; NTs – new towns; O/D – Origin – destination matrix or listing; IOB – details of those working in but resident outside, those working outside but resident in, and those both resident and working in area; Mode – travel-to-work modes specified.

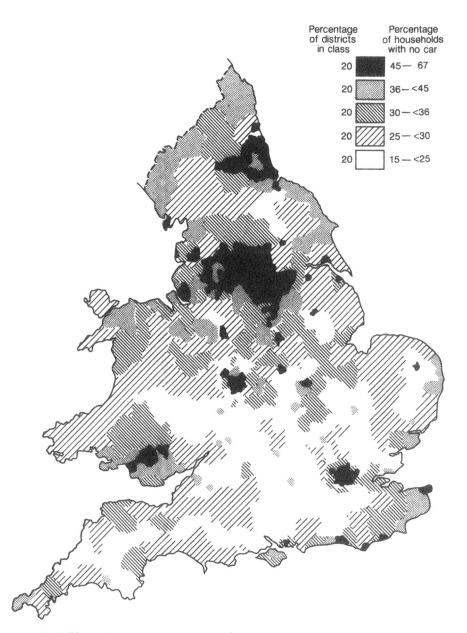

Fig. 19.6 The carless geography of 1981 (England and Wales by districts). (*Source:* Based on data from 1981 CoP.)

here – car ownership and journey to work (Channing 1984). In these respects, as in many others, census data are the most consistent available nation-wide, as well as the most detailed by locality, notwithstanding being up to more than a decade out of date, once publication delays are taken into account. As well as the three published census series covering the UK, further relevant data arise under the small area statistics option (Rhind 1983) and through unpublished tabulations at the normal 'published' scales. The precise types of data available vary among the national censuses, and from 1971 to 1981. Tables 19.2 and 19.3 summarize them.

Car ownership (strictly speaking, combined with van ownership)

These statistics derive from a 100 per cent count of census returns. As a 'stand-alone' they can be used both as a measure of affluence of local authority areas and as an input into personal mobility and transport policy analyses, given their implications for car parking provision, road and public transport investment. Hence the 1981 pattern in England and Wales (Fig. 19.6) underscores the relative deprivation of inner cities, the town–country distinction and the likely problems felt in certain remote localities (Thanet, the Welsh Valleys and Furness). Geeson (1981) illustrates how such small area car ownership data can be harnessed to a household-based questionnaire survey of differential levels and patterns of personal mobility in urban and rural areas.

Travel to work

These data, other than for Northern Ireland, derive from sampling only 10 per cent of returns, but are otherwise a richer data source. Most tabulations fall under three headings: the O/D pattern of daily commuters across local authority boundaries; the balance/surplus/deficit of local authorities with respect to homes and workplaces; and the profile of transport media used locally for commuting. While not the only regular form of household travel (cf. shopping and school-run trips, for example) such flows are significant in defining the area of urban influence of major centres (O'Connor 1980) and the socio-economic complexion of local areas. Not surprisingly, they also feature prominently when boundaries are being redrawn, whether for electoral or labour market administration purposes (Smart 1974). A range of other geographical uses of these data has also been explored in the UK (e.g. Hoare 1984; Humphrys n.d).

Housing and dwelling surveys

Two one-off interview surveys were run separately in all Welsh and 84 English (urban) districts as part of the Callaghan government's housing programme a decade ago (DOE 1978, 1980; Welsh Office 1981). These generated tables on 'car/van availability' and off-street parking by district (England) and county (Wales). The former were trumped by the fuller census coverage three years later, but the latter are not otherwise available at this spatial scale in Great Britain.

The National Travel Survey (NTS)

In the words of its sponsoring body 'this report is about personal travel. The only comprehensive national source of travel information, which links different kinds of travel with the characteristics of travellers and their families' (DTp 1988: 1). Since its inception in 1965 the *NTS* has appeared erratically (in 1972/73, 1975/76, 1978/79 and 1985/86), but publication has become speedier and fuller following earlier criticism.

Cooperating households, derived from a geographically stratified sample (there were slightly over 10 000 in 1985/86 – representing a response rate of 76%), are interviewed about personal details and vehicle ownership. All journeys over 50 yards (including on foot) are also recorded in a 7-day travel diary, which is checked on a return interview. These diaries are staggered throughout a full year. Only internal (British) travel is included, and Northern Ireland is excluded altogether. The essence of the data collected is summarized in Table 19.4 and further details of the survey and its changes over time reported elsewhere (Charles 1987; DTp 1988; Sando 1984; Sheriff 1982).

Table 19.4 *National Travel Survey* data (as of 1985/86)

1.	**Area details**: Region, population density, type of area (urban), concessionary fare schemes in force?
2.	**Household samples**: Tenure, facilities, accommodation, size and demographic structure, employment and income details, transport access and licence holding
3.	**Individuals in household**: Gender, age, movement status, employment details, vehicle access, frequency and problems of travel, licences and travel passes
4.	**Vehicle(s) owned**: Type, costs, mileage annually
5.	**Travel diary**: (a) Journey – timing, purpose, length, time, main mode (b) Segments of journeys – mode, length, time, ticket details, parking details, vehicle occupants

Despite this wealth of information its use in regional and subregional research has been limited, mirroring the lack of interest of the *NTS* in movement patterns measured by conventional locational markers, rather than by length, duration, purpose and mode of travel. A few regional tabulations appeared in earlier reports but with a caveat concerning interregional response rate bias (DoE 1975; viii). The latest *NTS* only published spatial disaggregations by 'type of area' (an urban–rural discriminant). Unpublished regional disaggregations may be available from the *NTS* database, which researchers are invited to discuss with the DTp (DTp 1988: 4). To the best of the author's knowledge, no published geographical work has yet exploited this option.

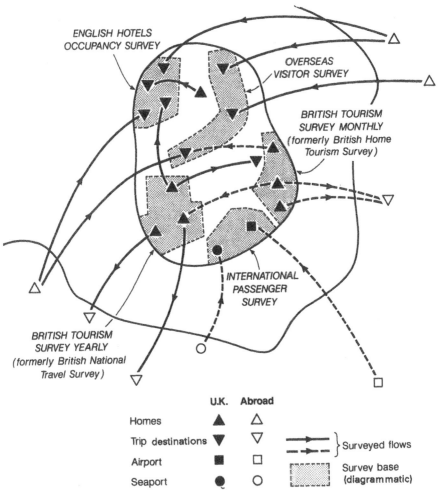

Fig. 19.7 Coverage domains of major surveys of tourist flows

19.4 Tourism

Tourism statistics can be multi-media or media-indifferent, and either household-, destination- or transit-based. This, and the purpose-specific nature of tourist travel, justifies treating them separately. Unfortunately, some otherwise key sources are expensive or available only to trade subscribers. Five useful sources are considered here, and depicted diagrammatically in Fig. 19.7.

The International Passenger Survey (IPS)

This is an annual survey dating from 1964, based (in 1987) on 165 000 randomly selected sea- and airport interviews with overseas visitors and Britons travelling abroad. All travel purposes are covered, although, in practice, most interviewees are holidaymakers. End-of-trip respondents supply details of expenditure incurred and length of stay, while all respondents identify their country of trip origin (foreigners) or destination (Britons). Various permutations of these variables are published, including a coarse-grained tabulation of nights spent by overseas visitors in four British regions. The IPS is published annually in the DTI's *Business Monitor: Overseas Travel and Tourism* series.

The British Tourism Survey – Yearly (BTS–Y)

This very detailed annual (November) sample survey began life in 1960 as the *British National Travel Survey*, changing to its present name in 1985. A home-based random sample of British resident adults provides details of all holiday trips of four or more nights in the previous year. Only subscribers can access the complete results, its database and that of the next survey, below, but disaggregations by holiday location (overseas country and domestic tourist board (TB) region) and region of residence appear in the British Tourist Authority (BTA)'s summary, *The British on Holiday*.

The British Tourism Survey – Monthly (BTS–M)

This is a monthly stablemate of the *BTS–Y* but covering all 'one-night-plus' trips for all purposes, and now published in the annual *The British Tourism Market* by the three national tourism agencies. A standard regional breakdown for trip origins and a regional TB one for destinations is provided, though unfortunately not an O/D cross-analysis. Affluent researchers able to afford the £100 per day consultation fee can use detailed data files held in the BTA Library. Its previous incarnation (between 1972 and 1985) was as the *British Home Tourism Survey*.

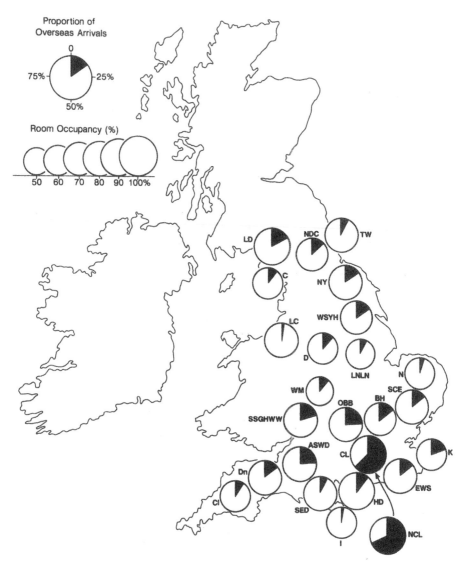

Fig. 19.8 Hotel room occupancy and overseas visitor levels: LD = Lake District National Park; C = rest of Cumbria; NDC = Northumberland, Durham and Cleveland; TW = Tyne and Wear; LC = Lancashire and Cheshire; NY = North Yorkshire; WSYH = West and South Yorkshire and Humberside; WM = West Midlands; SSGHWW = Salop, Staffs., Gloucs., Hereford and Worcs.; D = Derbyshire; LNLN = Lincs., Notts., Leics. and Northants; OBB = Oxford, Bucks. and Berks.; BH = Beds. and Herts.; SCE = Suffolk, Cambs. and Essex; N = Norfolk; CL = central London; NCL = non-central London; Dn = Devon; Cl = Cornwall; ASWD = Avon, Somerset, Wilts. and part of Dorset; I = Isle of Wight; SED = south-east Dorset; HD = Hampshire and part of Dorset; EWS = E. and W. Sussex; K = Kent. (*Source:* Based on data in *English Hotel Occupancy Survey, September 1987*: 10–11.)

The English Hotel Occupancy Survey

This is also produced by the BTA, and reports the monthly bed- and room-occupancy rates from a reporting panel of hotels. Duration of stay, time of week and the domestic/overseas origin of visitors are also noted. The sample is designed to be representative by hotel size with results available by 12 TB regions, 27 subregions (see Fig. 19.8) and other locational discriminants (seaside–inland; town size). For small areas, however, the number and representativeness of the reporting hotels become a problem. Jeffrey (1985) and Jeffrey and Hubbard (1988) offer spatial and time-series analyses of these data, but drawing on the normally confidential individual returns from hotels. The Northern Ireland Tourist Board (NITB) produces a similar monthly survey disaggregated for five subdivisions of the Province.

The Overseas Visitor Survey

This is organized annually by the BTA and the national tourist agencies. It provides information on overseas visitors beyond that in the IPS, which it uses to fix sampling locations and quotas. Certain questions each year are confidential to sponsors while the published geographical tables by visitor origin and visited region are inevitably coarse, given the small sample size.

UK Hotel Groups Directory

This provides a useful supplement to the familiar TB accommodation guides by showing the location(s) of hotels of over 200 companies, from some one-hotel firms to Trusthouse Forte's 221 (in 1988). Separate county listings show companies active there, and their numbers of hotels.

In addition to these individual sources useful compendia of tourist statistics reside in the *Digest of Tourist Statistics*, a supposedly annual but recently erratic publication from the BTA. For each TB region *Tourism Fact Sheets* (BTA and the English TBs) also collate many of the above sources. In both series origins, purpose, numbers and expenditure of visitors are summarized, along with regional tourist accommodation and the major visitor attractions. Northern Ireland is separately served by the NITB's annual *Tourism in Northern Ireland* reports which cover both holidaymaking by NI residents and the pattern of inward tourism. These draw on its equivalent of the *IPS* (the *Northern Ireland Passenger Survey*) for departing air and sea travellers, and the *Northern Ireland Holiday Survey*, an amalgam of the *BST–Y* and *BST–M*. The Scottish TB has conducted its equivalent of the *BST–M* since 1983, and this *National Survey of Tourism in Scotland* has significantly increased the trawl of data on tourism to Scotland and by Scots. It is reported in a series of *Factsheets* produced by, and available from, the board.

A wealth of locality-specific tourist reports also exist from the TBs and other sources, many available in the BTA Library. Those interested in British tourism in a wider context will find Pearce's (1987) review of international tourism statistics useful, along with the annual cross-national compendia in *Tourism Policy and International Tourism* and the one-off overview of Bar-On (1989). Finally, Wanhill (1988) provides further discussion of the sources above and evaluates British tourist statistics in the light of European experience and future needs.

19.5 Urban transportation studies

After about 1960 many metropolitan- and city-region scale transportation studies were commissioned, often harnessed to grandiose schemes for urban motorway investments. In many, the basic premise was that inter-areal transportation flows at this scale were a function of the disposition of land use, so allowing future transport demands to be predicted from likely or planned land use changes (Blunden and Black 1984). Dinosaur-like in their size and resource requirements, these studies proved equally unable to respond to changed circumstances (in political climate and planning philosophy), and became extinct in the 1970s.

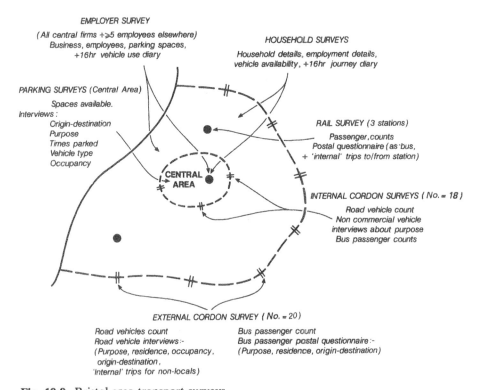

Fig. 19.9 Bristol area transport surveys

Other writers have reviewed their genesis, structure and demise (Bruton 1970; Plowden 1967; Starkie 1976) but their interest here lies in the wealth of transport-related data gathered from a family of contemporaneous surveys covering, in aggregate, many cities and city regions throughout the UK (see Starkie 1976: 11). Unfortunately, a comprehensive listing of them is hard to come by; the central register once kept by the Department of the Environment (DoE) now seems defunct.

As an example, consider Bristol's *Land Use Transportation Study* (Jamieson Mackay and Partners 1975). This was average in size, and its suite of surveys, shown in Fig. 19.9, gives a flavour of the wealth of information lodged in voluminous 'Reports of survey' and 'Technical appendices'. Of course, datedness and the detailed geographical formatting adopted by study teams restrict their research possibilities two decades on, but not necessarily their value as context-setters, question-raisers and guides to surveying practices.

19.6 Conclusion

This chapter has been something of a catalogue of names and contents but, as will be clear by now, much geographically-formatted information on the UK's transport, trade and tourism is scattered throughout a variety of sources, rather than being consolidated in a few all-important ones as applies with some other chapters of this volume. Research in these fields is equally diffuse, but, as in any other social science field, its practitioners must often compromise between problems they most wish to investigate and those they can most easily investigate, given the statistical information to hand. Where constraints of time and money apply – as for those engaged on school or college projects – the latter will tend to dominate, and research designs cut to the cloth of existing data series. But elsewhere the former may have the upper hand, and the future utility of the sources reviewed here must then be seen against shifts in academic interest towards socially relevant topics and larger spatial scales (international/global) (Rimmer 1988b). In some ways, too, these sources have taken a step backwards in recent years: the absence of any later sophisticated O/D freight matrices to compare with those of the 1960s is regrettable, as is the loss of interest of the Census of Production in interregional transport cost patterns (cf. Chisholm and O'Sullivan 1973; Chisholm 1987; Edwards 1975 – but see also Ch. 4).

However, there are also important and encouraging trends to report. Significant improvements have been made at the interface between data gatherers and data users: transport statistics in particular are much more user-friendly than before. Richer, fuller digests are now regularly published. As well as the *TSGB* newly upgraded annual series of *Welsh* and *Scottish Transport Statistics*, are not only valuable compendia but also include some information not otherwise easy to access on these regions and their subdivisions. Computer storage and retrieval capabilities have also dramatically increased the opportunities for 'special runs' (Akinbolu 1987). Many survey agencies positively encourage such enquiries, though there is obviously a cost to pay for services eventually rendered. The *TSGB* records names and phone numbers of contact staff associated with published tables. A Transport Statistics Users' Group furthers the interests of 'outside' customers, while the greater co-ordination of *all* official statistics now in hand through the Central Statistical Office (CSO) is another hopeful sign.

Such developments alone may not cause more researchers to interest themselves in the UK's transport, trade and tourism but they should at least make life that much easier for those who do.

Acknowledgements

The author acknowledges the helpful guidance of many librarians in the compilation of this chapter, especially those of the DTP, CAA and BTA, and also the wise comments of Professor Michael Chisholm on an earlier draft.

Appendix 19.1   Independently published statistical data series referred
to in the text

ABC Air Cargo Guide (ABC International, Dunstable)
ABC Air Travel Atlas (ABC International, Dunstable)
ABC World Airways Guide (ABC International, Dunstable)
Annual Digest of Port Statistics (NPC, London)
Annual Statement of the Overseas Trade of the United Kingdom (HM Customs and
   Excise, HMSO, London)*
Basic Road Statistics (British Road Federation Ltd, London)
British Airports Authority Annual Report and Accounts (BAA, London)
The British on Holiday (BTA, London)*
The British Rail Board Annual Report and Accounts (BR Board, London)
The British Tourism Market (BTA, English and Welsh Tourist Boards, London,
   Edinburgh and Cardiff)
British Waterways Board Annual Report and Accounts (British Waterways Board,
   London)
Bus and Coach Statistics, Great Britain (DTp, London)
Census (of England and Wales) (OPCS, HMSO, London)
Census (of Scotland) (Registrar-General Scotland, HMSO, Edinburgh)
Digest of Tourist Statistics (BTA, London)
Direction of Trade Statistics (IMF, Washington DC)
English Hotel Occupancy Survey (BTA, London)
Family Expenditure Survey (DE, HMSO, London)
Heavy Goods Vehicles in Great Britain (DTp, London)
Highway and Transportation Statistics (CIPFA, London)
International Road Haulage by United Kingdom Registered Vehicles (DTp, HMSO,
   London)
Kelly's Business Directory (Reed International, East Grinstead)
Kompass (Reed International, East Grinstead)
The Little Red Book (Ian Allen, Shepperton)
Lloyd's Loading Lists (Lloyd's, London)
Local Road Maintenance Expenditure, England (DTp, London)
National Bus Company Annual Report (NBC, London)*
National Road Maintenance Condition Survey (DTp, London)
National Travel Survey (DTp, HMSO, London)
The Northern Ireland Census (Registrar-General Northern Ireland, HMSO,
   Belfast)
Operating Costs of Bus Undertakings (Association of District Councils, London)*
Overseas Trade Statistics of the United Kingdom (*Business Monitor* MM 20, DTI,
   HMSO, London)
Overseas Travel and Tourism (*Business Monitor* MA 6, DTI, HMSO, London)
Overseas Visitor Survey (BTA and English Tourist Board market research,
   London)
Port Statistics (DTp and BPF, London)
Quarterly Statistical Abstract of the United Kingdom Ports Industry (BPA,
   London)*
Regional Trends (CSO, HMSO, London)
Road Accident Statistics: English Regions (DTp, London)
Road Accidents: Scotland (Scottish Office, Edinburgh)
Road Accidents: Wales (Welsh Office, Cardiff)
Road Lengths in Great Britain (DTp, London)
Scottish Transport Statistics (SDD, Edinburgh)
Statistics of Trade through United Kingdom Ports (HM Customs and Excise,
   HMSO, London)*

Tourism Fact Sheets (BTA and regional tourist boards, London)
Tourism in Northern Ireland (Northern Ireland Tourist Board, Belfast)
Tourism Policy and International Tourism in OECD Countries (OECD, Paris)
Traffic by Flight Stage (ICAD, Montreal)
The Transport of Goods by Road in Great Britain (DTp, London)
Transport Statistics Great Britain (DTp, HMSO, London)
UK Airlines (CAA, London)
UK Airports (CAA, London)
UK Hotel Groups Directory (Hotel and Catering Research Centre, Huddersfield
   Polytechnic, Huddersfield)
UK Trade Data (BPF, London)*
Welsh Transport Statistics (Welsh Office, Cardiff)
Who's Who in the Bus Industry (Vandell Publishing Ltd, Milton Keynes)

---

* Series terminated

Appendix 19.2   Useful Libraries

British Tourist Authority (book for an appointment–reading space severely limited)
Civil Aviation Authority Library
Department of Transport Library (access by appointment only)

References

Akinbolu J 1987 Databases in the Directorate of Statistics. *Transport Statistics
   Great Britain* **1976–86**: 6–7
Aldcroft D H 1981 Rail transport. In Maunder W F (ed.) *Reviews of UK Statistical
   Sources* XIV Pergamon, Oxford pp 1–124
Anon 1981 Transport costs in Scottish manufacturing industries. *Scottish Economic
   Bulletin* **22**: 27–37
Baker S K 1988 *Rail Atlas of Great Britain and Ireland* Oxford Publishing Co,
   Oxford
Bar-On R 1989 *Travel and Tourism Data: A Comprehensive Research Handbook
   on the World Travel Industry* Euromonitor, London
Bassett K, Hoare A G 1987 Sanctions against South Africa: some regional and
   local dimensions of the national debate. *Area* **19**: 303–12
Baxter R E 1979 Ports and inland waterways. In Maunder W F (ed.) *Reviews of
   UK Statistical Sources* X Pergamon, Oxford pp 1–110
Blunden W R, Black J A 1984 *The Land-use Transportation System* Pergamon,
   Sydney
British Railways Board 1963 *The Reshaping of British Railways* British Railways
   Board, London
Brown A H 1981 *Commuter Travel Trends in London and South East 1966–79 –
   and Associated Factors* DTp, London
Bruton M J 1970 *Introduction to Transportation Planning* Hutchinson, London
CAA 1980 *Passengers at the London Area Airports in 1978* (CAP 430) CAA,
   London
CAA 1984 *Passengers at Aberdeen, Edinburgh, Glasgow and Prestwick Airports in
   1982* (CAP 497) CAA, London

Channing V M 1984 Transport data from the 1981 Census (England and Wales). *Transport Statistics Great Britain* **1973–83**: 1–3

Charles G S 1987 The National Travel Survey. *Transport Statistics Great Britain* **1976–86**: 3–4

Chisholm M 1971 Freight transport costs, industrial location and regional development. In Chisholm M, Manners G (eds) *Spatial Policy Problems of the British Economy* Cambridge University Press, Cambridge pp 213–44

Chisholm M 1985 Accessibility and regional development in Britain: some questions arising from data on freight flows. *Environment and Planning A* **17**: 963–80

Chisholm M 1987 Regional variations in transport costs in Britain, with special reference to Scotland. *Transactions of the Institute of British Geographers* new series **12**: 303–14

Chisholm M, O'Sullivan P 1973 *Freight Flows and Spatial Aspects of the British Economy* Cambridge University Press, Cambridge

Cloke P J (ed.) 1984 *Wheels within Wales: Rural Transport and Accessibility Issues in the Principality* St David's University College, Lampeter

Cook W R 1967 Transport decisions of certain firms in the Black Country. *Journal of Transport Economics and Policy* **1**: 325–44

DE 1989 *Family Expenditure Survey 1987* HMSO, London

DoE 1975 *National Travel Survey 1972/73* HMSO, London

DoE 1978/80 *National Dwelling and Housing Survey: Phases II and III* HMSO, London

DtP 1983 *Railway Finances: Report of a Committee Chaired by Sir David Serpell* 2 vols HMSO, London

DTp 1986 *Crime on the London Underground* HMSO, London

DTp 1988 *National Travel Survey: 1985/86 Report: Part 2 A Technical Guide* HMSO, London

DTp 1989 *Origins, Destinations and Transport of UK International Trade 1986* Directorate of Statistics, DTp, London

DTp/MDS 1988 *Waterborne Freight in the United Kingdom 1986* DTp/MDS, Chester

DTp/NPC 1980 *Inland Origins and Destinations of UK International Trade* DTp/NPC, London

Edwards S L 1975 Regional variations in freight costs. *Journal of Transport Economics and Policy* **9**: 1–12

Freeman M, Aldcroft D 1985 *The Atlas of British Railway History* Croom Helm, Beckenham

Geeson A J 1981 Geographical aspects of car ownership and availability in the Bristol area. Unpublished PhD thesis, University of Bristol

Halsall D A (ed.) 1982 *Transport for Recreation* Institute of British Geographers Transport Geography Study Group, Edge Hill College of Higher Education

Hay A 1981 Transport geography. In Wrigley N, Bennett R J (eds) *Quantitative Geography* Routledge and Kegan Paul, London pp 366–73

Hayter R 1986 The export dynamics of firms in traditional industries during recession. *Environment and Planning A* **18**: 729–50

Hoare A G 1975 Some aspects of rural transport. *Journal of Transport Economics and Policy* **9**: 1–13

Hoare A G 1977 The geography of British exports. *Environment and Planning A* **9**: 121–36

Hoare A G 1983 *The Location of Industry in Britain* Cambridge University Press, Cambridge

Hoare A G 1984 Space and society in Northern Ireland: the geography of journey to work. *Environment and Planning A* **16**: 289–304

Hoare A G 1985 Great Britain and her exports: an exploratory regional analysis. *Tijdschrift voor Economische en Sociale Geografie* **76**: 9–21

Hoare A G 1986 British ports and their export hinterlands: a rapidly changing geography. *Geografiska Annler* **68B**: 29–40

Hoare A G 1988 Geographical aspects of British overseas trade: a framework and a review. *Environment and Planning A* **20**: 1345–64

Humphrys G n.d. Employment and travel to work. In Carter H (ed.) *National Atlas of Wales* Oxford University Press, Oxford section 6.1

Jamieson Mackay and Partners 1975 *Bristol Area Land Use Transportation Study* 3 vols Jamieson Mackay and Partners, London

Jeffrey D 1985 Trends and fluctuations in visitor flows to Yorkshire and Humberside hotels: an analysis of bed occupancy rates 1982–84. *Regional Studies* **19**: 509–22

Jeffrey D, Hubbard N J 1988 Foreign tourism, the hotel industry and regional economic performance. *Regional Studies* **22**: 319–30

Jones D W 1987 Traffic speeds in London. *Transport Statistics Great Britain 1976–86*: 7–10

Jowett A 1989 *Jowett's Railway Atlas of Great Britain and Ireland* Patrick Stephens, Wellingborough

Lane R, Powell T J, Prestwood-Smith P 1971 *Analytical Transport Planning* Duckworth, London

Lickorish L J 1975 Tourism. In Maunder W F (ed.) *Reviews of UK Statistical Sources* IV Heinemann, London

Martech Consultants Ltd 1964 *Britain's Foreign Trade* Port of London Authority, London

Maultby A J 1984 Personal travel statistics. *Transport Statistics Great Britain 1973–83*: 10–13

Mort D 1981 Sea transport. In Maunder W F (ed.) *Reviews of UK statistical sources* XIV Pergamon, Oxford pp 125–268

Moseley M J, Harman R G, Coles O B, Spencer M B 1977 *Rural Transport and Accessibility* 2 vols Centre for East Anglian Studies, University of East Anglia, Norwich

Moyes A n.d. Rail, sea and air transport. In Carter H (ed.) *National Atlas of Wales* Oxford University Press, Oxford section 7.2

Munby D L 1978 Road passenger transport. In Maunder W F (ed.) *Reviews of UK Statistical Sources* VII, no. 12 Pergamon, Oxford

Nutley S D 1980 Accessibility, mobility and transport-related welfare: the case of rural Wales. *Geoforum* **11**: 335–52

Nutley S D 1982 The extent of public transport decline in rural Wales. *Cambria* **9**: 27–48

Nutley S D 1983 *Transport Policy, Appraisal and Personal Accessibility in Rural Wales* Geo Books, Norwich

Nutley S D 1984 Planning for rural accessibility provision: welfare, economy and equity. *Environment and Planning A* **16**: 357–76

O'Connor K 1980 The analysis of journey to work patterns in human geography. *Progress in Human Geography* **4**: 475–99

Overson C R J 1982 Commuting in London and the South East. *Transport Statistics Great Britain 1971–81*: 4–5

Pearce D 1987 *Tourism Today: A Geographical Analysis* Longman, Harlow

Phillips C 1979 Civil aviation. In Maunder W F (ed.) *Reviews of UK Statistical Sources* X Pergamon, Oxford pp 111–302

Plowden S P C 1967 Transportation studies examined. *Journal of Transport Economics and Policy* **1**: 5–27

Powell K S 1985 Travel demand. In England J R, Hudson K I, Masters R I,

Powell K S, Shortridge J D (eds) *Information Systems for Policy Planning in Local Government* Longman, Harlow pp 264–82

Rhind D (ed.) 1983 *A Census User's Handbook* Methuen, London

Rimmer P J 1985 Transport geography. *Progress in Human Geography* **9**: 270–5

Rimmer P J 1988a Transport geography: a bit of a caveat here and there. *Australian Geographical Studies* **26**: 172–83

Rimmer P J 1988b Transport geography. *Progress in Human Geography* **12**: 270–81

Rubra N 1975 *Transport and Communications* Unit 13, Open University 2nd level course in statistical sources, Open University Press, Milton Keynes

Sando F 1984 The role of survey methods in transport research. In Banister D, Hall P (eds) *Transport and Public Policy Planning* Mansell, London pp 244–51

Schrach-Szmigiel C 1983 Trading areas of the United Kingdom ports. *Geografiska Annaler* **55B**: 71–82

Sheriff P J 1982 National Travel Survey. *Transport Statistics Great Britain* **1971–81**: 1–4

Sly F 1986 *British Rail Origin and Destination Survey: London and South East 1981–82* DTp Statistics Bulletin 86(42), GSS, London

Smart M W 1974 Labour market areas: uses and definitions. *Progress in Planning* **2**: 239–353

Smith J A, Gant R L 1982 The elderly's travel in the Cotswolds. In Warner A M (ed.) *Geographical Perspectives on the Elderly* Wiley, Chichester pp 326–36

Sowerbutts D 1987 The development of waterborne freight transport statistics. *Transport Statistics Great Britain* **1976–86**: 11–14

Starkie D N M 1976 *Transportation Planning, Policy and Analysis* Pergamon, Oxford

Wanhill S 1988 *Tourism statistics to 2000*. Paper delivered at conference on current issues in services research, Department of Hospitality Management, Dorset Institute, Poole

Watson A H 1978 Road goods transport. In Maunder W F (ed.) *Reviews of UK Statistical Sources* VII, no. 13 Pergamon, Oxford

Welsh Office 1981 *Welsh Housing and Dwelling Survey 1978–79* Welsh Office, Cardiff

White M T 1985 Road network-based information systems. In England J R, Hudson K I, Masters R J, Powell K S, Shortridge J D (eds) *Information Systems for Policy Planning in Local Government* Longman, Harlow pp 283–302

Wignall C J 1983 *Complete British Railway Maps and Gazetteer from 1830–1981* Oxford Publishing Co, Oxford

Williams A F 1981 Aims and achievements of transport geography. In Whitelegg J (ed.) *The Spirit and Purpose of Transport Geography* Transport Geography Study Group, University of Lancaster pp 5–31

Annex 1

Useful addresses

Below are listed the addresses of most of the organizations mentioned in the text, and a few others, which may be able to help with enquiries about particular sources of information. Organizations concerned with specific areas or regions are, for the most part, omitted.

ABC International
World Timetable Centre, Church Street, Dunstable

Anbar
62 Toller Lane, Bradford BD8 9BY

Article Number Association
6 Catherine Street, London WC2B 5JJ

Association of British Mining Equipment Companies
Royal Victoria Hotel, Station Road, Sheffield S4 7YE

The Association for Geographic Information
12 Great George Street, Parliament Square, London SW1P 3AD

Audits of Great Britain
West Gate, London W5 1UA

Automobile Association
Fanum House, Basingstoke, Hampshire RG21 2EA

BAA PLC
130 Wilton Road, London SW1V 1LQ

Bernard Thorpe
19–24 St George Street, London W1A 2AR

BKS Surveys Ltd
Ballycairn Road, Coleraine, County Londonderry BT51 5HZ

British Aggregate Construction Material Industries
156 Buckingham Palace Road, London SW1W 9TR

British Air Survey Association
c/o Cartographical Services (Southampton) Ltd, Landford Manor, Landford, Salisbury SP5 2EW

British Chambers of Commerce
Sovereign House, 212 Shaftesbury Avenue, London WC2H 8EW

British Coal Corporation
Hobart House, Grosvenor Place, London SW1X 7AE

British Coal Opencast Executive
200 Lichfield Lane, Berry Hill, Mansfield, Nottinghamshire NG18 4RG

British Council of Shopping Centres
College of Estate Management, Whiteknights, Reading RG6 2AW

British Gas Corporation
Rivermill House, 152 Grosvenor Road, London SW1V 3JL

British Library Business Information Service
25 Southampton Buildings, London WC2A 1AW

British Ports Federation
Victoria House, Vernon Place, London WC1B 4LL

British Railways Board
24 Eversholt Street, London NW1 1DZ

British Road Federation
194–202 Old Kent Road, London SE1 5TT

British Telecom Database
Walpole House, 18–20 Bond Street, Ealing, London W5 5AA

British Tourist Authority
Thames Tower, Blacks Road, Hammersmith, London W6 9EL

British Urban Development
32 Queen Anne's Gate, London SW1H 9AB

British Urban and Regional Information Systems Association
School for Advanced Urban Studies, University of Bristol, Rodney Lodge, Grange Road, Bristol BS8 4EA

British Waterways Board
Melbury House, Melbury Terrace, London NW1 6JX

Business Statistics Office, see Central Statistics Office

CACI Ltd
Regent House, 89 Kingsway, London WC2B 6RH

Cambridge Econometrics
21 St Andrew's Street, Cambridge CB2 3AX

Cartographic Services (Southampton) Ltd
Landford Manor, Landord, Salisbury, Wiltshire SP5 2EW

Central Statistical Office
  Great George Street, London SW1P 3AQ
  National accounts, press publications and publicity

  Cardiff Road, Newport, Gwent NP9 1XG
  Annual Census of Production, distribution and service statistics, press
  publications and publicity (Room D114)

  Millbank Tower, London SW1P 4QU
  Overseas direct investment, acquisitions and mergers

Centre for Business Research
Manchester Business School, Booth Street West, Manchester M15 6PB

Chartered Institute of Public Finance and Accounting
3 Robert Street, London WC2N 6BH

Chas E Goad Ltd
18a Salisbury Square, Old Hatfield, Hertfordshire AL9 5BE

Chelmer Population and Housing Model
Population and Housing Research Group, Anglia Polytechnic,
Victoria Road South, Chelmsford, Essex CM1 1LL

Chief Valuer Scotland
Meldrum House, 15 Drumsheugh Gardens, Edinburgh 3

Civil Aviation Authority
CAA House, 45–49 Kingsway, London WC2B 6TE

Clyde Surveys Ltd
Clyde House, Reform Road, Maidenhead, Berkshire SL6 8BU

Coalfield Communities Campaign
9 Regent Street, Barnsley S70 2EG

Committee for Aerial Photography
University of Cambridge, Mond Building, Free School Lane, Cambridge

Companies House
  Companies House, Crown Way, Cardiff CF4 3UZ
  Companies House, 55–71 City Road, London EC1Y 1BB
  Companies House, 102 George Street, Edinburgh EH2 3DJ

  Satellite offices:
  Birmingham Public Library, Chamberlain Square, Birmingham B3 3HQ
  21 Bothwell Street, Glasgow G2 6NR

25 Queen's Street, Leeds LS1 2TW
75 Mosley Street, Manchester M2 2HR

Registrar of Companies, IDB House, 65 Chichester Street, Belfast BT1 4JX

Confederation of British Industry
103 New Oxford Street, London WC1A 1DU

Corporate Intelligence Group
51 Doughty Street, London WC1N 2LS

Countryside Commission
John Dower House, Crescent Place, Cheltenham,
Gloucestershire GL50 3RA

Countryside Commission for Scotland
Battleby, Redgorton, Perth PH1 3EW

County Planning Officers Society
St Edmund House, Rope Walk, Ipswich IP4 1LZ

Crofters Commission
4/6 Castle Wynd, Inverness IV2 3EQ

Crown Estate Commissioners
13–16 Carlton House Terrace, London SW1Y 5AH

Customs and Excise, Board of
New King's Beam House, 22 Upper Ground, London SE1 9PL

Data Consultancy
7 Southern Court, South Street, Reading RG1 4QS

Debenham Tewson and Chinnocks
44 Brook Street, London W1A 4NA

Department of Agriculture and Fisheries for Scotland
Scottish Office, New St Andrew's House, Edinburgh EH12 7AT

Department of Agriculture in Northern Ireland
Room 817, Dunonald House, Belfast BT4 3SB

Department of Employment
  Headquarters Buildings, East Lane, Halton, Runcorn WA7 2DN
  Employment and hours

  Caxton House, Tothill Street, London WS1H 9NF
  Unemployment and vacancies, redundancy statistics,
  Labour Force Survey, Employment Intelligence Units, VAT statistics

  Exchange House, 60 Exchange Road, Watford, Herts WD1 7HH
  Employment Census

Training, Enterprise and Education Division
Moorfoot, Sheffield S1 4PQ

Department of Energy
1 Palace Street, London SW1E 5HE

Thames House South, Millbank, London SW1P 4QJ
Library

Department of the Environment
2 Marsham Street, London SW1 3EB

Department of the Environment (Northern Ireland)
83 Ladas Drive, Belfast BT6 9FT

Department of the Registers of Scotland
Meadowbank House, 153 London Road, Edinburgh EH8 7AU

Department of Trade and Industry
20 Victoria Street, London SW1H 0NF
Overseas trade

Ashdown House, 123 Victoria Street London SW1E 6RB
Specific industries

Kingsgate House, 66–74 Victoria Street, London SW1E 6SW
Invest in Britain Bureau

Department of Transport
2 Marsham Street, London SW1P 3EB
Library

Romney House, 43 Marsham Street, London SW1P 3PY

Publication Sales Unit, Building 1, Victoria Road,
South Ruislip, Middlesex HA4 0N2

Dun and Bradstreet Ltd
26–32 Clifton Street, London EC2P 2LY

Earth Observation Satellite Company
141 London Road, Branksome Chambers, Branksomewood Road,
Fleet, Surrey GU15 3JY

Economic and Social Research Council
Polaris House, North Star Avenue, Swindon, Wilts SN2 1UJ

Economic and Social Research Council Data Archive
University of Essex, Wivenhoe Park, Colchester, Essex CO4 3SQ

Economist Intelligence Unit
27 St James's Place, London SW1A 1NT

Edward Erdman
6 Grosvenor Street, London W1X 0AD

Engineering Industry and Training Board
54 Clarendon Road, Watford, Herts WD1 1LB

English Industrial Estates Corporation
St George's House, Kingsway, Team Valley, Gateshead,
Tyne and Wear NE11 0NA

Extel Financial Ltd
Fitzroy House, 13–17 Epworth Street, London EC2A 4DL

Factory Inspectorate, see Health and Safety Executive

Federation of Small Mines of Great Britain
29 King Street, Newcastle, Staffordshire ST5 1ER

Forestry Commission
231 Corstorphine Road, Edinburgh EH12 7AT

Fraser of Allander Institute, Strathclyde Business School, University of
Strathclyde, Curran Building, 100 Cathedral Street, Glasgow G4 0LN

General Register Office (Scotland)
Ladywell House, Ladywell Road, Edinburgh EH12 7TF

General Register Office (Northern Ireland)
   Oxford House, 49–55 Chichester Street, Belfast BT1 4LH

   Census Office, 5A Frederick Street, Belfast BT1 2LW

Geosurvey International Ltd
Geosurvey House, Orchard Lane, East Molesey, Surrey KT8 0BT

Goad Ltd, see Chas E Goad Ltd

Healey and Baker
29 St George Street, London W1A 3BG

Health and Safety Executive (HM Factory Inspectorate)
1 Chepstow Place, Westbourne Grove, London W2 4TF

Highlands and Islands Development Board
20 Bridge Street, Inverness IV1 1QR

Hillier Parker May and Rowden
77 Grosvenor Street, London W1A 2BT

HM Land Registry (administration only)
Lincoln's Inn Fields, London WC2

HMSO Publications Centre
PO Box 276, London SW8 5DT

Hunting Technical Services Ltd
Elstree Way, Borehamwood, Hertfordshire WD6 1SB

ICC Business Publications Ltd
Field House, 72 Oldfield Road, Hampton, Middlesex TW12 2HQ

ID Communication Services Ltd
Finlay House, 142 High Street, Berkhamstead HP4 3AT

Incomes Data Services Ltd
193 St John Street, London EC1V 4LS

The Industrial Marketing Research Association
11 Bird Street, Lichfield, Staffs WS13 6PW

Inland Revenue Valuation Office
Chief Valuer's Office, Carey Street, London WC2A 2JE

Institute for Employment Research
University of Warwick, Coventry CV4 7AL

Institute for Retail Studies
University of Stirling, Stirling FK9 4LA

Institute of Grocery Distribution
Letchmore Heath, Watford WD2 8DQ

Institute of Terrestrial Ecology
Merlewood Research Station, Grange-over-Sands, Cumbria LA11 6JU

Inter Company Comparisons, see ICC Business Publications Ltd

Invest in Britain Bureau, see Department of Trade and Industry

Investment Property Databank Ltd
7/8 Greenland Place, London NW1 0AP

JAS Photographic
92–94 Church Road, Mitcham, Surrey CR4 3TD

Jones Lang Wootton
22 Hanover Square, London W1A 2BN

Jordan and Sons Ltd
Jordan House, Brunswick Place, London N1 6EE

Key Note Publications Ltd, see ICC Business Publications

King & Co Ltd
7 Stratford Place, London W1N 9AE

King's College London
Department of Geography, University of London, Strand,
London WC2R 2LS

LAMSAC
Vincent House, Vincent Square, London SW1P 2NB

Local Authorities Research and Intelligence Association
c/o Dr L Worrall, Policy Unit, Wrekin Council, PO Box 213,
Telford TF3 4LD

Locate in Scotland, see Scottish Development Agency

Logica PLC
64 Newman Street, London W1A 4SE

London Research Centre
81 Black Prince Road, London SE1 7SZ

London School of Economics
Department of Geography, Houghton Street, London WC2A 2AE

Low Pay Unit Researchers' and Campaigners' Unit
9 Upper Berkeley Street, London W1

Macaulay Land Use Research Institute
Craigiebuckler, Aberdeen AB9 2QJ

McCarthy Information Ltd
Manor House, Ash Walk, Warminster, Wiltshire BA12 8BY

Management Horizons
Ryde House, 391 Richmond Road, Twickenham, London TW1 2EF

Market Location Ltd
1 Warwick Street, Royal Leamington Spa, Warwickshire CV32 5LW

The Market Research Society
175 Oxford Street, London W1R 1TA

Ministry of Agriculture, Fisheries and Food,
  Whitehall Place West, London SW1A 2HH
  General enquiries

  Ergon House, c/o Nobel House, 17 Smith Square, London SW1P 3HX
  Fisheries

  Government Buildings, Epsom Road, Guildford GU1 2LD
  Agricultural Census

Mintel
18–19 Long Acre, London EC1A 9HE

National Economic Development Office
Millbank Tower, Millbank, London SW1P 4QX

National House Building Council
Chiltern Avenue, Amersham, Buckinghamshire HB6 5AP

National Library for Scotland
George IV Bridge 1, Edinburgh EH1 1EW

National On-line Manpower Information System
Unit 3P, Mountjoy Research Centre, University of Durham, Durham DH1 3SW

National Remote Sensing Centre
Royal Aircraft Establishment, Farnborough, Hants GU14 6TD

National Union of Mineworkers
Holly Street, Sheffield S1 2ST

Natural Environment Research Council
Polaris House, North Star Avenue, Swindon SN2 1EU

Nature Conservancy Council (now changed to English Nature,
Nature Conservancy for Scotland, and Countryside Council for Wales)
English Nature is based at:
Northminster House, Peterborough PE1 1UA

Northern Ireland Departments
  Stormont, Belfast BT4 3SW
  Economic statistics, transport statistics

  Dundonald House, Upper Newtownwards Road, Belfast BT4 3SF
  Agricultural statistics

  Netherleigh House, Massey Avenue, Belfast BT4 2JS
  Employment statistics, production and trade statistics

  IDB House, 64 Chichester Street, Belfast BT4 2JS
  Industrial Development Board; Valuation and Lands Office

  Oxford House, 49–55 Chichester Street, Belfast BT1 4HF
  Population statistics

  River House, High Street, Belfast BT1
  Land Registry

Northern Ireland Economic Research Centre
46–48 University Road, Belfast BT7 1NJ

Northern Ireland Electricity Service
120 Malone Road, Belfast BT9 5HT

Northern Ireland Tourist Board
48 High Street, Belfast BT1 2DS

North of Scotland Hydro-Electric Board (now Scottish Hydro-Electric plc)
16 Rothesay Terrace, Edinburgh EH3 7SE

Office of Population Censuses and Surveys
  Census Division, St Catherine's House, 10 Kingsway,
  London WC2B 6JP
  Census of Population, OPCS Reference Library

Titchfield, Fareham, Hampshire PO15 5RR
Supply of special and unpublished census statistics

Ordnance Survey
Romsey Road, Maybush, Southampton SO9 4DH

Ordnance Survey Northern Ireland
Colby House, Stranmillis, Belfast

Oxford Institute of Retail Management
Templeton College, Oxford OX1 5NY

PA Cambridge Economic Consultants
62–64 Hills Road, Cambridge CB2 1LA

Pinpoint Analysis Ltd
Mercury House, 117 Waterloo Road, London SE1 8UL

Public Record Office
   Chancery Lane, London WC2A 1LR
   Ruskin Avenue, Kew, Richmond, Surrey TW9 4DU

RAC Motoring Services Ltd
   89 Pall Mall, London SW1
   Library

   RAC House, PO Box 100, Landsdowne Road, South Croydon,
   Surrey CR9 2AJ

Red Deer Commission
Knowsley, 82 Fairfield Road, Inverness IV3 5LH

Regional Studies Association
Wharfdale Projects, 15 Micawber Street, London N1 17TB

Register of Sasines
Registers of Scotland, Meadowbank House, 153 London Road,
Edinburgh EH8 7AU

Remote Sensing Society
c/o Department of Geography, University of Nottingham, University Park,
Nottingham NG7 2RD

Retail Consortium
Bedford House, 69/79 Fulham High Street, London SW6 3JW

The Reward Group
Reward House, Diamond Way, Stone Business Park, Stone,
Staffordshire, ST15 0SD

Richard Ellis
Berkeley Square House, London W1X 6AN

Royal Geographical Society
1 Kensington Gore, London SW7 2AR

Royal Institution of Chartered Surveyors
12 Great George Street, London SW1P 3AD

Royal Scottish Geographical Society
10 Randolph Crescent, Edinburgh EH3 7TU

The Royal Statistical Society
25 Enford Street, London W1H 2BH

The Royal Town Planning Institute
26 Portland Place, London W1N 4BE

Rural Development Commission
11 Cowley Street, London SW1P 3NA

Sales and Marketing Information Ltd
1 Warwick Street, Leamington Spa, Warwickshire CV32 5LW

Sales Performance Analysis
1 Warwick Street, Leamington Spa, Warwickshire CV32 5LW

Sand and Gravel Association
1 Bamber Court, 2 Bramber Road, London W14 9PB

Scottish Office
New St Andrews House, St James's Centre, Edinburgh EH1 3SX
General enquiries, road/rail transport, planning/environment, library and
publication sales, Central Register of Air Photography, Scottish Office
Environment Department (previously Scottish Development Department

Pentland House, 47 Robbs Loan, Edinburgh EH14 1UE
Agricultural and fisheries statistics

Alhamara House, 45 Waterloo Street, Glasgow G5 6AS
Scottish economic matters

Scottish Development Agency (now Scottish Enterprise)
120 Bothwell Street, Glasgow G2 7JP

Scottish Record Office
HM General Register House, Princes Street, Edinburgh EH1 3YY

Scottish Tourist Board
23 Ravelton Avenue, Edinburgh EH14 1UE

Sea Fish Industry Authority
10 Young Street, Edinburgh EH2 4JQ

Small Business Research Trust
School of Management, The Open University, Walton Hall, Milton Keynes
MK7 6AA

Social Science Forum
61 Richborne Terrace, London SW8 1AT

South of Scotland Electricity Board (now Scottish Power plc)
Cathcart House, Spean Street, Glasgow G44 4BE

SPA Database Marketing
Blackburn House, London Road, Coventry CV3 4AP

SPOT Image
18 Avenue Edouard-Belin, F31055, Toulouse, France

Statistics Users' Council
Lancaster House, More Lane, Esher, Surrey KT10 8AP

TMS
Oxford House, 182 Upper Richmond Road, London SW15 2SH

Training, Enterprise and Education Division,
see Department of Employment

Transport and Environmental Studies
177 Arlington Road, London NW1 7EY

Transport and Road Research Laboratory
Old Workingham Road, Crowthorne, Berkshire RG11 6AU

UK Atomic Energy Authority
11 Charles II Street, London SW1Y 4QP

UK Science Park Association
44 Four Oaks Road, Sutton Coldfield, West Midlands B74 2TL

Unemployment Unit
9 Poland Street, London W1V 3DG

Union of Democratic Mineworkers
The Sycamores, Moor Road, Bentwood, Nottinghamshire NG6 8VE

Unit for Retail Planning Information Ltd, see Data Consultancy

Urban and Economic Development Group (URBED)
3 Stanford Street, London SE1 9NT

Wales Tourist Board
2 Fitzgerald Road, Cardiff CF2 1UY

Water Authorities' Association
1 Queen Anne's Gate, London SW1H 9BT

Welsh Development Agency
  Pearl House, Greyfriars Road, Cardiff CF1 3XX

  Welsh Development International (previously WINvest)

Welsh Office

Economic and Statistical Services Division, Crown Building, Cathays Park, Cardiff CF1 3NQ

Central Register of Air Photographs for Wales, Cathays Park, Cardiff CF1 3NW

Index